Solid Mechanics and Its Applications

Volume 223

Aims and Scope of the Series

The fundamental questions arising in mechanics are: *Why*?, *How*?, and *How much*? The aim of this series is to provide lucid accounts written by authoritative researchers giving vision and insight in answering these questions on the subject of mechanics as it relates to solids.

The scope of the series covers the entire spectrum of solid mechanics. Thus it includes the foundation of mechanics; variational formulations; computational mechanics; statics, kinematics and dynamics of rigid and elastic bodies: vibrations of solids and structures; dynamical systems and chaos; the theories of elasticity, plasticity and viscoelasticity; composite materials; rods, beams, shells and membranes; structural control and stability; soils, rocks and geomechanics; fracture; tribology; experimental mechanics; biomechanics and machine design.

The median level of presentation is to the first year graduate student. Some texts are monographs defining the current state of the field; others are accessible to final year undergraduates; but essentially the emphasis is on readability and clarity.

More information about this series at http://www.springer.com/series/6557

Gérard A. Maugin

Continuum Mechanics through the Ages - From the Renaissance to the Twentieth Century

From Hydraulics to Plasticity

Springer

Gérard A. Maugin
Institut Jean Le Rond d'Alembert
Université Pierre et Marie Curie
Paris Cedex 05
France

ISSN 0925-0042 ISSN 2214-7764 (electronic)
Solid Mechanics and Its Applications
ISBN 978-3-319-26591-9 ISBN 978-3-319-26593-3 (eBook)
DOI 10.1007/978-3-319-26593-3

Library of Congress Control Number: 2015954983

Springer Cham Heidelberg New York Dordrecht London

Printed on acid-free paper

Springer International Publishing AG Switzerland is part of Springer Science+Business Media
(www.springer.com)

Preface

In 1800 (An VIII of the First French Republic), the civil engineer Gaspard-Clair-François-Marie RICHE, Baron de PRONY (1755–1839), published a book entitled "*Mécanique philosophique*" ("Philosophical Mechanics"). This title bears the whole spirit of the approach to science and more generally knowledge that was expanded during the Enlightenment and its acceptance by contemporaries of the French Revolution and early nineteenth century. I very much like this title which in fact is explained by the subtitle (in translation) "Analysis of various parts of the Science of Equilibrium and Motion." It is in this state of mind that I have expanded some aspects of the historical developments of continuum mechanics in the immediate post-Newtonian era till the second half of the twentieth century in two previous volumes. The first one[1] published in 2013 was neatly characterized by my own experience in the field with several national institutions that I visited and/or worked with from time to time (e.g., in the USA, UK, Germany, Italy, Poland, Japan—and obviously France), and by my friendly relationship with research leaders of various nationalities whom I practically all knew personally, being aware of both their strength and pettiness, but always emphasizing the former. The choice of studied groups and schools as also a selection of particular fields was strictly personal since I had decided to speak only about what I knew best, although sometimes superficially and not without some unavoidable misinterpretations and factual errors. This personal and nonobjective vision was naturally criticized by some readers. Most of the time, these people complained that I had not treated at all, or not developed enough, their own field of research—this was the case of finite-strain elasto-plasticity and phase-transformations in deformable solids, although this was touched upon but cursorily—for which I thought that some other scientists would be much more competent than me. Obviously, with a more than

[1]Continuum mechanics through the twentieth century: A concise historical perspective, SMIA Volume 196, Springer, Dordrecht, 2013.

natural humane bias, some complained about not being cited. To these last people I sincerely apologize. A well-balanced exercise in citation is not easy and is a dangerous exercise when one deals with his own contemporaries. I did not take any chance with the second volume[2] devoted to the eighteenth and nineteenth centuries where all contributors could not complain to me orally or via e-mail, save perhaps in some of my nightmares. Still my choice of subjects and main actors was strictly personal, although greatly influenced by some famous predecessors such as J.L. Lagrange, A. Barré de Saint-Venant, P. Duhem, I. Todhunter and K. Pearsons, R.T. Whittaker, M. Jouguet, and more recently, R. Dugas,[3] I. Szabò,[4] G.A. Tokaty,[5] S.P. Timoshenko,[6] and naturally C.A. Truesdell.[7]

The reader will have noticed that with growing age I am going back in the past. But this has a technical limit due to my lack of knowledge of some "dead" languages. Of course, I could deal with ancient Greek having at home all necessary help for the reading and interpretation of the primary sources. My improved reading of Latin would require the help of a priest but I am not a religious person. Furthermore, the reader must realize that I am not concerned with a total history of mechanics—for which Dugas (see Footnote 3) and Szabó (see Footnote 4) provided beautiful but not completely satisfactory attempts—but only with that part called *continuum* mechanics. This poses the fundamental question of the never-ending debate between the molecular-particular discrete vision and the continuum one. This matter touches upon both history and philosophy and a return to the ancient Greeks may be needed at this very point. But the subject remained of actuality during all centuries in particular with the early developments of continuum mechanics with Poisson, Navier, Cauchy, Piola, etc. In this volume, I consider that continuum mechanics starts with hydrostatics, the notion of *pressure,* and applications to *hydraulics* while *modern* continuum mechanics starts with hydrodynamics, three-dimensional elasticity, and the notion of stress tensor. Hydraulics brings to the foreground the important role of some experimentalists, starting with the Renaissance but also especially in the eighteenth century. Although a theoretician, I have made all efforts to deal with this aspect with sympathy.

---

[2]Continuum mechanics through the eighteenth and nineteenth centuries: Historical perspective from John Bernoulli (1727) to Ernst Hellinger (1914), SMIA Volume 214, Springer, Dordrecht, 2014.

[3]Dugas R. (1950). History of Mechanics, Editions du Griffon, Neuchatel, Switzerland [Dover reprint, New York, 1988] (This exists in both French and English editions).

[4]Szabo I., (1977). Geschichte der mechanischen Prinzipien, Birkhauser, Basel.

[5]Tokaty G.A., (1971). A history and philosophy of fluid mechanics. Foulis, Henley-on-Thames [Dover reprint, New York, 1994].

[6]Timoshenko S.P., (1953). History of the Strength of Materials, McGraw Hill, New York [Dover reprint, New York, 1983].

[7]Truesdell C.A., (1968). Essay in the history of mechanics, Springer, New York.

The main purpose of the present volume is to fill in the gaps left in the previous two volumes by completing some details of the genesis and birth period and introducing some important and original fields that I previously neglected. The same format of interrelated essays has been kept. These essays may be read separately although many authors, including me, prefer a global orderly reading of their books. Subject matters that were previously neglected include: a deeper attention to hydraulics, the question raised by porosity in solids, the theory of mixtures and reacting media, and the influence of fast flows and of the birth of aeronautics in fluid mechanics (with Reynolds, Prandtl, von Kármán). This brings me closer to my initial interest as alumnus of the school of Aeronautics in France. Also, in my historical approach, I always try to avoid any "precursoritis" and duly examine the (at the time) contemporary reaction to the then most recent advances. This can be achieved by carefully perusing the lecture notes and treatises published by famous scientists—albeit not necessarily the most creative ones in the field of continuum mechanics—because such works reflect both the state of the art at the time of their writing and what the author tries to input from his own viewpoint with, usually, a deeply thought appraisal. I already applied this strategy with the treatise of Paul Appell in France and the encyclopedic article by Ernst Hellinger in Germany in my second volume. Here, this is applied to the lecture notes and some collected technical works of Pierre Duhem on hydrodynamics and elasticity, and the lecture notes of a course given by Henri Poincaré on elasticity and an introduction course on continuum mechanics delivered by David Hilbert. To some readers, this may seem to grant too much importance to the "Belle époque." Not only is the viewpoint of two such giants of mathematics as Poincaré and Hilbert of intrinsic interest, but the period at which the lectures were given was a critical one for the whole of physics. It is salient to see whether the burgeoning new physics had any influence on a mature science then thought to have stabilized with magisterial treatises by H. Lamb and A.E.H. Love, respectively, in fluid mechanics and elasticity. Furthermore, the synthetic works of Appell, Hellinger, Duhem, Poincaré, and Hilbert in fact provide the most valuable documentary basis on which a revival of continuum mechanics and its formalization by Truesdell et al. was built in the second half of the twentieth century. We are dutifully following the advice of Rabbi Rashi (Eleventh century) of Troyes in Burgundy: "Ask your master his sources." This is already what Duhem did when examining Leonardo and his possible sources of inspiration in pre-Renaissance times. This is complemented by essays on the special behaviors of viscoelasticity of solids and plasticity—only superficially mentioned in the previous volumes—on fracture—so important in continuum dynamics—the role of geometry as a cornerstone of the field, and a kind of sociohistorical appraisal of the seminal contributions by our direct masters in the second half of the twentieth century.

Acknowledgments

I take this opportunity to thank the members of the team at Springer (in particular, Nathalie Jacobs and Cynthia Feenstra) for their kind and fruitful cooperation. From the scientific viewpoint, my heartfelt thanks go to some members of the *Institut Jean Le Rond d'Alembert* (Université de Paris 6)—among them, Drs. Joël Frelat, Jérôme Hoepfner, Pierre-Yves Lagrée, and Martine Rousseau—for their critical but constructive reading of my manuscripts.

Gérard A. Maugin

Contents

Chapter 1
Particles/Molecules Versus Continuum: The Never-Ending Debate

Abstract Corpuscular tiny bodies ("particles") interacting through empty space via at-a-distance forces and a continuum transmitting forces by contact are traditionally presented as two alternate pictures of the physical mechanical world, and this from the early inception of physics by the ancient Greek "philosophers of Nature". But for more than two millenaries endeavours by the most illustrious physicists and mathematicians have been to reconcile these two visions or, for the least, to deduce the second one from the first thought to be more "physical". This contribution focuses more precisely on such a move from Newton to the early twentieth century with special attention to mechanicians of fluids and solids such as Poisson, Navier, Cauchy, Piola, Green, and others, who created the modern theory of continuum mechanics, essentially through its special branch known as the theory of elasticity. The latter provides the best illustration of the various strategies applied by great scientists in their harsh competition. The nineteenth century French engineers-scientists take the largest share in this endeavour. The subject matter, however, remains an unsettled one, all the more that recent powerful techniques of computation favour a return to the corpuscular vision.

1.1 Old Times, New Ideas

The dichotomy between an atomic vision of matter and a continuum vision is as old as the ancient Greeks. The word atom or $\alpha\tau o\mu o\nu$ = *atomon* in Greek means "uncuttable" or "indivisible". Apart from "atoms" the world is a void. A logical contradiction already appears at this level since "a void" means that there exists nothing so that there *cannot exist* a void. Notwithstanding this contradiction, the idea of such an atomic world is attributed to fifth-century (BC) Greek philosophers such as Leucippus, one of his students Democritus, and then Epicurius and Lucretius. Heraclitus (cf. Fig. 1.1) is often cited as the best supporter of this description of Nature made of atoms and voids. An altogether differing vision was introduced by Aristotle about 330 BC: he entertained the idea of continuity and the absence of

G.A. Maugin, *Continuum Mechanics through the Ages - From the Renaissance to the Twentieth Century*, Solid Mechanics and Its Applications 223,
DOI 10.1007/978-3-319-26593-3_1

Fig. 1.1 Heraclitus

Fig. 1.2 Galileo Galilei (1564–1642)

voids. The latter's vision was supported for a long time, especially with the Scholastics who fully endorsed Aristotle's curriculum in the "philosophy of Nature" (physics). However, it seems that Galileo Galilei (1564–1642; Fig. 1.2), one of the first "modern" physicists, advocated the atomic vision. But it is really with the seventeenth century and the development of an embryonic mechanics of matter by physicists such as Pierre Gassendi (1592–1655; Fig. 1.3), René Descartes (1596–1650; Fig. 1.4) and Robert Boyle (1627–1691; Fig. 1.5), that we witness a true rebirth of atomism, but with some alteration. For instance, Descartes claims that

Fig. 1.3 Pierre Gassendi (1592–1655)

Fig. 1.4 René Descartes (1596–1650)

void does not exist, so that Nature is made of "molecules" or "particles" (the name we now give to "atoms") that interact directly with one another by contact. This, in turn, explains the "mechanics" of Descartes and his theory of the impetus. The great Newton (1643–1727) duly implemented another means of interaction between distant particles: *attraction*. This is Newton's concept of *at-a-distance force*, perhaps a somewhat mysterious notion (according to Leibniz, d'Alembert, etc.;

Fig. 1.5 Robert Boyle (1627–1691)

cf. Maugin 1980, 2012), but a very efficient one that requires no further explanation: Newton "frames no hypotheses". This notion is justified by the immediate success of Newton's mechanics in accounting for phenomena both on earth and in the heaven. This was not without criticism, even among "Newtonians", for whom attraction inevitably yields collapse. Another mechanism must be at play to avoid this catastrophe. Thus, Roger Joseph Boscovich (1711–1787; Fig. 1.6), a polymath Jesuit priest of Croatian origin, proposed that if attraction is certainly active at large distances between two particles, the interaction becomes of the *repulsive* type at short distances. Excuse the play of words: this is an "attractive" concept that will recur from time to time in mechanics and theoretical physics.[1]

The introduction of action-at-a-distance and attraction by Newton became extremely popular but his theory of light required a phenomenon seen as the propagation of "granules" ("atoms" of light). Although mentored by d'Alembert in Paris, Pierre Simon (de) Laplace (1749–1827; Fig. 1.7), was a Newtonian disciple in mechanics, so that he favored the notion of attraction as shown by his theory of capillarity proposed in 1805 (and published in Laplace, 1819). There Laplace states that capillary action is due to "*... les lois dans lesquelles l'attraction n'est sensible qu'à des distances insensibles;* ..." (… the laws in which attraction is sensible [significant] only at insensible [infinitesimal] distances …). But he did much more than that by showing, together with Poisson and then Gauss, that in Newton's theory, the *potential* of attraction—from which the attraction force can be derived—satisfies a partial differential equation (the celebrated second-order "Laplace"

[1]John Archilbald Wheeler (1911–2008) always cited Boscovich as one of his heroes during his lectures on general relativity as witnessed by the writer in 1968–69.

Fig. 1.6 Roger Boscovich (1711–1787)

Fig. 1.7 Pierre Simon Laplace (1749–1827)

equation). This is related to the mathematical field of analysis. This in turn is the application domain of the notion of *field*, itself practically synonymous with a continuum hypothesis because singular fields, although of great interest, are more rarely considered and more difficult to deal with than continuous ones. Thus,

Fig. 1.8 Pierre Simon Girard (1765–1836)

Laplace (1805) in fact created a link between the typically nonlocal theory of attraction and a continuum vision. From then on, there will be recurring attempts to derive a continuum description from a corpuscular one by means of some smoothing and limiting procedure.

At this point it is necessary to make a pause and look at the situation in the field of theoretical and applied mechanics as of the closing of the eighteenth century.

On the one hand, Euler and Lagrange have set the bases for the continuum mechanics of perfect fluids in both Eulerian and Lagrangian formats but we are still missing a true general three-dimensional theory of continuum mechanics in the absence of the notion of stress tensor. However, in addition to pressure, the notion of shear force has been introduced by Euler in the study of the deformation of some structural elements. The study of the mechanics of such elements has given rise to the *strength of materials* (called as such after the formidable book on the theory of elasticity by Pierre Simon Girard (1765–1836; Fig. 1.8) entitled *Traité de la résistance des solides* in Girard 1798). Girard was formed at the Royal School of Ponts et Chaussées (ERPC) and became a powerful figure in the administration of the corresponding Corps of engineers in the early nineteenth century. According to Todhunter and Pearson (1886, p. 74), this "first historical review in the field contained the total knowledge of that day on the subject", and also "an admirable introduction".

On the other hand, a true craze for corpuscular theories of media in all the physics of the period developed in the expert hands of Coulomb, Laplace, and others, to be taken over by a group of remarkable French scientists including Ampère, Poisson, Cauchy, Navier and Fresnel in the years 1810–1830. This was to bring the looked for progress in the three-dimensional theory of elasticity, what is examined next, not forgetting that all these engineers-scientists, excellent as they were, had also strong personalities and sometimes an overdeveloped ego that was going to cause a lot of priority disputes and reciprocal accusations of plagiarism.

1.2 Three-Dimensional Elasticity in the Early Nineteenth Century

1.2.1 Poisson on Elastic Surfaces (1812)

Simeon Denis Poisson (1781–1840; Fig. 1.9), a graduate from the Ecole Polytechnique in 1800, became a professor of mathematics at this institution. His scientific interests were multisided. However, as a disciple of Laplace, he paid specific attention to Boscovich's ideas and became one of the most active propagandists for a theory of elasticity based on a molecular structure. His first contribution in this line was an analysis of deformation of elastic bodies in the investigation of the bending of plates. In this analysis presented in a memoir (Poisson 1812), the author considers a plate as a system of particles distributed in the middle plane of the plate (cf. Timoshenko 1953, p. 104). Unfortunately, as remarked by Timoshenko, this modelling yields a mechanical structural element that will resist extension but not bending. So that it strictly applies to an ideally flexible membrane, but not to a plate. We deal in greater detail with this general type of approach with the next step taken by Navier.

Fig. 1.9 Siméon Denis Poisson (1781–1840)

1.2.2 Navier's Corpuscular Theory (1820)

C.L.M.H. Navier (1785–1836; Fig. 1.10), a member of the Corps of Engineers of Ponts et Chaussées (Formation: Ecole Polytechnique followed by Ecole des Ponts et Chaussées), is one of the greatest elasticians (but also fluid dynamicists) of the nineteenth century. He presented to the Paris Academy of Sciences on May 14, 1821—but published in Navier (1824) with a long abstract published in 1823 after publication of an abstract by Cauchy in the same bulletin (see below)—, a memoir on the laws of equilibrium and motion of elastic solid bodies based on the idea of a corpuscular description. Navier's strategy in his molecular theory of elasticity is very well explained by Timoshenko (1953, pp. 105–107) that we shall paraphrase.

Navier assumes that in the body there exist two systems of forces, one, ΣF is such that forces balance each other and represent molecular forces acting when external forces are absent. The other system of forces, ΣF_1, balances the external forces, such as that due to gravity. They are assumed to be proportional to changes $r_1 - r$ of the distances between the particles and to act along the lines connecting them. Let u, v, w the components of the displacement of a particle $P(x, y, z)$ and $u + \Delta u, v + \Delta v, w + \Delta w$ the corresponding displacements of an adjacent particle $P_1(x + \Delta x, y + \Delta y, z + \Delta z)$. Then the change of distance between these particles is given by

$$r_1 - r = \alpha\,\Delta u + \beta\,\Delta v + \gamma\,\Delta w, \tag{1.1}$$

Fig. 1.10 C.L.M.H. Navier (1785–1836)

where (α, β, γ) are the cosines of the angles which the direction r makes with the coordinate axes x, y, z. The corresponding force F_1 is then given by an expression of the type

$$F_1 = f(r)(\alpha\,\Delta u + \beta\,\Delta v + \gamma\,\Delta w), \tag{1.2}$$

where $f(r)$ is a rapidly decreasing function with increasing r. Then Navier expands $\Delta u, \Delta v, \Delta w$ in series of the form

$$\Delta u = \frac{\partial u}{\partial x}\Delta x + \frac{\partial u}{\partial y}\Delta y + \frac{\partial u}{\partial z}\Delta z + \frac{1}{2}\frac{\partial^2 u}{\partial x^2}\Delta x^2 + \frac{\partial^2 u}{\partial x \partial y}\Delta x\,\Delta y + \cdots, \tag{1.3}$$

and limits the expansion at the second order. Substituting these in the Eq. (1.2) above and setting $\Delta x = r\alpha, \Delta y = r\beta, \Delta z = r\gamma$, yields for the x component of the force F_1:

$$\alpha F_1 = rf(r)\left[\alpha^3\frac{\partial u}{\partial x} + \alpha^2\beta\left(\frac{\partial u}{\partial y} + \frac{\partial v}{\partial x}\right) + \cdots\right] + r^2 f(r)\left[\frac{\alpha^4}{2}\frac{\partial^2 u}{\partial x^2} + \alpha^3\beta\frac{\partial^2 u}{\partial x\,\partial y} + \cdots + \frac{\alpha\gamma^3}{2}\frac{\partial^2 w}{\partial z^2}\right] \tag{1.4}$$

To compute the total force acting in the x-direction of the displaced particle $P(x,y,z)$, a summation over expressions such as (1.4) is performed for all the particles within the *sphere of action*—or region of molecular activity to be defined by a characteristic distance—of the particle P. Assuming that the molecular forces are independent of the direction of r, i.e., that the body is *isotropic*, terms containing cosines in odd powers cancel out. It is then remarkable that the evaluation of those integrals in the last contribution in (1.4) can be reduced to the calculation of a single integral of the form

$$C = \frac{2\pi}{15}\int_0^\infty r^4 f(r)dr. \tag{1.5}$$

If this is known then we can evaluate $\Sigma\alpha F_1$ and the other two projections on the axes. For example,

$$\Sigma\alpha F_1 = C\left(3\frac{\partial^2 u}{\partial x^2} + \frac{\partial^2 u}{\partial y^2} + \frac{\partial^2 u}{\partial z^2} + 2\frac{\partial^2 v}{\partial x \partial y} + 2\frac{\partial^2 w}{\partial x \partial z}\right). \tag{1.6}$$

With a more modern notation for the cubic dilatation and the Laplacian operator

$$\theta = \frac{\partial u}{\partial x} + \frac{\partial v}{\partial y} + \frac{\partial w}{\partial z}, \quad \nabla^2 = \frac{\partial^2}{\partial x^2} + \frac{\partial^2}{\partial y^2} + \frac{\partial^2}{\partial z^2}, \tag{1.7}$$

the equilibrium equations can be expressed in the following manner:

$$C\left(\nabla^2 u + 2\frac{\partial\theta}{\partial x}\right) + X = 0, \ldots, \tag{1.8}$$

where X is the x-component of the externally applied force. All this is apparently fine but only *one* constant C is involved! Navier also obtained the associated "natural" boundary conditions by invoking the principle of virtual displacements. They also involve the constant C only. This, as we know now, is a defect in this first approach by Navier.

1.2.3 Fresnel's Corpuscular Approach to Wave Optics (1822)

Augustin Fresnel (1788–1827; Fig. 1.11) is another member of the Corps of Ponts et Chaussées. He specialized in physical optics. He is well known for his invention of a system of lenses that much improved the light beam signal produced in lighthouses, and for the discovery of the transverse nature of light vibrations—as opposed to the longitudinal nature of sound waves—in his studies on double refraction. In order to obtain a model that allows for such a propagation Fresnel considered early in 1822 (presented to the Academy in January and March 1822) a

Fig. 1.11 Augustin Fresnel (1788–1827)

corpuscular medium (not to be mistaken for Newton's view of granular light waves). With Fresnel it is the propagation medium that is granular to start with, but light propagates by means of waves i.e., ondulatory signals, hence the possible occurrence of interferences. The concept of such a medium of propagation may have been inspired by Navier's work of 1820. Fresnel had to evaluate the molecular forces that act upon a single slightly displaced particle of the medium from the other particles around it. This he discussed with Cauchy in 1822.[2]

1.2.4 Cauchy's First Theory (1822–3; 1828)

The long contribution of Navier presented in 1820 and the presentation of his own ideas by Fresnel kindled the interest of Augustin Louis Cauchy (1789–1857) for the construction of a three-dimensional theory of elasticity. But this he achieved in a very original move, eschewing any corpuscular approach. As a matter of fact, he generalized Euler's notion of normal pressure acting on a surface element by a generally obliquely applied "traction" and was thus led to the notion of "stress".[3] This is obtained completely independently of any material behaviour, whether elastic or inelastic, whether solid or fluid, a generality clearly emphasized by the title of his contribution (Cauchy 1823, 1828a), a theory that we call CAUCHY-1. Cauchy presented succinctly his ideas orally at the Paris Academy on September 30, 1822. He published a well written long abstract in 1823, but the corresponding lengthy memoir was published only in 1828 (Cauchy 1828a). It is of interest to note in the list of attendants to the meeting of the Paris Academy on September 30, 1822 (see its reproduction in p. 49 in Maugin 2014a) the names of Fourier, Laplace, Ampère, Cauchy, Girard, and Poisson—who all belonged to the same scientific environment, if not to the same formation or Corps. They all were interested in the contemporary research in theoretical mechanics and may all have been called at one time or another as experts on the papers of Poisson, Navier, Fresnel and Cauchy. We cannot speak of a clique but certainly of a kind of "consanguinity"; but "comrades" in a Corps may have become unforgivable opponents on scientific matters. The postponement of the publication of Cauchy's memoir may have been due to some quarrels of priority—in particular with Navier—but we must also note that Cauchy kept very active and so busy in his research in mathematics (theory of integrals, complex analysis, theory of residues) during the period 1822–1828.[4]

While in the corpuscular approach the equilibrium equation—e.g., Eq. (1.8)—emerges directly without the need for a constitutive equation, in CAUCHY-1 the system must be closed, in the case of elasticity, by such a relation that one must

---

[2]Cf. Cauchy's remark in the 1823 abstract of Cauchy translated in Maugin (2014a, pp. 50–52).

[3]Of course, not called this way by Cauchy, who rather uses the plural "pressures" to denote the six components of this new physical object.

[4]On this activity see Belhoste (1991).

postulate between the six strains—of which the theory was also essentially due to Cauchy—and the stresses. The first such proposal by Cauchy for a linear isotropic theory was wrong although astutely designed. Cauchy thought of relating proportionally the so-called principal strains (dilatations) and stresses (pressures) forgetting that the cubic dilatation should also be involved, so that he obtained a representation of isotropic linear elasticity with one elasticity coefficient only. This he corrected in his belated memoir of 1828. For linear elastic crystals he would later on introduce a general linear homogeneous relationship (in modern language, a linear application of R^6 onto R^6) which, on account of the symmetry for stresses, would involve at most twenty one independent elasticity coefficients.

1.2.5 Poisson's Memoir (1827–1828) and Cauchy's Second Theory (1828, 1833)

Finally, Poisson announced on October 01, 1827, but published in 1828, his extremely long memoir on his corpuscular theory (see also Poisson's (1833) course on mechanics). Poisson there proved the existence of two types of elastic waves in isotropic bodies, longitudinal and transverse waves. It seems that Cauchy responded to this by working out by himself a theory of elasticity (called CAUCHY-2) based on a corpuscular vision. He may have been encouraged to do so with the hope to generalize Fresnel's work on birefringence of transparent media. Fresnel had died in 1827. But this looks strange enough and awkward to modern (continuum) mechanicians like us because its looks like a real step backward from a conceptual viewpoint. Poisson's publication caused a priority dispute with Navier, but Cauchy succeeded in remaining somewhat outside this dispute. Of course one dream of Cauchy was to go to the case of anisotropic media by carefully examining the equilibrium and motion of a system of material points (Cauchy 1828b, c) envisaging (in agreement with Boscovich) that interparticle interactions $f(r)$ be of the attractive *or repulsive* type and proportional to the two masses of the interacting particles. In the case of isotropy, he found that the problem would be solved by the computation of two coefficients G and R such that (cf. Szabó 1977, pp. 398–399; with a reproduction of p. 199 from Cauchy 1828b).

$$G = \pm\frac{2\pi}{3}\int_0^\infty r^3 f(r)dr, \quad R = \pm\frac{2\pi}{15}\int_0^\infty \left[r^4 f'(r) - r^3 f(r)\right]dr, \tag{1.9}$$

to be compared to (1.5) above. In arriving at this result, Cauchy had to implement some hypotheses on the fast decrease with distance of the function $f(r)$, so that he could neglect infinitesimal quantities of the second order in the infinite series expansion that he introduced to approximate the finite differences involved in the displacement of the considered molecules. He also had to make some additional assumptions regarding the distribution of molecules in the system. Anyway, on account of definitions (1.7) and redefinition of the scalar coefficients, this whole manipulation—certainly likely to be subjected to serious criticism from a purist's viewpoint—yields instead of (1.8) the equilibrium equation of modern linear isotropic elasticity as the vectorial equation

$$\mu\nabla^2\mathbf{u} + (\lambda + \mu)\nabla\theta + \mathbf{X} = \mathbf{0}, \tag{1.10}$$

where λ and μ are Lamé's coefficients. Love (1892, pp. 616–618 of 1944 edition) describes well Cauchy's use of the corpuscular description to arrive at the stress-strain relations in a crystalline elastic material.

As rightly emphasized by Belhoste (1991, p. 99), following Laplace in gravitation, Coulomb in electrostatics, Ampère in electrodynamics, Fresnel in optics, and now Navier and Poisson in elasticity, the molecular model had become "so much a part of thinking of that era", that even Cauchy, who had many other successes to his glory, could not resist the temptation to work within this framework, all the more that he expected to improve on Fresnel's theory of light waves. In fact, Cauchy had obtained in passing equilibrium equations for cubic symmetry and uniaxial crystals. He also applied his theory to specific structural elements such as anisotropic plates and rods, and even to perfect fluids. In the last case he wanted to prove that the equality of pressure in all directions was indeed a characteristic property of fluids within this molecular framework. Cauchy's application to light waves started in 1828. In this line he applied his theory of anisotropic elasticity to the phenomenon of double refraction, using Huygens' Principle where appropriate for spherical waves. It seems that by 1830 Cauchy had chosen the molecular theory over his own general and definitive continuum theory of 1823–1828, in fact the only one universally acknowledged nowadays by both applied mathematicians and mechanical engineers. Practically all students—and professors—ignore Cauchy's epochal choice and his strong involvement in the theory of the propagation of light. This was to become outdated with the rejection of the existence of "ether" as a possible physical substrate of this propagation late in the nineteenth century.

1.2.6 Piola's Original Works (1836, 1845)

Gabrio Piola (1794–1850) is an Italian mathematician who was an enthusiastic disciple of Joseph Louis Lagrange (1736–1813). He is, therefore, an ardent supporter of variational formulations in the Euler-Lagrange tradition. He is the author of generally lengthy memoirs that have often been unjustly overlooked. But we must admit that these papers are difficult to read, in reason both of the obsolete mathematical terms and the somewhat (now) antiquated Italian language. We focus attention on his memoirs of 1836 and 1845.

One line followed by Piola is the a priori consideration of the motion of an ensemble of points in interaction, following Poisson (and also the second theory of continua of Cauchy), and then passing to a limit providing equations for a continuum, with an appropriate definition of what will later on be called the first Piola-Kirchhoff stress tensor. This is developed at length in Piola (1836) frequently referring to French mathematicians and mechanicians (Cauchy, Laplace, Poisson, Legendre, Lacroix, and of course Lagrange, hardly a Frenchman to him).[5] To make a long story short, Piola (1836) considered a system of identical point particles of unit mass that we can label (α). Each one is initially at position denoted by (a, b, c) with label (α) and after motion at position (x, y, z) with label (α). In modern notation this would yield the change of position as $x_i^{(\alpha)}$ or $\mathbf{x}^{(\alpha)}$ function of $X_K^{(\alpha)}$ or $\mathbf{X}^{(\alpha)}$, in Cartesian tensor notation and intrinsic notation, respectively. Thus the kinematic description may be said to be *referential*. With externally applied force $\mathbf{f}^{(\alpha)}$, and a model of interactions between particles (called "molecules") whose exploitation is somewhat obscure, Piola is able to write a variational formulation for motion of the following type (Piola 1836, p. 173, Eq. (15))

$$\sum_i \sum_\alpha \left(\frac{d^2 x_i^{(\alpha)}}{dt^2} - f_i^{(\alpha)} \right) \delta x_i^{(\alpha)} + \sum_i \sum_{\alpha,\beta} \phi(S_{\alpha,\beta}) \delta S_{\alpha,\beta}|_i = 0 \tag{1.11}$$

where the S's—whose details are irrelevant—depend on the relative distances between particles, hence on the x_i. For arbitrary variations of the x_i's this formally yields equations of motion of individual particles in the form (Piola 1836, p. 189, Eq. (41))

$$f_i^{(\alpha)} - \frac{d^2 x_i^{(\alpha)}}{dt^2} + I_i^{(\alpha)} = 0, \quad \alpha = 1, 2, \ldots \tag{1.12}$$

where $I_i^{(\alpha)}$ is the interaction force with other particles that we do not elaborate further. The "tour de force" of Piola rests in the approximation of these interaction terms (pp. 175–200) and passing to some kind of continuum limit that brings the generic local equation of motion to the vectorial form (Piola 1836, p. 201, Eq. (56); here the initial density is arbitrarily taken as $\rho_0 = 1$)

$$\mathbf{f} - \frac{\partial^2 \mathbf{x}}{\partial t^2} + div_X \mathbf{T} = \mathbf{0}, \tag{1.13}$$

where $\mathbf{T}$ is an object with nine independent components (for it has no symmetries) and the modern symbol div_X means the divergence operator with respect to the referential coordinates (a, b, c) i.e., X^K. Obtaining (1.13) involves the neglect of supposedly small terms. Equation (1.13) can also be written as

[5]This we examined in detail in a previous work (cf. Maugin 2014a, Chap. 4) that we closely follow here.

$$f_i - \frac{\partial^2 x_i}{\partial t^2} + \frac{\partial}{\partial X_K} T_{Ki} = 0. \tag{1.14}$$

The change of position and its inverse (assuming invertibility in agreement with Lagrange) can be noted (Piola 1836, p. 202, Eqs. (58) and (59))

$$(x, y, z) \; functions \; of \; (a, b, c) \; and \; time \; t \tag{1.15}$$

and

$$(a, b, c) \; functions \; of \; (x, y, z) \; and \; time \; t \tag{1.16}$$

or in modern notation

$$\mathbf{x} = \bar{\mathbf{x}}(\mathbf{X}, t) \; and \; \mathbf{X} = \bar{\mathbf{X}}(\mathbf{x}, t). \tag{1.17}$$

Let J denote the Jacobian determinant of the first of these transformations (this is denoted H by Piola, p. 204), i.e.,

$$J = \det \mathbf{F}, \quad \mathbf{F} = \left\{ F_{iK} = \frac{\partial \bar{x}_i}{\partial X_K} \right\}. \tag{1.18}$$

For a referential mass density ρ_0 not set equal to one ρ_0 has to be introduced and (1.13) has to be rewritten as

$$\rho_0 \left(\mathbf{f} - \frac{\partial^2 \mathbf{x}}{\partial t^2} \right) + div_X \mathbf{T} = \mathbf{0}. \tag{1.19}$$

In order to compare his equation of motion with the formulation obtained by Cauchy (1828a) and Poisson (1829) in the actual configuration, Piola needs to do some work since he must pass to the spatial parametrization of the Eulerian type in terms of the actual position (x, y, z) or $\mathbf{x} = \{x_i; i = 1, 2, 3\}$. He shows that he can introduce a geometrical object $\boldsymbol{\sigma}$ (noted $\boldsymbol{K}$ by Piola) such that (cf. Piola 1836, p. 204, Eq. (60))

$$\sigma_{ij} = J^{-1} \frac{\partial \bar{x}_i}{\partial X_K} T_{Kj} \quad or \quad \underline{\sigma} = J^{-1} \mathbf{F} \mathbf{T}. \tag{1.20}$$

Reciprocally, (Piola 1836, p. 205, Eq. (63))

$$\mathbf{T} = J \mathbf{F}^{-1} \cdot \underline{\sigma} \quad or \quad T_{Ki} = J \frac{\partial \bar{X}_K}{\partial x_j} \sigma_{ji}. \tag{1.21}$$

These are the celebrated *Piola transformations*. With the mass conservation equation written as

$$\rho_0 = J\rho, \tag{1.22}$$

where ρ is the actual density at (x, y, z), Piola finally shows that Eq. (1.19) above renders the equation of motion (p. 212, Eq. (74))

$$\rho\left(\mathbf{f} - \frac{d^2\mathbf{x}}{dt^2}\right) + div_x\underline{\sigma} = \mathbf{0}. \tag{1.23}$$

This Piola identifies with the equation obtained by Cauchy (1828a, p. 166) or Poisson (1829, VIII, p. 387; X, p. 578). Accordingly, $\boldsymbol{\sigma}$ is none other than the *Cauchy stress tensor* for any continuum, whether solid or fluid, while **T** deserves to be called the *Piola stress* (*first Piola-Kirchhoff stress* in modern jargon). Equilibrium is obtained by making the acceleration term vanish in Eq. (1.23)—cf. Piola (1836, p. 215, Eq. (79)). Piola had thus succeeded to deduce the continuum Eq. (1.23) from a starting point that involved a corpuscular vision, this quite independently of other authors, and thus not a trivial achievement.

What is most original in Piola's strategy is that he has formulated what we now call the "Piola format" of the basic equations of continuum mechanics. His "format" involves two configurations with a preference for the referential one for the space parametrization. It is sometimes called the *material* formulation (cf. Maugin 2011) since **X** refers directly to the material "points" that belong to the "material manifold". The only inconvenience is the appearance of geometrical objects such as **F** and **T** that have two "feet" in different configurations and will later on be called *two-point tensor fields*—i.e., tensors depending on two "points"—by Einstein or, here precisely *double vectors*. But it must be understood that all computations are effected by Piola with all explicit scalar components of the introduced objects since he has no notion of a tensor (only introduced in the 1880s by Waldemar Voigt).

Another important innovative point brought in by Piola is that, independently of any microscopic concept, stresses can be viewed as a priori co-factors of virtual variations of strain components in a formulation of the principle of virtual work (Piola 1848). This idea may have found its origin in Piola's treatment of rigid bodies by accounting for a mathematical constraint that requires the introduction of a Lagrange multiplier (in fact a second order tensor in modern language). The introduced tensor Lagrange multipliers can be interpreted as "reaction internal forces" needed to maintain the rigidity of the body (Piola 1833). These internal forces are undetermined for a rigid body. In the case of a generally deformable body, the principle of virtual work can be written as (in modern notation)

$$\int_{V_0} \rho_0 \ddot{\mathbf{x}} \cdot \delta\mathbf{x}\, dV_0 = \int_{V_0} \rho_0 \mathbf{f} \cdot \delta\mathbf{x}\, dV_0 - \int_{V_0} S_{KL}\delta E_{KL} dV_0 + \int_{\partial V_0} \mathbf{t} \cdot \delta\mathbf{x}\, dS_0, \tag{1.24}$$

for a description in the referential configuration. Here E_{KL} are the Cartesian components of the finite strain defined by

$$E_{KL} = \frac{1}{2}\left[\sum_{i=1}^{3} F_{iK}F_{iL} - \delta_{KL}\right], \quad F_{iK} = \frac{\partial \bar{x}_i}{\partial X_K}, \tag{1.25}$$

while S_{KL} is the introduced cofactor, now called the second Piola-Kirchhoff stress. It is related to the Cauchy stress σ by the full Piola transformation

$$S_{KL} = J\frac{\partial X_K}{\partial x_i}\sigma_{ij}\frac{\partial X_L}{\partial x_j} \quad or \quad \mathbf{S} = J\mathbf{F}^{-1}\underline{\sigma}\mathbf{F}^{-T}, \tag{1.26}$$

so that (1.24) can also be replaced by the following formulation in the actual configuration:

$$\int_V \rho\ddot{\mathbf{x}}\cdot\delta\mathbf{x}dV = \int_V \rho\mathbf{f}\cdot\delta\mathbf{x}\,dV - \int_V \sigma_{ij}\delta e_{ij}dV + \int_{\partial V} \mathbf{t}\cdot\delta\mathbf{x}\,dS \tag{1.27}$$

Formulations of the type (1.24) and (1.27) will be adopted by most mechanicians on the Continent (France, Germany, and Italy). This is the case of Clebsch (1862), Kirchhoff (1876), Barré de Saint-Venant (1883), Poincaré (1892), Appell (1900), Hilbert (1906) and Hellinger (1914), and most Italian authors (e.g., Beltrami 1880–1882; Betti 1874). It is also intimately related to the renascent formulation of the principle of virtual power by Germain (1973). Advantages of such formulation are the easy rendering of associated natural boundary conditions, and generalization to all kinds of complex media including media with a mechanical or physical (e.g., electromagnetic) microstructure and the allied complex kinematics (cf. Maugin 1980, 2010).

1.2.7 Green's Energy Argument (1839)

The expression of the work of internal forces in (1.24) or (1.27) is tantamount to introducing a potential energy density for elasticity W per unit of referential volume, $W = W(\mathbf{E})$, such that

$$\delta\int_{V_0} W(\mathbf{E})dV_0 = \int_{V_0} \delta W(\mathbf{E})dV_0 = \int_{V_0} \mathbf{S}{:}\delta\mathbf{E}dV_0 \tag{1.28}$$

and

$$\mathbf{S} = \frac{\partial W}{\partial \mathbf{E}} \quad or \quad S_{KL} = \frac{\partial W}{\partial E_{KL}}. \tag{1.29}$$

This was proposed by George Green (1793–1841) in his study of the wave propagation in transparent elastic bodies and of the laws of reflection and refraction of light (Green 1839). This was applauded by W. Thomson (the future Lord Kelvin), an ardent propagandist of all ideas related to energy. With a formulation such as (1.27)–(1.28) all traces of a microscopic underlying description have disappeared if one postulates a reasonable expression for the energy density. The Cosserat brothers (Cosserat and Cosserat 1909) will exactly follow this line with a more complicated kinematics—such as involving internal degrees of freedom—of the considered medium.

1.2.8 Other Works: Lamé, Clapeyron, etc.

The corpuscular approach to elasticity was also pursued and exploited by Gabriel Lamé (1795–1870) and Benoît Clapeyron (1799–1864)—both members of the Corps of Mining engineers (formation: Polytechnique followed by School of Mines)—cf. Lamé (1852); Lamé and Clapeyron (1831).

It is more surprising that as late as 1892 Henri Poincaré (1892)—also originally a member of the Corps of Mining Engineers—in his lectures on elasticity avoids Cauchy's first theory, and bases his presentation on the fundamentals of the corpuscular theory. Indeed, in his Chapter II he introduces the notion of *elastic forces* by distinguishing between theories based on molecular hypotheses and those based on thermodynamics (in the manner of George Green) where the intimate constitution of matter is disregarded. Nowhere is Cauchy's fundamental notion of stresses (what we call CAUCHY-1) introduced. Poincaré here prefers to start with the *molecular approach* expanded by Poisson, Navier and CAUCHY-2. That is, bodies are made of molecules that are very small compared to their separating distances. They are acted upon by both "internal" and "external" forces. Internal forces are like Newtonian forces between molecules and are ultimately responsible for the macroscopically observed elasticity of the body. They are of the central type, and according to Lamé (1852), vanish at natural equilibrium (in the absence of external forces) with a vanishing attraction. Conservation of energy then implies that internal forces derive from a potential (p. 30), to be called the "force function". External forces may act on all molecules, while at the surfaces they may act on superficial molecules (whatever these be) only. In modern standard continuum mechanics the former are recognized as "body" forces and the latter as applied "tractions". Poincaré cites the example of a gas in a container. Like for Poincaré's predecessors, passing to the case of true deformations calls for an exercise in the Taylor expansion of the force functions in terms of the small displacements. The possibility of the existence of mathematical constraints connecting molecules is envisaged and taken care of by the introduction of Lagrange multipliers. This is more original. Approximation of the expansion of the force function requires the introduction of a small radius of molecular "activity" (interaction), yielding the neglect of far-distance mutual interactions. This, as usual, is the most disputable part of the procedure, in particular when dealing with molecules placed at the boundary of a body. The case of isotropic bodies indeed involves so-called isotropic polynomials[6] that must be linearly independent. The number of remaining elasticity coefficients in this approach remains a discussed matter as already discussed by Lamé (1852) and Clebsch (1862).

The equilibrium equations are then introduced in Chapter III by means of the principle of virtual velocities. The resulting bulk equations and boundary conditions are finally obtained (pp. 58–59) with identification of what are called "pressures" (in modern jargon, the components of a—Cauchy—*stress tensor*). It is only at this point (p. 62) that Poincaré considers as an alternate approach the energy viewpoint of Green (1839). But he improves on Green's presentation by using a thermodynamic argument involving entropy and free energy that, obviously, Green could not know.

We surmise that the choice of approach by Poincaré is based on his will to pursue the line opened by Fresnel and Cauchy regarding the propagation of light waves in the so-called "ether". As a matter of fact, Poincaré shows more interest in the application of elasticity to this field of physics than to potential mechanical-engineering applications. This is corroborated by his enthusiastic development of the mathematical theory of light (Poincaré 1889). But we must also account for the general state of physics at the end of the nineteenth century and the ideas that permeate the whole medium at that critical time (see next Section). An interesting aside is that Poisson continued in the late 1830s his pursuit of an ideal corpuscular model of elasticity. In this endeavour (Poisson 1842), he proposed to replace the points of his corpuscular approach by tiny rigid bodies capable of rotation as well as of translational motion. This prefigures some development in microstructured bodies such as by Cosserat and Cosserat (1909). Poisson's idea was worked out in detail by one of Poincaré's contemporaries, Voigt (1887, 1910), the inventor of the word "tensor" in its current meaning.

1.3 Action at a Distance, Electromagnetism and Crystal Dynamics

As mentioned before in this contribution, the possible complementarity between a corpuscular vision and a continuum description exploiting nice partial differential equations has been kept alive for a long time. James Clerk Maxwell (1831–1897) has described the fluctuating opinion of physicists about this matter in his deeply thought contribution to "Action at a distance" (Maxwell 1873). He was well placed to discuss this matter as he had shown in electromagnetism that a ponderomotive force (seen as an *at-a-distance force*) could also be expressed in the form of a

[6]The theory of isotropic polynomials (rotationally invariant quantities) was expanded by Poincaré in his lectures

contact action by means of the divergence of an appropriately defined stress (the "Maxwell stress"), i.e., symbolically,[7]

$$\mathbf{f}^{em} = div_x \underline{\sigma}^{em}. \tag{1.30}$$

This casts a doubt on the most appropriate physical interpretation. Here the notion of stress appears as *secondary*, being introduced via an identity. Note that this was already the case in the transformation from Eq. (1.12) into (1.13) in Piola's theory. But more was going to come with the dynamics of crystalline lattices.

In the third edition of his well known book on the mathematical theory of elasticity, A.E.H. Love (1863–1940) added a Note B on the notion of *stress*. There he clearly set forth the criteria that a corpuscular theory of elasticity must satisfy: First it ought to "provide a foundation for the notion of stress". Second, "it ought to lead to Hooke's law" as a paradigmatic linear constitutive equation. Third, "it ought to lead to the existence of a strain-energy function". Fourth, as a more technical detail, it ought to avoid the occurrence of the so-called Cauchy's relations[8] between elastic constants. Theories briefly presented in previous pages have to pass these four tests. We let the reader decide on which of the theories pass them.

He also clearly defined what the basic constituents of these theories are (pp. 617–618): e.g., centres of force endowed with the property of mass, with interparticle forces being directed along the line joining any two pairs of particles, forces being reciprocal, and decreasing markedly with distance, and perhaps vanishing beyond "a radius of the sphere of molecular activity". With these criteria and definitions, Love found that Cauchy's second theory (CAUCHY-2) and Poisson's one are none other than the ancestors[9] of the modern theory of *crystal lattices*, as developed by Max Born (1882–1970) and von Kármán (1881–1963) in the early 1900s (e.g., Born and von Kármán 1912; Born 1915; Born and Huang 1954). Remember that a crystal lattice can be defined by an ordered arrangement of "atomic" mass points with vector positions

$$x_i = x_{0i} + \sum_{\alpha=1}^{3} n_\alpha a_{\alpha i}, \tag{1.31}$$

where x_{0i} are the coordinates of one of the particle, the n_α are positive or negative integers, and the $a_{\alpha i}$ are the components of fixed constant vectors. This applies well to crystallized matter, but the isotropic case is recovered by supposing that in these materials the "crystallization is confused so that no part at all large compared with

[7]This we developed at length in our treatise on the electromagnetism of deformable solids (Maugin 1988, Chap. 3). In relation with these developments we note that Lorentz's (1916) theory of "electrons"—worked out in the late nineteenth century—is to start with a theory of point-wise electric charges interacting through their mutual distances in an otherwise vacuum space, and was therefore much to the taste of Poincaré (Cf. Fig. 1.12).

[8]These spurious relations reduce artificially the number of independent constants.

[9]This is not acknowledged in modern texts on lattice dynamics, e.g., Maradudin et al. (1963).

Fig. 1.12 Henrik A. Lorentz

molecular dimensions, forms a single homogeneous assemblage". In other words, in crystallized matter, the elements (particles) do form a homogeneous assemblage.

Love explains in great detail the two steps of the derivation, one concerning first the equilibrium of a particle, and the second accounting for the changes in the involved quantities when the body undergoes a small strain accompanied by a small rotation. Equilibrium equations of a particle can thus be obtained. On evaluating the coefficients of these strains in the equations, he then shows that they can indeed be written as the derivatives of a potential, from which one concludes with the existence of a strain-energy function. According to Love this should have been more or less obvious because "in a system of attracting and repelling particles, when the force between two particles is function of the distance between them, there must be a potential energy function, which depends on these distances only" (Love 1892, p. 627). This was already explicitly stated by Poincaré in his lectures where he pointed out that conservation of energy implies that internal forces $f(r)$ derive from a potential to be called the "force function" (Poincaré 1892, p. 30). Anyway, it is also proved that the spurious Cauchy's relations are not a necessary consequence of the considered structure theory which, therefore, satisfies the criteria he had enounced. We can conclude this point with the following remark. In a simple introduction to the lattice dynamics of Born and von Kármán, the interaction between neighboring particles is represented by restoring forces (derivatives of a potential) of the linear spring type. We can say that this is phenomenology at its best with the aphorism that "the cause of the effect is the effect", since macroscopic elasticity is deduced by an approximation procedure in which the microscopic departure point uses the concept of springiness. This may look quite naïve and for the least opportunistic, but remarkably enough, this proved to be a fruitful basis for devising more complex realistic models of continua in the late twentieth century (see, e.g., Askar 1986).

1.4 Conclusion

In spite of Cauchy's genial idea of his introduction of the stress concept in 1823–1828, the approach that considers first a corpuscular structure and at-a-distance forces survived through the whole nineteenth century. More than that, it has found its way in the twentieth and twenty-first centuries essentially through statistical mechanics and new computational techniques that eschew the notion of continuum. In the first line, we must mention the original kinetic theory of gases created by J.C. Maxwell and L. Boltzmann where the notions of distribution function and probability density appear for the first time. A model of interactions via collisions may be necessary to close the system in the presence of dissipation. This requires the study of a special equation known as the Boltzmann equation, and of its various "moments" so as to arrive at the notion of stress albeit as a secondary one. But closer to the already exposed notion of corpuscular description—but still with statistical considerations—we mention the statistical theory of liquid solutions by the physico-chemist John G. Kirkwood (1907–1959). This scientist really created a statistical physics yielding a liquid state theory in which the properties of liquids are obtained in terms of interactions between the constituent molecules (Kirkwood 1936). This approach involves a molecular distribution function. Along the same line we note the recent works of A. Ian Murdoch synthesized in Murdoch (2012).

In a second line, one should not forget the dream of computing all dynamical behaviors of materials by completely eschewing a continuum view and solving simultaneously the Newtonian equations of motion of an extremely large ensemble of individual particles, all interacting via known potentials of force. This ideal Laplacian view, but with more or less involved potentials, is practically becoming true with the huge computational possibilities offered by modern computers. This has been proved to be spectacular with, for instance, the simulation of the propagation of cracks and of other structural defects. This molecular dynamics approach (cf. Rapaport 1995) poses the fundamental question whether the very concept of continuum should not be rejected to the dark ages of phenomenological physics. But quid of the admirable concept of stress introduced by Cauchy in 1823–1828, which in the end remains the essential ingredient to decide on whether something will resist a load or start to flow in a more or less dramatic way? From a purely epistemological viewpoint there is no need for a continuum theory to be substantiated by an underlying corpuscular theory. This is well accepted by contemporary applied mathematicians—great amateurs of partial differential equations—and engineers alike without special pondering of the matter. We can think with Pierre Duhem that the continuum theory provides a good representation of phenomena and that its efficiency is what counts above all other argument. It does not provide any explanation, avoiding any unnecessary microscopic modelling. But what we estimate to be a true physicist cannot help but compare the two approaches, corpuscular and continuous, and try to reconcile these two in some irresistible move. This is the justification for the existence of the present contribution.

References

Appell P (1900) Traité de mécanique rationnelle, vol 3, 1st edn. Gauthier-Villars, Paris

Askar A (1986) Lattice dynamical foundations of continuum theories. World Scientific, Singapore

Barré de Saint-Venant AJC (1883) Théorie des corps élastiques (French translation from the German with many comments and additions of "Clebsch A (1862) Theorie der Elastizität fester Körper. Leipzig" (Translation with the help of A. A. Flamant); Reprint by Johnson Reprint Corp. New York, 1966

Betti E (1874) Teoria dell'elasticità. Soldaini, Pisa

Belhoste B (1991) Augustin-Louis cauchy: a biography. Springer, New York (English translation from the French original "Cauchy, 1789–1857". Editions Belin, Paris)

Beltrami E (1880–1882) Sulle equazioni generali della elasticità. In: Opere matematiche, vol 3, pp 383, 407. Hoepli, Milano (1902–1920)

Born M (1915) Dynamik der Kristallgitter. Teubner, Leipzig and Berlin

Born M, von Kármán T (1912) On fluctuations in spatial grids. Physik Zeit 13:297–309 [Also: On the distribution of natural vibrations of spatial lattices, Zeit Phys 14:65–71 (1913)]

Born M, Huang K (1954) Dynamical theory of crystal lattices. Oxford University Press, London

Cauchy AL (1823) Recherches sur l'équilibre et le mouvement intérieur des corps solides ou fluides, élastiques ou non élastiques. Bull Soc Filomat Paris, January issue, 9–13 [This is a kind of long abstract of a work that will materialize in the next listed publication]

Cauchy AL (1828a) Sur les équations qui expriment l'équilibre ou les lois du mouvement intérieur d'un corps solide élastique ou non élastique. Exercices de mathématiques, vol 2, pp 160–187, Sept 1828 (This presents in print the ideas originally submitted to the Paris Academy of Sciences on September 30, 1822)

Cauchy AL (1828b) In: Exercices de Mathématiques, vol 3, Sept 1828, pp 188–212, and «De la pression ou tension dans un système de points matériels», Oct 1828, pp 213–236

Cauchy AL (1828c) Sur l'équilibre et le mouvement d'un système de points matériels sollicités par des forces d'attraction ou de répulsion mutuelle. In: Exercises, vol 3, or Œuvres complètes de Cauchy, 2nd series, vol VIII, p 227

Clebsch RFA (1862) Theorie der Elastizität fester Körper. Teubner, Leipzig (French translation with many additions and comments by A. Barré de Saint-Venant, Paris, 1886)

Cosserat E, Cosserat F (1909) Théorie des corps déformables. Hermann, Paris (Facsimile reprint by Gabay, Paris, 2009)

Germain P (1973) La méthode des puissances virtuelles en mécanique des milieux continus-I Théorie du second gradient. J Mécanique (Paris) 12:223–274

Girard PS (1798) Traité de la résistance des solides. Firmin Didot, Paris, An VI of First French Republic, 238 p

Green G (1839) On the laws of reflection and refraction of light at the common surface of two non-crystallized media. Trans Cambridge Phil Soc 7:245–269

Hellinger E (1914) Die allgemein Ansätze der Mechanik der Kontinua. In: Klein F, Wagner K (eds) Enz. Math. Wiss., vol 4, Part 4. Springer, Berlin, pp 602–694

Hilbert D (1906–1907) Vorlesung über die Mechanik der Kontinua (Lectures on the mechanics of continua), Gehalten in Wintersemester 1906–907, zu Göttingen (Mimeographed, 239 pages; with an introduction (in German) by J. Stickforth, TU Braunschweig, 1970)

Kirchhoff G (1876) Vorlesungen über mathematische Physik: Mechanik (known as "Lectures on Mechanics"). Teubner, Leipzig [available on the site Gallica of the Bibliothèque Nationale de France (BNF)]

Kirkwood JG (1936) Statistical mechanics of liquid solutions. Chem Rev 19:275–307

Lamé G (1852) Leçons sur la théorie mathématique de l'élasticité des corps solides. Paris, (2nd edn, 1866; Facsimile reprint, Gabay, Paris, 2006)

Lamé G, Clapeyron E (1831) Mémoire sur l'équilibre intérieur des corps solides Homogènes. Journal de Crelle (J. für reine und angewandt Mathematik), vol VII (Also Mémoires de Savants étrangers, Acad Sci Paris, vol IV, pp 463–562, 1833; available on the website of the BNF, Paris)

Laplace PS (1805) Traité de mécanique céleste, vol 5. Paris (An XIII of the First Republic), Chez Courcier
Laplace PS (1819) Considérations sur la théorie des phénomènes capillaires. J Physique 89:259–264 (Also: Ann Chim et Phys 12, 1819)
Lorentz HA (1916) The theory of electrons. Teubner, Leipzig and Berlin
Love AEH (1892) A Treatise on the mathematical theory of elasticity. Two volumes. Cambridge University Press, UK (1944, 4th edn, Reprint in one volume. Dover, New York)
Maradudin AA, Montroll EW, Weiss GH (1963) Theory of lattice dynamics in the harmonic approximation. Academic Press, New York
Maugin GA (1980) The method of virtual power in continuum mechanics: application to coupled fields. Acta Mech 35:1–70
Maugin GA (1988) Continuum mechanics of electromagnetic solids. North-Holland, Amsterdam
Maugin GA (2011) Configurational forces. CRC/Chapman & Hall/Taylor and Francis, Boca Raton/New York
Maugin GA (2012) The principle of virtual power: from eliminating metaphysical forces to providing an efficient modelling tool. Cont Mech Thermodynam 25:127–146 (Special issue in the honor of G. Del Piero)
Maugin GA (2014a) Continuum mechanics through the eighteenth and nineteenth centuries: historical perspectives from John Bernoulli (1727) to Ernst Hellinger (1914). Springer, Dordrecht
Maugin GA (2014b) Poincaré and Hilbert on continuum mechanics, Preprint UPMC, Paris (Published as Chap. 6 in the present book)
Maxwell JC (1873) Action at a distance. In: Niven WN (ed) Proceedings of the Royal Institution of Great Britain, vol VII, 11 p (1876). Reprinted in Maxwell's Scientific papers, vol 2, pp 311–322, 1890. Reprint by Dover, New York, 1963
Murdoch IA (2012) Physical foundations of continuum mechanics. Cambridge University Press, UK
Navier CLMH (1824) Mémoire sur les lois de l'équilibre et du mouvement des corps solides élastiques, Presented to the Academy on May 14, 1821 (but printed in the Mém Inst Natl. 1824, with an excerpt published in Bull Soc Philomatique Paris, 1823, p. 177)
Piola G (1833) La meccanica de'corpi naturalmente estesi tratta col calculo delle variazioni. Opusculi matematici e fisici di diversi autori. Giusti, Milan, pp 201–236
Piola G (1836) Nuova analisi per tutte le questioni della meccanica moleculare. Mem Mat Fis Soc Ital Modena 21(1835):155–321
Piola G (1848) Intorno alle equazioni fondamentali del movimento di corpi qualsivoglioni considerati secondo la naturale loro forma e costituva. Mem Mat Fiz Soc Ital Modena 24(1):1–186
Poincaré H (1889) In: Blondin J, Lamotte M, Hurmuzescu D (eds) Leçons sur la théorie mathématique de la lumière. George Carré, Paris
Poincaré H (1892) Leçons sur la théorie de l'élasticité, Georges Carré, Paris (Facsimile reprint, Gabay, 1990; also University of Michigan, Ann Arbor, 2012)
Poisson SD (1812) In: Mémoires Acad for 1812, pp 167–226
Poisson SD (1829) Mémoire sur l'équilibre et le mouvement de corps élastiques. Mémoires de l'Institut de France, vol VIII:357–570 (and vol X, p 578)
Poisson SD (1833) Traité de mécanique, vol II. Bachelier, Paris
Poisson SD (1842) Mémoire sur l'équilibre et le mouvement des corps cristallins. In: Mém. de l'Académie, vol 18. Paris (submitted in 1839 but published posthumously in 1842)
Rapaport DC (1995) The art of molecular dynamics simulation. Cambridge University Press, NewYork
Szabó I (1977) Geschichte der mechanischen Prinzipien. Birkhauser, Basel
Timoshenko SP (1953) History of the strength of materials, McGraw Hill, New York (Dover reprint, New York, 1983)

Todhunter I (1886) A history of the theory of elasticity and the strength of materials from Galileo to the present time, vol 1, Cambridge University Press, UK (edited and published posthumously in 1886 by Karl Pearson)
Voigt W (1887) Theorie des Lichts für bewegte Mediem. Göttinger Nachrichten 8:177–238
Voigt W (1910) Lehrbuch der Kristallphysik. Leipzig and Berlin (see Chap. VII, Sect. III]

Chapter 2
Hydraulics: The Importance of Observations and Experiments

Abstract Before the expansion of analysis, studies of fluid mechanics were necessarily confined to experiments although a complete body of technical applications of hydraulics had developed starting with the most ancient civilizations around the Mediterranean Sea and in Asia. Greeks, Egyptians and Romans, in particular, conceived all kinds of elementary machines but also well designed systems of distribution of water, whether in the fields or in the supply of cities. The true experimental aspects concerning this matter had to await the birth of the modern scientific spirit of the Renaissance period. The notion of pressure, the most elementary form of stress, appears with experiments performed by Torricelli, Pascal, and Mariotte. With these scientists and Newton, the first formulas appear, some just reporting pure empirical results, others founded on some rational reasoning. In the eighteenth century one witnesses parallel developments in the works of mathematicians (mostly, Clairaut, the Bernoullis, d'Alembert, Euler and Lagrange) and in the careful experiments of a group of gifted engineers (e.g., Borda, Bossut, du Buat,..). The role played by viscosity, but still in laminar flows, will be best captured by Poiseuille, Hagen, and others in experiments, and obviously by Navier, Saint-Venant and Stokes in analysis, before the consideration of turbulence. Accordingly, this contribution places the emphasis on experiments that were decisive in the perused evolution of continuum mechanics, from ancient times to the nineteenth century.

2.1 Introduction

Our personal vision of continuum mechanics is essentially theoretical and mathematical. It must be realized that before the works of our grand predecessors (the Bernoullis, d'Alembert, Euler, Lagrange, Cauchy, etc.) in the eighteenth century and early nineteenth century,[1] continuum mechanics may be identified in

[1]The reader may consult our book (Maugin 2014) for the history of that period.

G.A. Maugin, *Continuum Mechanics through the Ages - From the Renaissance to the Twentieth Century*, Solid Mechanics and Its Applications 223,
DOI 10.1007/978-3-319-26593-3_2

experimental developments and engineering material realizations, for instance, in so far as the statics and flow of some fluids are concerned. The first natural fluid of this kind was obviously **water**, from which the word **hydraulics** naturally follows from the combination of two Greek words, "hydra" (for water) and "aulos" (for pipe). This sounds very much like the flow of water in conduits, channels, and rivers that all ancient Egyptians, Mesopotamians, Persians, Greeks, Romans and other people around the globe had to consider of necessity in practical applications. The choice of water as the main object of study was particularly salient because it is a good example of a practically incompressible fluid that can transmit without much loss what will become known as the *pressure*, a force per unit area, and obviously the simplest form of our modern notion of stress. Thus hydraulics in its most elementary form certainly is the backbone of the fruitful development of notions that provided the basic ingredients in the theory of hydrodynamics and then continuum mechanics. Per force, very few relevant mathematical expressions were formulated before the eighteenth century, i.e. before sufficient expansion of mathematical analysis.

In accordance with its basic etymology, hydraulics is usually characterized as dealing with the physical behavior of water at rest or in motion, while the *Encyclopaedia Britannica* is more severe in speaking about "the practical applications of liquids in motion" and thus puts the emphasis on an applied field that for a long time was considered a part of civil engineering. Engineering and experimental sides cannot be overlooked as they provided a support for all inclusive ideas that developed in time between the ancient Greeks and the eighteenth century. Here we need not go back in detail to these ancient times for which there exist nice histories and reviews (e.g., Mays 2008). All the reader will retain are expressions like running water, water power, pumps, hydraulic machines, water supply to cities, etc., and the role of luminaries such as Archimedes (287–212 BC) in hydrostatics. What is really surprising to our modern scientific mind is that contrary to our common view of the development of a science, hydraulics first went through technological applications, then much later through a phase of experiments, and finally reached a theoretical level in hydrodynamics or, more generally, fluid dynamics. Thus, after a rapid perusal of technological developments where one cannot totally forget the material realizations of the Romans and of the Hellenistic period of which we inherited, one then usually jumps directly to the Renaissance period with true experiments that mark the birth of the practice of modern science.

2.2 Ancient Times: Hydraulic Technology

In ancient times, with the development of sedentary civilization and the associated agriculture, and the advent of some kind of urbanism, needs appeared to control the distribution of water to the fields in the form of organized irrigation, as also to bring water to cities. In both cases water had to be directed from some natural spring or a river to a specific place, usually at a lower level, and to bring it in the best conditions of delivery without too much loss on the way. Greeks and especially Romans—whom we consider the inventors of civil engineering—developed a special taste for the related technology, having determined the best conditions to have a descending flow, creating machines if necessary to take water to a higher elevation, and possibly devising true mega hydraulic projects for which the remnant Roman aqueducts in Italy, Spain and France provide the most spectacular examples. Cisterns and other means of water collecting and containment, the design of drainage channels, the construct of public fountains and the invention of animated fountains all are illustrations of this vivid development. Among the most impressive oldest inventions we cannot avoid citing the water screw by Archimedes (287–212 BC) in Syracuse (the founder of hydrostatics with his principle of buoyancy), and the force pump (or water lifting device) by Ctesibius or Ktesibios (circa 270 BC) in Ptolemaic Alexandria, Egypt, and later adapted by Roman engineers. Ctesibius is also supposed to have invented instruments using cleverly the properties of flowing water such as in hydraulic clocks and some musical instruments. All these are well documented in the text of Mays (2008) to whom we refer the reader. Of course we cannot say that these technical achievements came out of the blue. Although we cannot identify here any true scientific method, we must assume that some technical solutions—which in modern terms would involve the notions of pressure, flux through orifices, and certainly a bit of friction—were found only after a long series of trials and errors, a kind of archaic precursor of the scientific method. The Greek, Hellenistic, and Roman engineers greatly influenced those of the Renaissance (cf. Gille 1964, First two chapters), in particular via Vitruvius' teaching (cf. Vitruvius ca 25 BC).

2.3 The Renaissance Experimentalists-Thinkers: Leonardo, Stevin, Galileo Galilei

With the Renaissance period that we grossly delineate by the dates 1453 (fall of Constantinople)—or the invention of printing by Gutenberg circa 1450—and the introduction of the Baroque style in arts or the successful career of Shakespeare in drama or still the death of Galileo Galilei in 1642, we witness a renewed interest in the antic culture and a characterization by the reading and re-evaluation of fundamental texts. This applies well to both science and technology. For us, the symmetric, perfectly geometrically proportionate, representation (cf. Fig. 2.1) of Vitruvius' man by Leonardo da Vinci (1452–1519) is an ideal that directly connects with the preceding section. Indeed, Leonardo is an unavoidable figure including in our present subject of interest, hydraulics. At this point, we must mention the thorough studies of Duhem (1906–1912) on Leonardo. By studying in depth so many original texts from the Middle Ages (from the masters at the Sorbonne or in Oxford, etc.) and the immediate Pre-renaissance Italian scientists, Duhem tried, and sometimes seems to have succeeded, to trace what may have influenced Leonardo,

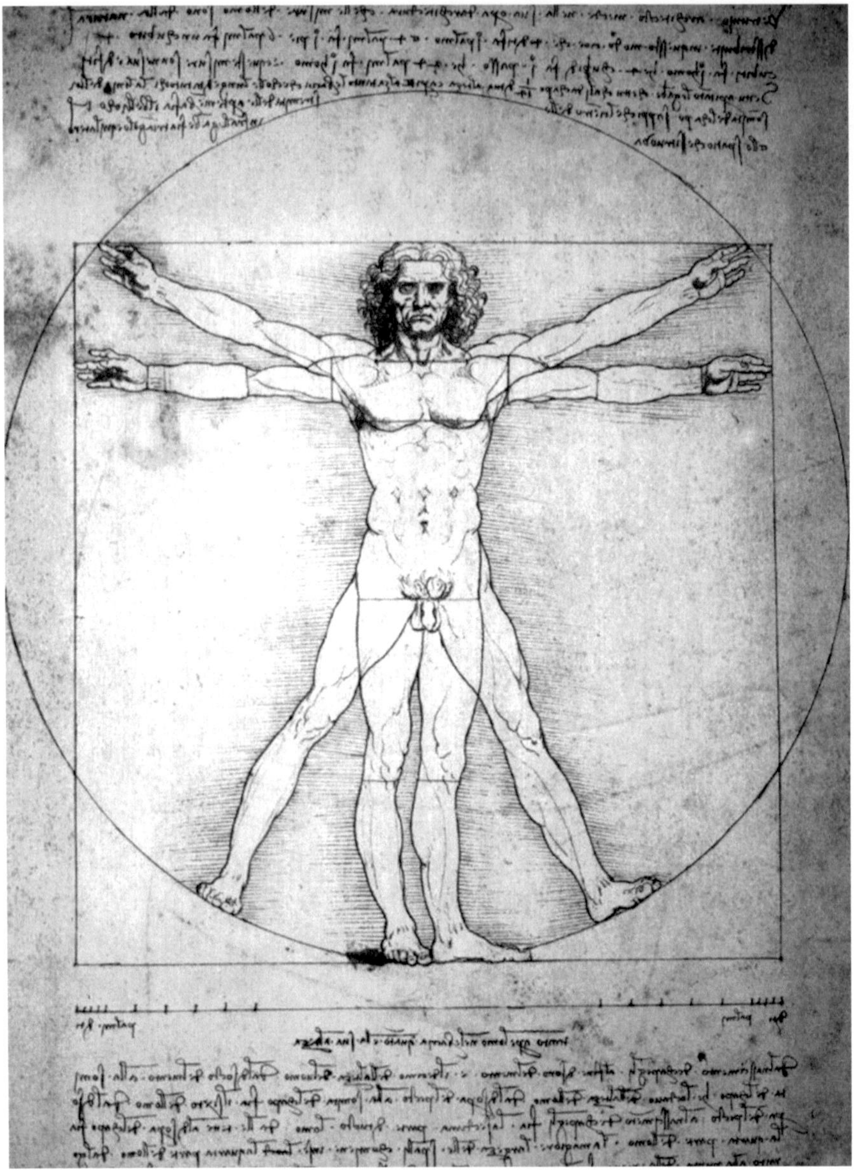

Fig. 2.1 Vitruvius' man by Leonardo da Vinci

among his predecessors in the Middle Ages, and what he may have read, although Leonardo himself does not provide any clues. This has remained a discussed matter among historians of science (including A. Koiré, Th. Kuhn, etc., who are in favour of a clear discontinuous evolution of science such as in the change of paradigm, while Duhem leans toward a greater continuity, from what his thesis follows).

In the present case, Leonardo is not a scientist in the modern sense of the word, but he is a great observer of Nature and a true engineer's mind. We can say with Truesdell (1956, p. 4), that Leonardo is "an observer of the undisturbed nature". His graphic talents are here extremely useful because a good accurate picture can replace many words. Whether his drawings represent a report on a true (well designed) experiment or just a visual observation is not clear, although the second avenue is more likely. But Leonardo, like some Japanese artists of woodprints (e.g., Hokusai) in the eighteenth and nineteenth centuries, captures very well the intricacy of some flows with vortices and fantastic waves. But this is too involved, and one must start with some more quiet situations, as a matter of fact, *hydrostatics*. In so far as theoretical thinking is concerned, Leonardo seems to be somewhat limited, but we must attribute to him some kind of recognition of a *principle of continuity* in a simple form (cf. Truesdell 1956, p. 6). His explanation for the fact that fluids such as water at the top of mountains or blood in the human head can go all this way up is erroneous (cf. comments by Dugas 1950, p. 80). But his vision of the flux of a fluid is more appropriate according to Duhem (1906–1912, First volume, p. 198). In effect, in his "Del moto e misura dell'acqua" Leonardo states that "all motion of water of uniform width and surface will be faster in one place than in another one all the more that this water is less deep in one case than in the other". He also sketches out a theory of hydraulic pumps that, in some sense, prefigures Pascal's findings. So much for Leonardo's accomplishments that we do not want to belittle in any way.

We may consider that Leonardo is at the turning point of the birth of a true modern science. For pedagogical purpose this birth is often attributed to a few individuals such as Francis Bacon (1581–1626), a developer of an empirical theory of knowledge with a specification of rules in the experimental method (See his "Novum Organum", 1620), and Galileo Galilei (1564–1642), a true mathematician with a taste for experiments. But Simon Stevin (1548–1620), a Flemish all round scientist (mathematician, mechanician, engineer and physicist, much concerned with the basic principles of mechanics and the non existence of perpetuum mobile—perpetual motion), is also to be considered. Stevin brought to the science of hydrostatics two fundamental ideas. One is the *principle of solidification* according to which a solid body of any shape but with the same density as water can stay in equilibrium in water in any position without altering in any amount the pressure field in the rest of the fluid. This helped him to determine the pressure exerted on each element of the base of the container by solidifying thus the whole fluid save a fine channel leading from the open surface to this element. As a result the pressure is independent of the shape of the container; it depends only on the weight of the column that fills up the channel. Another consequence of this is the so-called *hydrostatic paradox*. According to this paradox, the total force—resultant sum of pressures—that a fluid can exert on the bottom of a vase can be considerably larger than the weight of the same fluid. Another result of Stevin deals with the resultant of pressures acting on an inclined plane. He used here a reasoning that involves an astute limit procedure by cutting the surface of this plane by a succession of small horizontal slices and then increasing indefinitely the number of slices. He finally related Archimedes' principle to the impossibility of a perpetual motion. He is a direct predecessor of Galileo Galilei although his publications in Flemish—a language of which he was an ardent propagandist—may have been missed by Galileo.

Galileo Galilei is a true mathematician in that he does provide formulas based on mathematical reasoning. He is also a true experimentalist because he performed experiments that were purposefully designed. He is in quest of checking a hypothesis but may also arrive at an unexpected result. This is the quintessence of Francis Bacon's advised method. But regarding hydrostatics, he first borrows elements to Aristotles' mechanics. According to his "Discorso intorno alle cose che stanno in su l'acqua o che in quella si movimento" (Galileo 1612) one such principle is that "two weights moving with equal velocities are endowed with the same power or *momento* in all their operations" (Dugas 1950, p. 138), a statement in which we can see an application of the principle of virtual velocities as conceived by Aristotle. A second principle along the same line is that "the power of gravitation grows with the velocity of the moving object". Accordingly, absolutely equal weights moving at different velocities have unequal *momenti*: the most "powerful" one is that one that moves faster than the other, and this in proportion of its velocity compared to the velocity of the other one. From this one deduces, following Aristotle, that two weights of unequal magnitude can balance reciprocally one another and will be endowed with equal *momenti* each time that their "gravity" will be in the inverse ratio of the velocity of their motion. Galileo applies this reasoning to the siphon where a small quantity of water contained in a narrow vase can equilibrate a large mass of fluid in a wide vase as a small decrease in height in the second vase will correspond to a large level increase in the former. If we are to believe Duhem (1906–1912, Volume 2, p. 214), this prefigures Pascal's vision (see below). Galileo also brings the properties of floating bodies to an application of the principle of virtual velocities. It is noted by Dugas (p. 139) that Galileo still believed in the "horror of vacuum" ("resistenza del vacuo"). But he is pragmatic enough to accept the evidence that a newly developed pump cannot elevate water higher than a certain limit, thus in contradiction with this "horror of vacuum".

2.4 Seventeenth Century Experiments: Torricelli, Pascal, Mariotte

If we keep with our own chronological delineation of the Renaissance, then we must admit that Evangelisto Toricelli (1608–1647) worked at the final period of this prestigious epoch. Although interested in all things mechanical, Torricelli left an indelible trace in the history of hydraulics with his law of fluid flow ("efflux") through a small orifice placed at the bottom of a vase (container, tank). Exploiting an analogy with the fall of heavy objects, he observes that the coming out water cannot by itself rise again at an elevation larger than the one of the fluid in the vase;

while this elevation would be the same if there were no resistance at all in the system. As a conclusion he formulates—but does not prove—his famous law according to which the velocity of the flowing water at the orifice goes like the square root of the height of the fluid in the vase. Both Newton and Varignon (1725) will have to ponder this matter again. But Torricelli's law acted as a true motive power in further developments in hydraulics. Note that Torricelli improved on others' experiments by using mercury (known as quicksilver in his time) instead of water.

The next scientist in our gallery of distinguished "hydraulicians" certainly is Pascal. Blaise Pascal (1623–1662) is a French philosopher and religious thinker. His interests cover a wide spectrum both in humanities and sciences. He may be considered a somewhat unorthodox catholic intellectual,[2] but also an excellent geometer, an inventor (of a computing machine), and an amateur physicist who gave his name to a unit of pressure, the "pascal" (one newton per square meter). It is this last point that requires our attention. The story is told in all books for French teenager apprentice scientists in order to encourage them to engage in a scientific career. Pascal repeated Torricelli's experience in 1644 in Rouen, Normandy, with a friend called Petit. Then convinced that Nature does not have horror of vacuum, he projected in 1647 an experiment that he indeed conducted on September 19, 1648 at the Puy-de-Dôme mount in central France. This experiment proved that, in Pascal's words, "fluids weigh more given their height" since he observed a change in the level of mercury in a vertical tube depending on the altitude at which the experiment was conducted. The result was printed in October 1648 while Pascal's "Traité de l'équilibre des liqueurs" (Treatise on the equilibrium of liquors") was published in 1663. In other words, he demonstrated the variation of atmospheric pressure with altitude. A corollary of this result was the first well established definition of *pressure*. "A vessel full of water is a new machine to multiply a force to any desired degree" since a fluid in a small vertical tube connected to a much larger reservoir can balance the large amount of fluid in the corresponding height in the reservoir. We face here the observation that the cause of equilibrium in all examples treated by Pascal stems from the fact that the "matter that extends from the bottom of a vessel up to its open surface is indeed liquid". This shows the possibility of incompressible fluids such as water to transmit pressure without loss. This is usually referred to as Pascal's principle. It contains the basic idea for the conception of a *hydraulic press*, what was indeed industrially realized by Samuel Bramah (1748–1814), a British inventor, with the deposit of a patent for such a machine in

---

[2]Pascal is often presented as an opportunist and pragmatic—within the framework of apologetics—who considered a surprising probability reasoning (Pascal was interested in gambling games) and dared to bet on the existence of God (the celebrated "Pascal's wager"): In short, we can say that it is "safer" to believe in the existence of God than in his nonexistence; the odds are better this way since there is nothing to lose in making this bet. In truth this "bet" is best understood as an expression of the mathematical expectation, i.e. the product of gain and probability (a small, nonzero probability can produce a large expectation). The main writings of Pascal in both science and philosophy are collected in his "*Pensées*" (Thoughts). The author was offered a reprint copy of this thick book by his elder brother when he was in high school. No need to say that this was very difficult reading that went much beyond his understanding.

1795. Indeed, the press has two cylinders and pistons of different cross-sectional areas. If a force is exerted on the smaller piston, this will be translated into a large force on the larger piston. The difference in the two forces will be proportional to the difference in area of the two pistons (remember that pressure equals force per unit area). In effect, the cylinders act in a similar way that a lever is used to increase an exerted force. This sounds very much like another statement of the principle of virtual displacements. In modern technology one often meets the word "oil-hydraulics", what simply means that oil has replaced water in the process (although the Greek root for "water" has remained in the word!).

The Abbé Edmé Mariotte (1620–1684) appears to be a true experimentalist who dealt with solids, fluids and gases. He published on the laws of shocks of bodies (See his "Traité sur la percussion des corps") as well as on the flow of fluids (see his "Traité du mouvement des eaux" 1684). As a member of a committee in charge of checking Torricelli's law, he indeed checked this law, but missed the constriction of the emerging flow. This will be corrected by Newton in a second edition of his *Principia*. Mariotte also showed the importance of the deviation of the momentum in the experience of the impinging of a flow on a flat surface. He is a pioneer in considering size effects and introducing thus an embryonic form of a law of similitude in the study of the resistance of fluids where he formulates proportionality between resistance and the squared velocity. He also looked at the velocity of a flow in channels or rivers, and envisaged the action of friction in the flow of water in pipes. These achievements, of course, may appear modest compared to Mariotte's empirical determination of the famous Boyle-Mariotte law of the elasticity of ideal gases according to which the volume of such a gas varies like the inverse of the pressure at mass and temperature kept constant. Mariotte's result was obtained in 1676 but Robert Boyle (1627–1691) had proposed the same law earlier in 1662. This empirical result will later on be justified by the kinetic theory of gases.

Of course Isaac Newton (1643–1727) could not have left unexplored any branch of mechanics and physics of his time. No wonder, therefore, that he paid some attention to flows and undertook to validate Torricelli's law. In his interpretation of the free fall of water, each particle acquires its velocity in free fall so that Torricelli's law is verified. But his solution concerning a vertical vase with an orifice at its base contradicts the principles of hydrostatics (Dugas 1950, p. 216). It is Pierre Varignon (1654–1722), a French Jesuit and mathematician, better known for his proposal of composition of forces and his analytic definition of velocity and acceleration, who gave a more natural explanation of Torricelli's law (Varignon 1725). According to this explanation recalled by Dugas (1950, pp. 216–217), each fluid particle receives instantaneously a finite velocity at the exit of the vase through the effect of the column of water above it. Accounting for the flux, the so generated quantity of motion is proportional to the square of the velocity. But the weight of the fluid column is proportional to its height h, from which there follows the law $h \approx v^2$.

2.5 Eighteenth Century Theoreticians: Clairaut, Daniel Bernoulli, D'Alembert, Euler[3]

With the post-Newtonian and Leibnizian mathematical achievements, a new era starts that will harmoniously mix analysis and experiments, perhaps with an advantage for the former that we examine first.

Although much less known in continuum mechanics than in mathematics, Alexis Clairaut (1713–1765)[4] contributed to hydrostatics through his marked interest in geophysics and the equilibrium figure of the earth in particular (see his book: Clairaut 1743). Of necessity he had to envisage the equilibrium of water volumes under the action of the earth's gravity. The equilibrium will decide on whether the earth is a perfect sphere or is more or less flattened at the poles (in agreement with Newton) since there existed conflicting theories about the actual shape of the earth. In looking at this problem, Clairaut applied the strategy proposed by Stevin of the solidification of fluid volumes, considered fluid masses in motion (rotation) and practically introduced the notion of *potential* in mathematics (cf. Dugas 1950, p. 273). As a matter of fact, the general purpose of Clairaut's research was to find the laws of hydrostatics that are in accord with all hypotheses on gravity (work of 1734). Lagrange (1788, p. 128) emphatically said that Clairaut had "changed the face of hydrostatics to create out of it a new physical theory". The equilibrium principle enounced by Clairaut that provides a necessary and sufficient condition is that "a fluid mass can be in equilibrium only in so far as all parts that are contained in a channel of any form that traverses the whole mass, do not destruct each other" (Clairaut 1743, p. 1). From the application of this almost trivial principle, Clairaut deduces that the total weight of a column of fluid of constant section and density extending from the center of the earth to the equator or to a pole is none other than the "effort" on the column. This is obtained by integration over the height of the column. Applied to a short element of a channel this yields that the gradient of the "pressure" is balanced by the components of gravity. This result must be independent of the precise form of the channel, from which it follows that the spatial variation of pressure should be an exact differential. In a rotating mass, this transforms into the fact that the sum of the potential of gravitation—without specific hypothesis on its expression—and of the potential of the centrifugal force is a constant on a level surface. The vanishing of the variation of this potential yields the surface of equilibrium that is more than often perpendicular to the action of gravity as suggested earlier by Huygens. The spheroid itself will be such a level surface. On using the Newtonian gravitation law this indeed provides the flattening of the earth at the poles.

---

[3]The pioneering works on the foundations of continuum mechanics in the eighteenth century were surveyed in our book (Maugin 2014).

[4]For the life of Clairaut, see Brunet (1952). For Clairaut's original contribution to hydrostatics, see Passeron (1994) and the corresponding extract of 2009. The analysis of Dugas (1950, Chap. 7) is also very informative.

The expedition to Laponia led by Maupertuis—but supported by Clairaut—in 1736–1737 proved that a degree of an arc of meridian is larger at the north pole than in France, so that the hypothesis supported by Newton and Clairaut seems to be correct, but with a flattening of the order of 1/300 th only, i.e., much below Newton's estimate. This will be corroborated by more recent precise measurements. Clairaut, however, concluded from the discrepancy with Newton' proposal that the earth is probably inhomogeneous with a density increasing with distance from the centre. It is of interest to note in conclusion that Lagrange (1788, p. 129) finds in Clairaut's analysis the idea that pressure is the same in all directions, i.e., is isotropic although it is Euler who really formulated this essential property.

Practically contemporary of Clairaut we find Bernoulli (1700–1782) whose name remains for ever attached to a famous theorem which in modern standard notation and for a stationary flow reads

$$p + \frac{1}{2}\rho v^2 + \rho g h = const. \tag{2.1}$$

Here we have reinstated the pressure p that Bernoulli did not conceived as such. No need to emphasize the tremendous role played by this equation in all further developments in theoretical fluid mechanics, simple applications, and aeronautics. As noticed before (Maugin 2014, p. 10), this was not the only result of Daniel who produced a fundamental book on hydrodynamics written in 1734 (Hydrodynamica 1738). To the bewilderment of many, this scientific achievement by his own son caused a burst of envy and jealousy from the shameless John (Johann) Bernoulli (1667–1748). He hastily published a competing book with the title *Hydraulica* in (1739). He even anti-dated the date of writing this opus to 1732 to pretend to a (false) priority! But, to the credit of John, we must acknowledge that his book also had many merits. In particular, it presented the first successful use of the *balance of forces* to determine the motion of a deformable body. This was possible for John because he had recognized that "the fluid on each side of an infinitesimal slice pressed normally upon that slice, with a varying force which was itself a major unknown" (Truesdell 1968, p. 121). With this we are very close to the notion of *internal pressure* and a concrete view of *contiguity* of action in continuum mechanics in a line that both Euler in the period 1749–1752 and Cauchy in 1823–1828 will expand. Also, John was the first to practically give the above modern form to his son's theorem. As to Daniel, we must also note that he combined the notions of "pressure" and motion, being guided by the conservation of living forces in the sense of Leibniz. He also considered a motion by slices that are perpendicular to the direction of motion, all particles in the same slice having the same velocity that is inversely proportional to the considered section. This is very close to our modern view of the flow problem in a tube. Daniel again discussed the principle of living forces at the time of the death of his father (cf. D. Bernoulli 1748), a principle to which he was strongly committed.

The great Leonhard Euler (1707–1783), perhaps more than the Bernoullis, is to be credited for many definitive results in hydraulics and continuum mechanics in a general manner. This, of course, is the viewpoint of Truesdell (1984, pp. 212–217) whose real god in mechanics is none other than Euler. As emphasized earlier, *pressure* is the main constructive concept in hydraulics. Nobody did more to clarify this notion than Euler for whom, much influenced by the recent progress in the theory of hydraulics by the Bernoullis, father and son, *pressure* can be seen as the action "*from all sides* and *from neighbouring elements of fluid on an isolated element of fluid*" (a "particle"). In modern terms, it is *isotropic* and, with Euler, will be viewed as normal force acting on an element of surface. This means that Euler has reached the idea of pressure in a quasi-modern sense—while Lagrange attributed it to Clairaut-, and that the notion of *contiguity* is thus definitely formulated. Furthermore, pressure becomes a true *field* that depends on both space and time in the general case of dynamics. With the additional inception by (Euler 1757) of the balance equations in Eulerian format, we have then at hand the basic formulation of the field theory of perfect fluids. This brief discussion on fluids may make the reader believe that Euler had no notion of a tangential force. But this is not true because Euler himself dealt with this notion in a problem of solid mechanics—the elastica—that had been examined by James (Jacob) Bernoulli a long time before. The high value of Daniel Bernoulli's and Euler's seminal works in hydraulics is emphasized by Truesdell (1956). Daniel conducted some experiments in parallel with his mathematical approach, while Euler was for ever a pure theoretician.

In referring to Euler we have bypassed Jean Le Rond d'Alembert (1717–1783) who left a quantity of innovative works in the science of Mechanics in his young age. Perhaps not in the same class of mathematicians as Euler with whom he had many discussions and exchanges of correspondence, and certainly not as clear as Euler in his writing and proof reports, d'Alembert must be singled out for his originality and his general view of mechanics in a Leibnizian tradition. Thus in 1744, this gentleman, well educated in the best college in Paris, but mostly self-taught in mathematics, published his own book on the emerging fluid mechanics (D'Alembert 1744) after his publication of a celebrated Treatise on Dynamics in 1743. Furthermore, in 1752 d'Alembert obtained correct partial differential equations for axially symmetric and plane flows (of the type now called irrotational flows; cf. D'Alembert 1752). This is one of the first considerations of a two-dimensional motion of a continuum. He had already introduced for the first time the notion of *partial differential equations* in a previous work of 1743 on the mechanics of a heavy hanging rope. As noticed by Truesdell (1968, p. 228), d'Alembert does not speak of "pressure" but of "forces" that are viewed as "reversed accelerations". This fits well in d'Alembert's vision of reducing hydrodynamics to hydrostatics in accord with the general approach to mechanics he had given in his treatise of 1743. D'Alembert was also responsible for the introduction of the notion of *stream function* while Euler had introduced that of *velocity potential*. And we cannot avoid mentioning the d'Alembert paradox for the absence of drag on a cylinder placed in a perfect fluid flow. This was proved by d'Alembert in 1750. Of course, this theoretical result is contrary to common experience. A resolution of this paradox could be given only with the introduction of discontinuities in the flow field and the notion of wake.

In a general manner Truesdell has an exaggerate tendency to belittle the importance of d'Alembert's works as compared to Euler's. Thus he claims concerning d'Alembert's treatise of 1744 that (Truesdell 1968, p. 227) "this entry of a newcomer [he means d'Alembert] in the field added nothing to the subject". He repeated the same kind of argument in his review of the book by Rouse and Ince (1957) on the History of Hydraulics (Truesdell 1984, pp. 212–217). There he emphasized the inaccuracies introduced by Lagrange (1788) in his historical introduction on the history of fluid mechanics; he claims that Lagrange wrongly attributes successes to d'Alembert and also Clairaut, and doing so does not hesitate to introduce some innuendos (e.g., tactic fully making a remark that d'Alembert was the "patron of the career of Lagrange in France"; p. 213) or demonstrating (p. 215) that his translation of the Latin text of Daniel Bernoulli is much better than the one produced by others (e.g., Dugas—who certainly knew his Latin like all people of Dugas' calibre highly educated before the First World War—or Rouse and Ince). In spite of these harsh criticisms, we note that d'Alembert provided the bases on which Lagrange was going to build his grand scheme of mechanics.

Lagrange (1736–1813) is the last great theoretician of mechanics in the eighteenth century. His beautiful contributions to fluid mechanics need not be reported here again. Suffice it to remind the reader of the Lagrangian formulation of (perfect) fluid mechanics and his solutions of various wave problems in fluids such as the flow in a shallow channel of small slope and uniform width for which Lagrange has shown that the speed of propagation of a wave is proportional to the square root of the depth of the channel.

Toward the end of the eighteenth century we still note the contribution of Lazare Carnot (1753–1823), successful politician, military organizer, but also mechanician of value. He exploits the principle of solidification of Stevin and Clairaut in his consideration on hydraulics which chronologically precedes the treatise of Lagrange although published only in Carnot (1803). Carnot contributed to the proposal of the empirical Borda-Carnot equation (see below).

2.6 The True Experimentalists: Borda, Bossut, and Du Buat

We qualify Jean-Charles de Borda (1733–1799), Abbé Charles Bossut (1730–1814) and Pierre du Buat (1734–1809) of true experimentalists as these scientists devised special purposeful experiments and studied the influence of various parameters. With these scientists we arrive at the study of the resistance of fluids on their flow in real situations, what is obviously of great importance in applied hydraulics. With the new account of essential friction effects this contrasts with the theoretical studies of d'Alembert, Euler and Lagrange.

The "Chevalier" de Borda—as he is often called—in fact was a military naval engineer with a taste for mathematics and physics who fulfilled various state missions throughout the world as a navigator and hydrograph (cf. Mascart 2000). He demonstrated remarkable qualities as an experimentalist in fluid mechanics in works performed in the 1760s and 1770s. Among his successful studies we must record his study on the resistance offered by air to a moving material surface. This effect is a global one that cannot be obtained by integration over simple elements. The global resistance is carefully shown to be proportional to the square of the velocity and to the sine of the incidence angle (not to the squared sine as proposed earlier by Newton). In the case of water he checked the law of proportionality to the squared velocity. He observes that this resistance decreases with depth in water and that it may grow faster than the square of the velocity at the surface. To explain the phenomenon he exploits a theory of loss of living forces that is not entirely convincing. In all, his conclusion is that the theory put forward by Newton (who used the image of the shock of fluid particles on the obstacle) is not valid and should not be applied to the motion of boats. The considerations about the loss of living forces are of special interest because they followed from a divergence from the famous Bernoulli formula (2.1). First of all the idea is original. Second, the result is corroborated by experiments. We remind the reader that this well known problem concerns the flow of a fluid by an orifice standing for a sudden expansion or contraction of the flow, say in a horizontal pipe. Carnot revisited the problem, hence the often given name to the obtained formula, the *Borda-Carnot equation*. Using modern notation, this equation can be stated in the following simple form (for decreasing velocity):

$$\Delta E = \frac{1}{2}\xi\rho(v_1 - v_2)^2, \tag{2.2}$$

where ΔE is the loss of mechanical energy by the fluid—or dissipated kinetic energy-, ρ is the fluid density, v_1 and v_2 are the mean flow velocities before and after expansion, and ξ is an empirical nondimensional coefficient with value between zero and one (in fact equal to one for an abrupt and wide expansion). The energy quantity ΔE is none other that the variation in the "constant" in Daniel Bernoulli's equation (2.1) along a streamline. For an open channel flow we have

$$\Delta E = \rho g \Delta H, \quad H := h + v^2/(2g), \tag{2.3}$$

where $h = z + p/(\rho g)$ is called the total *head* in hydraulics, i.e., the free surface elevation above a reference height. In practice, the flow separates and one observes a *turbulent* recirculating zone after expansion. Separating recirculating zones are also observed at the entrance of a narrower pipe in the case of contraction. The flow is contracted between the separated flow areas and then re-expands to cover the full pipe section. The contraction coefficient for a sharp-edge contraction was experimentally determined by Julius L. Weisbach (1806–1871) in the nineteenth century (cf. Weisbach 1855). The original bold idea of Borda (1766) was to assimilate the hydraulic phenomenon to a shock with loss of living force, that is, a shock between *hard* bodies according to the terminology of the early eighteenth century (in the controversy between momentum and vis-viva as the conserved quantity in mechanics). Borda also contributed to the theory of hydraulic wheels (cf. Borda 1767, 1786).

The Abbé Charles Bossut is a personality who differs completely from that of Borda. We do not know of any adventures through the world by this quiet immovable priest. We noted in our book (Maugin 2013, p. 7) that he was "*a disciple of d'Alembert and a specialist of hydrodynamics* (cf. Bossut 1771), *but also an underestimated historian of mathematics* (cf. Bossut 1810), *and a remarkable pedagogue. His course of mathematics at the Military school of Mézières was first published in 1781. Its last edition* (1800) *was in seven volumes, of which two were devoted to differential and integral calculus in the Leibniz notation. This had a decisive influence on the rebirth of mathematical physics in the UK. He was a colleague of Laplace and Lagrange at the Paris Academy of Sciences, but not in the same class as these two mathematicians-mechanicians from the point of view of creativity*". He was a long time examiner (1796–1808) in mathematics at the Ecole Polytechnique after the creation of this school. However, the surprise here is provided by the outstanding qualities he demonstrated as a great experimentalist in a series of crucial experiments in fluid dynamics.

In truth, a team composed of d'Alembert, Nicolas Caritat de Condorcet (1743–1794)—another protégé of d'Alembert—and Abbé Bossut at the Paris Academy of Sciences was asked by the King's government to investigate the means of improving the navigation of ships. But the corresponding experiments were conducted, and the resulting report written, only by Bossut (cf. D'Alembert et al. 1777) and a few helpers. Bossut performed his experiments in the water basin available at the Royal Military School in Paris. The experiments involved as many as twelve different models of ships, and were repeated a total number of 300 times, 200 of these in a practically indefinite space and about 100 in an artificially realized channel in the basin, with depth and width adjustable at will. He also accounted for corrections due to the friction in the pulling apparatus due to a falling weight (of course no other type of motor was available at the time) and due to the part of the model in contact with air. We can say that the theoretically minded Bossut followed a remarkable experimental protocol. Among the conclusions reached we single out: (1) the resistance to a similar cross-sectional surface but at different speeds follows approximately the square law in velocity; (2) the resistance of fully submerged bodies is less than for those of partly submerged ones; For bodies equally submerged in water but with different widths, the resistance varies in the same direction as the surface; (3) the squared sine law (of Newton) is not well verified for small angles of incidence. In fact, Bossut assumed a tentative law involving a sine at a power n, and experimentally found n between 0.66 and 1.79. In the conclusion by Bossut, the resistance perpendicular to a plane surface in an indefinite fluid equals the weight of a column of fluid of which the section is that of the said surface and the length is the height determined by the speed at which the interaction with the fluid (for him a "shock") takes place. Clearly, Bossut is interested by the role played by the friction that the ship suffers along its length. This friction is rather small once the ship has been set in motion. A final remark is that the observed resistance is larger in a channel of restricted width so that he advised channels as deep and wide as possible for easier navigation, but for the limitation in cost. Bossut also advised on the construction of canals. It is said that the results of his experimental work influenced the design of ships by the American inventor Robert Fulton (1765–1815) when the later experimented with steam ships on the Seine river in Paris (Bossut, Carnot and Prony witnessed these well publicized events).

Pierre Du Buat was a French engineer specialist of hydraulics. Educated at the Ecole Royale du Génie de Mézières (Royal School of Engineering in Mézières, the forerunner of the Ecole Polytechnique), he obtained his diploma in 1750. He is the author of a celebrated book on the Principles of Hydraulics (Du Buat 1779; second edition 1786) that will remain a kind of "bible" in the field until the publication of the course of Bélanger (1790–1874) delivered at the Ecole des Ponts et Chaussées (1850s). Du Buat may be considered a follower of d'Alembert and Abbé Bossut but he is more applied than them. Barré de Saint-Venant (1865) expressed his admiration for him (See Fig. 2.13 herein after). He proved to be as good an experimentalist as Bossut. His research on the drag exerted on ships is documented in his book. The major step taken by du Buat, both experimentally and intellectually, is that he wants to account for friction and viscosity. According to his vision, water resistance is due to the friction of the fluid on solid walls. This allowed him to propose a corrected formula for the flux of fluid in terms of: (i) the width of the river, (ii) its depth, and (iii) the slope of the bed of the river. This also included the acceleration of gravity and a proportionality coefficient that depends on the roughness of the banks. This formulation marks a deep progress in sophistication compared to formulas proposed earlier by hydraulicians such as Antoine Chézy (1718–1798) who gave a simple equation for the mean velocity of the flow. With Du Buat, viscosity is viewed only as intervening indirectly in the "delay" caused by the presence of walls. His experiments are conducted in artificially constructed wooden canals or metallic or glass tubes. He introduced the original and fruitful notion of *mean radius* for such conducts. Like some of his predecessors, he interpreted the mechanical interaction with walls through the notion of shocks of fluid particles with the walls especially when considering the sinuosity of the river. As a matter of fact, he envisaged all accidents that a river can meet during its flow all the way to the sea. Concerning the flow around a cylindrical object, he noted the presence of an overshoot of pressure at its head and a lowering of pressure and a suction effect at the rear. He multiplied the number of experiments by varying the various parameters, and he wisely concluded that much more was needed on the experimental side. With such a conclusion we can say that du Buat was remarkably honest and a true experimentalist, indeed.[5]

Two remarks are in order at this stage of our presentation. First, in the line of Newton, many of the discussed scientists view the interaction of water with obstacles as shocks of particles with these obstacles. But, second, the great progress

[5]There is some irony that the etymology of the name "Buat" brings us back to water. Indeed, a possible etymology is that "buat" stands for a water conduit, or a "washing place" ("lavoir" in old French) and that even in modern French "buandrie" means a covered place or special room devoted to washing household linen and clothes.

is that this interpretation results in *friction* and a new era will expand with account of viscosity in one way or another.

We can conclude this section that dealt mostly with the eighteenth century by mentioning that some more practical books had a strong influence on the civil engineering applications of hydraulics. This is the case of the treatise by the engineer Bélidor (1697–1761) where all developments since the ancient Greeks and the Romans are taken into account (Bélidor 1737–1753). This met great success and was translated into foreign languages.

2.7 The Role of Viscosity: Poiseuille, Hagen

At all times scientists and engineers have noticed the negative influence of friction on the motion of objects in contact with other objects (and the associated feeling of expanded heat) or on the flow of fluids in conducts. They had the feeling that something was lost. Was it momentum, energy or else; anyway, notions that were not well defined? But with experiments in hydraulics conducted at the end of the eighteenth century, the actuality of the problem was pregnant and more thought was needed about it. The French military engineer Charles Augustin de Coulomb (1736–1806), a pioneer in geotechnical engineering, may be considered the creator of the *science of friction* in the late eighteenth century. Of course a correct inclusion of the phenomenon of friction in irreversible thermodynamics had to await the second part of the twentieth century to find a satisfactory formulation. But Coulomb, a gifted and careful experimentalist, provided the corner stone on which the theory could be built. This endeavour was undertaken by Coulomb after the publication of a competition prize offered by the Paris Academy of Sciences s in 1781 with subject: "To produce new and large scale experiments applicable to machines used in the Navy, e.g., pulleys, capstans, and inclined planes". Coulomb won the prize. He realized a very thorough series of experiments by varying all kinds of parameters such as the nature of surfaces in contact (wood, metal), with or without lubricant, the roughness of these surfaces, the pressure exerted on these surfaces, their size, the duration of contact, the more or less large speeds of the contacting objects, and the humidity and dryness of the apparatus. Rarely had such complete studies been made before. Among the reached conclusions we note: the proportionality of friction to the pressures; heterogeneous surfaces yield markedly different results; and friction increases with the velocity (with an arithmetic progression in friction for a geometric progression in velocities. In order to pull a weight P on a horizontal plane Coulomb finds that one must expand a force (in our notation)

$$T = A + Pa, \tag{2.4}$$

where A is a small constant that depends on the "coherence" of the surfaces in contact, and a is a friction coefficient that depends on the nature of the contacting surfaces. Considering an inclined plane (as is the case during the launching of a new ship) he finds that the force needed to hold the ship is given by the following formula that generalizes (2.4):

$$T = \frac{A + P(a\cos n + \sin n)}{\cos m + a\sin m}, \tag{2.5}$$

where n is the angle made by the plane with the horizontal, and m is the angle made by the pulling force with the inclined plane. T is minimum for $a = \tan m$. Equations (2.4) and (2.5) contain the main message left by Coulomb.

In his long memoir where he collects his results of research for the period (1838–1844) Poiseuille (1846) says that "I began my investigations because progress in physiology demands knowledge of the laws of motion of the blood…in small-diameter pipes". Poiseuille (1797–1869) received first the standard engineering-science education of the day at the Polytechnique School, but he decided to shift to medical studies to devote his whole life to the transposition of physical laws to the laws of physiology. As a first success in his doctoral thesis in medicine (1828) he proposed the first scientific means of measuring the blood pressure with a manometer using mercury to be called a hemodynamometer.[6] However, his most well known research results are those that pertain to his delicate studies of the flow of liquids in tubes of very-small diameters (i.e., capillaries) and are now summarized in the well-known Poiseuille law. A detailed history of this law and a description of the apparatus used by Poiseuille are to be found in Sutera and Skalak (1993). This law is often called the Hagen-Poiseuille law—especially in Germany—although Hagen's (1839) studies are much less extensive and accurate than those by Poiseuille. There was no contact between these two scientists and they published in different languages, but their researches were indeed contemporary. Hagen (1797–1884) was a Prussian hydraulician with an administrative career as a construction official (cf. Uhlemann 2009).

Although viscosity was duly introduced by Navier in his landmark work on fluid mechanics in 1823, Poiseuille does not use the expression "coefficient of viscosity", but his work indeed relates to the flow of viscous fluids, of which blood truly is a good representative. Using a modern notation that was obviously foreign to Poiseuille, Poiseuille's law for the velocity field of the *laminar* flow of a liquid in a circular tube of radius R is given in cylindrical coordinates by the following expression:

$$v(r,\theta,z) = v(r) = v_{\max}\left(1 - \frac{r^2}{R^2}\right), \tag{2.6}$$

[6]Poiseuille's perduring influence of this technique was such that we are still giving the blood pressure in units of height of mercury.

where the maximal velocity is given in terms of the pressure gradient in the flow direction, the radius R and the viscosity coefficient η by

$$v_{\max} = \frac{R^2}{4\eta}\left|\frac{dp}{dz}\right|. \tag{2.7}$$

The proof of this equation requires assuming that the velocity of the fluid vanishes at the boundary $r = R$, i.e., a no-slip condition. Equation (2.6) is a neat progress compared to what existed before (like in formulas proposed by Chézy—see above—and de Prony), when only a *mean* velocity across the cross section of a tube was introduced. Here the velocity profile is parabolic with a maximum at the centre of the tube. A similar computation can be effected for the flow between two flat fixed walls. But the problem of the flow motion caused by the shearing effect of a moving top plane is another famous problem called the Couette flow.

The original work of Poiseuille did not provide an equation such as (2.6). Like that of Hagen, it purported to obtain the expression of a pressure drop along a pipe as a function of the diameter D or radius R of this pipe of length L. If ΔP is the pressure drop and Q is the volumetric flow rate, then Poiseuille's result in his notation can be written as

$$Q = \frac{K'' D^4 \Delta P}{L} \quad or \quad \frac{\pi R^4}{8\eta}\frac{|\Delta P|}{L}, \tag{2.8}$$

what is compatible with (2.6) and (2.7). Poiseuille's constant K'' equals $\pi/128\,\eta$. The determination of this constant (in fact depending on temperature) by Poiseuille produced the dynamic viscosity of water to a remarkable accuracy[7] (cf. Sutera and Skalak 1993). The great Stokes could have easily deduced the result (2.6)–(2.7) if he had indeed implemented the no-slip condition at the boundary but he still had some doubt about this condition. This result was deduced from the Navier-Stokes equations by Eduard Hagenbach (1833–1910) from Basel in 1860. He is the one who chose to call it the Poiseuille law without reference to Hagen.[8]

---

[7]According to Bingham (1922), Poiseuille's value is accurate to 0.1 %.

[8]The priority question between Poiseuille and Hagen does not require much development although Hagen in 1869 claimed priority because of a paper he had published in 1839 (see Szabó 1977, pp. 269–273; Hagen 1869). But Poiseuille published continuously his results in notes in the Comptes Rendus of the Paris Academy of Sciences between 1838 and 1844, the final publication of his long memoir of 1846 being due to the delay in publication of contributions by non-members of the Academy (so called "savants étrangers") in a special series of memoirs.

2.8 Summary and Conclusion

More than two thousand years elapsed between the original works of Archimedes and the study of blood flow by Poiseuille. A wealth of practical applications of hydraulics, whether in irrigation or the supply of water to cities, was developed using only an empirical approach. The much awaited breakthrough came with the consideration of true experiments with the Renaissance and the next period of the seventeenth and eighteenth centuries. This is due to a few scientifically-minded engineers who acknowledged the protocol of a good experiment (That is: consider the main facts and design the experiment accordingly, identify the most relevant parameters and vary them as far as possible, do measurements, and propose simple mathematical relations, that would be justified later on by a true analysis when this technique had been constructed). Before analysis was used all reasoning presented by experimentalists was vague and most often erroneous, although some of these gifted people had the right intuition.

According to our survey, three basic notions seem to have emerged from these experimental studies. The first of these notions is *pressure*. The emergence of this quantity was facilitated by the a priori consideration of fluids such as water that are nearly incompressible (in modern terms). This led to the idea of making a fluid flow by means of an applied pressure. But this is related to the second notion that played an essential role, the notion of *gravity*. This is materialized by the fact that pressure in its most primitive form is measured by the height of a fluid in a vertical column opened to air, that is, it weight. That is clearly demonstrated by Torricelli's and Pascal's experiments. This relates to a static view of the phenomenon, called *hydrostatics*. In turn, this played a determining role in the study of flows under gravity and the equilibrium figures of the earth. In any case, all these experiments and considerations remain very *earthy*. It will take some time to arrive at a more abstract vision of pressure as *contiguity of action from all directions*, as proposed by Euler; that is, the first step towards the true notion of *contact action* in the framework of continuum mechanics.

It will also take some time to introduce the third notion of interest, *viscosity*. Of course, experimentalists had noticed the loss in the rate flow of fluids in experiments involving a pressure gradient (via a decrease in altitude along an inclined plane). But they experimented with negligibly viscous fluids (e.g., water) and the accuracy of measurements was insufficient. Kinematics enters the picture with the measurement of velocities. Early experimental research provided some velocity of exit from a container as proportional to the square root of the height of water in the container (Torricelli, Borda, du Buat). But one of the most beautiful formulas of physics was proposed by Daniel Bernoulli with his Eq. (2.1). This indeed relates two of our three ingredients, pressure and gravity, together with a velocity field, but for perfect (nonviscous) fluids also called inviscid fluids. With some advocated enthusiasm, we can declare that this equation is formidable. In particular, a straightforward application of it to the flow around an obstacle such as a wing profile provides a direct elementary proof of the lift effect on the wing, hence the theoretical basis of aeronautical flight. Viscosity was first imagined as friction on the bed of rivers or along the walls of a conduct of limited width. This will destroy the constancy of the right-hand side in Eq. (2.1). More precisely, in problems the effect of viscosity will essentially come from the boundary conditions in solving a true boundary-value problem of hydrodynamics while in experiments fluids such as oil or blood will offer a better object for its study. That is why our second favorite formula here is Poiseuille's Eqs. (2.6)–(2.7). It really relates to *hydrodynamics* although in a laminar regime (cf. Darrigol 2005, Jouguet 1924, Tokaty 1971), that is—to be simple—to slow enough flows with regular stream lines. The case of turbulent flows and high velocities will be studied in depth at the end of the nineteenth century, especially through the experimental (Reynolds' number) and theoretical (Reynolds' stress) efforts of Osborne Reynolds (1868–1912). Also, the flow of fluids through porous media of interest in geophysics and civil engineering will benefit from the works of Henry Darcy (1803–1858) and other people. But this is another story of which we shall examine the most important episodes in another contribution (Portraits of the main scientists involved in this chapter are given in Figs. 2.2, 2.3, 2.4, 2.5, 2.6, 2.7, 2.8, 2.9, 2.10, 2.11, 2.12, 2.13, 2.14, 2.15, 2.16, 2.17).

Fig. 2.2 Archimedes: By Domenico Fatti (1620) [Museum of Alte Meister, Dresden, Germany]

Fig. 2.3 Leonardo da vinci, self-portrait

Fig. 2.4 Simon Stevin (1548–1626)

Fig. 2.5 Galileo Galilei (1564–1642)

Fig. 2.6 Evangelisto Torricelli ((1608–1647) 62)

Fig. 2.7 Blaise Pascal (1623–1662)

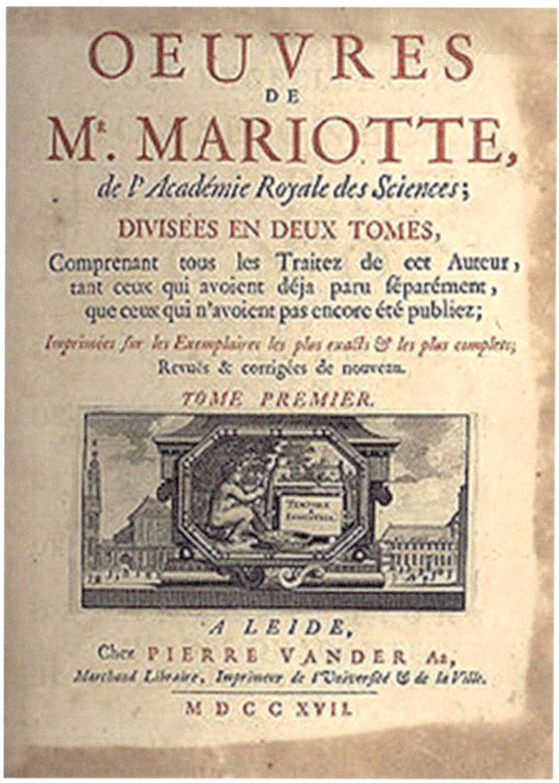

OEUVRES
DE
M^r. MARIOTTE,
de l'Académie Royale des Sciences;
DIVISÉES EN DEUX TOMES,
Comprenant tous les Traitez de cet Auteur,
tant ceux qui avoient déja paru séparément,
que ceux qui n'avoient pas encore été publiez;
Imprimées sur les Exemplaires les plus exacts & les plus complets;
Revûs & corrigées de nouveau.
TOME PREMIER.
A LEIDE,
Chez PIERRE VANDER Aa,
Marchand Libraire, Imprimeur de l'Université & de la Ville.
M D C C XVII.

Fig. 2.8 Works by Edmé Mariotte (1717)

Fig. 2.9 Edmé Mariotte (1620–1684)

Fig. 2.10 Alexis Clairaut (1713–1765)

Fig. 2.11 Chevalier de Borda (1733–1799)

Fig. 2.12 Abbé Charles Bossut (1730–1814)

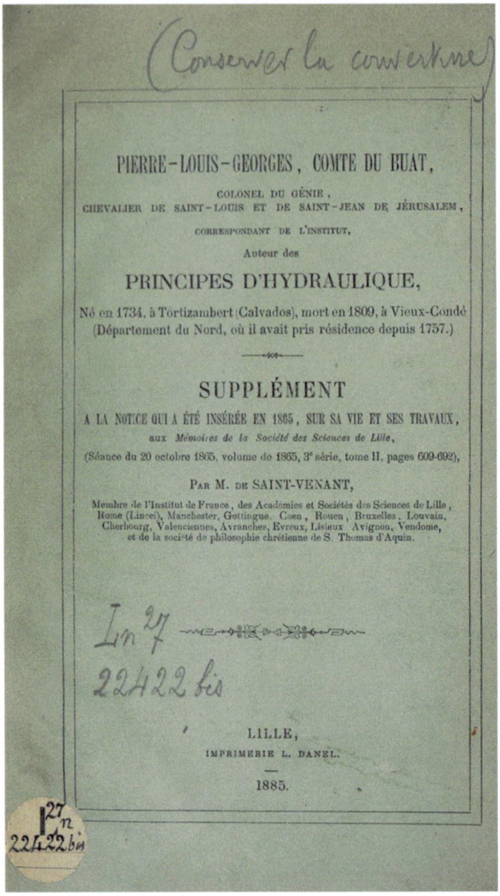

(Conserver la couverture)

PIERRE-LOUIS-GEORGES, COMTE DU BUAT,

COLONEL DU GÉNIE,
CHEVALIER DE SAINT-LOUIS ET DE SAINT-JEAN DE JÉRUSALEM,
CORRESPONDANT DE L'INSTITUT,

Auteur des

PRINCIPES D'HYDRAULIQUE,

Né en 1734, à Tortizambert (Calvados), mort en 1809, à Vieux-Condé (Département du Nord, où il avait pris résidence depuis 1757.)

SUPPLÉMENT

A LA NOTICE QUI A ÉTÉ INSÉRÉE EN 1865, SUR SA VIE ET SES TRAVAUX,
aux *Mémoires de la Société des Sciences de Lille*,
(Séance du 20 octobre 1865, volume de 1865, 3e série, tome II, pages 609-692),

PAR M. DE SAINT-VENANT,

Membre de l'Institut de France, des Académies et Sociétés des Sciences de Lille, Rome (Lincei), Manchester, Gottingue, Caen, Rouen, Bruxelles, Louvain, Cherbourg, Valenciennes, Avranches, Evreux, Lisieux, Avignon, Vendome, et de la société de philosophie chrétienne de S. Thomas d'Aquin.

LILLE,
IMPRIMERIE L. DANEL.

1885.

Fig. 2.13 On Du Buat (1734–1809) by Barré de Saint-Venant (1865, 1885)

Fig. 2.14 Poiseuille (1797–1869)

Fig. 2.15 Julius Weisbach (1806–1871)

Fig. 2.16 Gotthilf Hagen (1797–1884)

Fig. 2.17 Gotthilf Hagen (1797–1884)

References

Barré de Saint-Venant A(1865) Notice sur Louis-Joseph comte du Buat. Mémoires de la Société des Sciences, de l'agriculture et des arts de Lille, vol 1865, 3ème série, Tome II: 609–692, Imprimerie L. Danel, Lille [available on the site Gallica of the BNF]
Bélidor BF de (1737–1753) Architecture hydraulique. Four volumes, Paris
Bernoulli D (1738) Hydrodynamica. Basel
Bernoulli J (John) (1739) Hydraulica, nunc primum detecta ac demonstrata ex fundamentis pure mechanicis. Basel
Bingham CB (1922) Fluidity and plasticity. McGraw-Hill, New York
Borda JC de (1766) Mémoire sur l'écoulement des fluides par les orifices des vases. Mém Acad Sci. Paris, p. 147 on
Borda JC de (1767). Mémoire sur les roues hydrauliques. Mém Acad Sci, 1767 (1770), pp 270–287
Borda JC de (1786) Principes d'hydraulique. Chez Farrois, Paris
Bossut C (1771) Traité élémentaire d'hydrodynamique, Paris (improved in two volumes: Traité théorique et expérimental d'hydrodynamique, 1786–87, Paris)
Bossut C (1800) Cours de mathématiques. Seven volumes, Didot Editeur, Paris
Bossut C (1810) Histoire générale des mathématiques. Two volumes, Paris
Brunet P (1952) La vie et l'œuvre de Clairaut (1713–1765). Presses Universitaires de France, Paris
Clairaut A (1743) Traité de la figure de la terre tirée des principes de l'hydrodynamique. Chez Durand, Paris
Carnot L (1803) Principes généraux de l'équilibre et du mouvement. Paris
D'Alembert Jean Le Rond (1744) Traité de l'équilibre et du mouvement des fluides pour servir de suite au traité de dynamique. Chez David, Paris
D'Alembert Jean Le Rond (1752) Essai sur une nouvelle théorie de la résistance des Fluids. Chez Durand, Paris (Competition essay for the 1750 prize of the Academy of Berlin)
D'Alembert J. Le Rond, Condorcet N de, Bossut C (1777) Nouvelles expériences sur la résistance des fluides. Chez Jombert, Paris
Darrigol O (2005) Worlds of flows: a history of hydrodynamics from the Bernoullis to Prandtl. Oxford University Press, Oxford
Du Buat P (1779) Principes d'Hydraulique. Paris (Second edition, 1786; translated to German, 1791)
Duhem P (1906–1913) Etudes sur Léonard de Vinci. Three volumes. Hermann, Paris
Dugas R (1950) Histoire de la mécanique (History of Mechanics, 1956). Editions du Griffon, Neuchatel, Switzerland [Dover reprint, New York, 1988]
Euler L (1757) Principes généraux de l'état d'équilibre des fluides. Mémoires de l'Académie des Sciences de Berlin, 11:217–273 (Memoirs presented to this Academy in 1755)[Also: Euler's collected works II, Volume 12, pp 2–132]
Gille B (1964) Les ingénieurs de la Renaissance. Hermann, Paris
Hagen G (1839) Über d. Bewegung d. Wassers in engen zylindr. Röhren. In: Hdb. d. Wasserbaukunst III:1841–1863
Hagen G (1869) Ueber die Bewegung des Wassers in cylindrischen nahe horizontalen Lettungen. Math Abh Akad Wiss., Berlin, pp 1–29
Jouguet E (1924) Lectures de mécanique (La mécanique enseignée par les auteurs originaux). Gauthier-Villars, Paris [Facsimile reprint, Gabay, Paris, 2007]
Lagrange JL (1788) Mécanique analytique, First edition. Vve Desaint, Paris; Second edition, Vve Courcier, Paris, 1811 [Facsimile reprint of 4th edition with notes and comments par J. Bertrand and G. Darboux, Gabay, Paris] [English translation of the 2nd edition (1811), Eds. A. Boissonnade and V.N. Vagliente. Springer, Dordrecht, 1997]
Mariotte E (1684) Traité du mouvement des eaux et des autres corps fluides. Paris (Posthumous edition by P. De la Hire)
Mascart J (2000) La vie et les travaux du Chevalier Jean-Charles Borda [Reprint by Presses de l'Université de Paris-Sorbonne, Paris]

Maugin GA (2014) Continuum mechanics through the eighteenth and nineteenth Centuries. Springer, Dordrecht

Mays LW (2008) A very brief history of hydraulic technology during antiquity. Environ Fluid Mech 8:471–484

Passeron I (1994) Clairaut et la figure de la terre au XVIIIe siècle, cristallisation d'un nouveau style autour d'une pratique physico-mathématique. Doctoral Thesis, University of Paris 7, Paris. (Extract: Le principe d'équilibre de Clairaut en hydrostatique (1743), Open Archives: hal-00361469, Version 1, 15 Feb 2009]

Poiseuille JLM (1846) Recherches expérimentales sur le mouvement des liquides dans les tubes de très petit diamètre. Mém Acad Sci 9:433–564

Rouse H, Ince S (1957) History of hydraulics, Reprint of original papers by Iowa State University [Dover reprint, New York, 1963]

Sutera SP, Skalak R (1993) The history of Poiseuille's law. Annu Rev Fluid Mech 25:1–19

Szabó I (1977) Geschichte der mechanischen Prinzipien. Birkhäuser, Basel

Tokaty GA (1971) A History and philosophy of fluid mechanics [Dover reprint, New York, 1994]

Truesdell CA (1956) Experience, theory, and experiment (1955) In: Proceedings of 6th hydraulics conference, bulletin 36, pp 3–18, University of Iowa Studies in Engineering [Reprinted in Truesdell 1984, pp 3–20]

Truesdell CA (1968) Essays in the history of mechanics. Springer, New York, Berlin

Truesdell CA (1984) An idiot's fugitive essays on science. Springer-Verlag, New York

Uhlemann HJ (2009) Gotthilf Hagen (1797–1884). In: Mitteilungen DWhG, Nr. 14/April 2009, Anhang, S. 1–33

Varignon P (1725) Traité du mouvement et de la mesure des eaux courantes et jaillissantes. Pissot, Paris (posthumous edition by Abbé Pujol)

Vitruvius L (ca 25 BC). De Architectura [Original Latin text and French translation available at Gallica, site of the BNF, Paris]

Weisbach JK (1855) Die Eksperimental-Hydraulik. Engelhardt, Freiberg

Chapter 3
On Porous Media and Mixtures

Abstract Porous media have obviously attracted the attention of hydraulicians and specialists of geo-materials as soon as the early nineteenth century. The essential breakthrough was presented by Henry Darcy who proposed a simple law for the flux of water per area in sand on the basis of careful experiments. Here the notion of porosity is essential. But the importance of the concept of viscosity was not fully made explicit in spite of Poiseuille's pioneering work on flows in pipes (in truth, blood in arteries). Darcy's law was to prove to be the driving force in further progress in the field, so that the present contribution is akin to a saga of this law through a hundred and fifty years. Thus, chronologically, Fillunger, introduced the concept of uplift, Terzaghi produced a basic theory of consolidation of soils, Fillunger approached the flow in porous media in the form of an elementary theory of mixtures, and Biot proposed a first satisfactory theory of poroelaticity with interesting dynamic properties. Modern authors in the 1960s–1970s followed Truesdell in his rational theory of mixtures. More mathematically oriented researchers used the homogenization technique to justify Darcy's law and its generalizations. The thermo-mechanical continuum theories in finite strains are also evoked, as well as very briefly the importance of porosity in plastic behaviour.

3.1 Introduction

Very early it was realized when watering a plant that it took some time for water to get absorbed by the ground, although it was naturally questioned whether this happened through some chemical process or water was making its way through interstices in the ground seen as some kind of granular material. Bulging of the ground would indicate that water was indeed occupying some volume and the observed fact that water at some point could not be accepted anymore hinted at a phenomenon of saturation. Also, the flow through aquifers was more or less slow or altogether hindered. These were properties that would call for a scientific explanation in due time. It seems that the idea of porous media developed quite naturally. The great Leonhard Euler (1762)

G.A. Maugin, *Continuum Mechanics through the Ages - From the Renaissance to the Twentieth Century*, Solid Mechanics and Its Applications 223,
DOI 10.1007/978-3-319-26593-3_3

himself "attributed the elasticity of a solid to a certain subtle matter in closed pores" (cf. De Boer 1996, 2000, p. 24). Pores indeed are thought as voids (in fact, air) that in some cases become filled up with a liquid. The relationship with geomaterials and soil mechanics is obvious, although we cannot escape from the vision of a spongy material or, now, that of the spectacular metallic foams. In summary, in modern practice, a porous medium is seen as a material body made of a more or less deformable solid matrix with its voids filled with fluids. These voids are interconnected so that fluids can flow through the medium; that is, the medium is permeable.

No theoretical progress could be achieved before the inception of the notion of *viscosity*, the beautiful experiments of Poiseuille and Hagen, and the introduction of well justified continuum equations governing viscous fluids by Navier, Stokes and Barré de Saint-Venant (cf; Maugin 2014a). A breakthrough was reached with the experimental results of Darcy, their justification on the basis of the Navier-Stokes equations, and their improvements by Brinkman and Forchheimer. This really took place in the expansion of studies in hydraulics and applications in civil engineering. It will take some time to mathematicians to correctly deduce these laws by means of some homogenization procedure, and to physicists and specialists of continuum mechanics to incorporate them in the framework of a good thermodynamic description, to include the theory of porous media within the more general frame of a recently developed theory of mixtures, and to assess the importance of porosity in plasticity. The role played by mechanicians such as Truesdell, Biot and Bowen, and engineers such as Fillunger and von Terzaghi (the creator of modern soil mechanics) must be emphasized.

Most recent developments are exposed in a number of books (Bear 1972; Rajagopal and Tao 1995; Coussy 1995, 2010; De Boer 2000; Ehlers 2010; Wilmanski 1998). The present essay provides some historical perspective with an emphasis on the main breakthroughs since the seminal works of Poiseuille and Darcy. It logically complements our previous contribution devoted to the beginnings of hydraulics as an engineering science (Maugin 2014b).

3.2 Reminder: Poiseuille and Blood Flow

In extremely careful experimental works conducted in the period 1838–1844, J.L. M. Poiseuille (1797–1869), an alumnus from Ecole Polytechnique but also a Doctor in medicine, fulfilled an important part of his programme to devote his whole life to the transposition of physical laws to the laws of physiology. This he did particularly by investigating the laws of motion of the blood in small-diameter pipes (Poiseuille 1846). Although viscosity was duly introduced by Navier in his landmark work on fluid mechanics in 1823, Poiseuille does not use the expression "coefficient of viscosity", but his work indeed relates to the flow of viscous fluids, of which blood truly is a good representative. The original result he obtained relates the expression of a pressure drop along a pipe as a function of the diameter D or radius R of this pipe of length L. If ΔP is the pressure drop and Q is the volumetric flow rate, then Poiseuille's result can be written in his notation as (cf. Maugin 2014b)

$$Q = \frac{K'' D^4 \Delta P}{L} \quad \text{or} \quad \frac{\pi R^4}{8\mu} \frac{|\Delta P|}{L}, \tag{3.1}$$

where the second expression is a modern rewriting involving the viscosity μ. Poiseuille's constant K'' equals $\pi/128\,\mu$. The determination of this constant (in fact depending on temperature) by Poiseuille produced the dynamic viscosity of water to a remarkable accuracy.

In modern textbooks Poiseuille's solution of the *laminar* flow of a liquid in the circular tube is given in the following more mathematical form in cylindrical coordinates

$$v(r, \theta, z) = v(r) = v_{\max}\left(1 - \frac{r^2}{R^2}\right), \tag{3.2}$$

where the maximal velocity is given in terms of the pressure gradient in the flow direction, the radius R and the viscosity coefficient μ by

$$v_{\max} = \frac{R^2}{4\mu}\left|\frac{dp}{dz}\right|. \tag{3.3}$$

The proof of this equation requires assuming that the velocity of the fluid vanishes at the boundary $r = R$, i.e., a *no-slip condition*. Equation (3.2) is a neat progress compared to what existed before (like in formulas proposed by Chézy and De Prony), when only a *mean* velocity across the cross section of a tube was introduced. Result (3.2)–(3.3) was deduced from the Navier-Stokes equations by Eduard Hagenbach (1833–1910) from Basel.

With Poiseuille's result, we have a first true manifestation of viscosity, although not mentioned as such by its author. From then on viscosity will have to be accounted for in the study of flows in various configurations. The case of turbulent flows and high velocities will be studied in depth at the end of the nineteenth century, especially through the experimental (Reynolds' number) and theoretical (Reynolds' stress) efforts of Osborne Reynolds (1868–1912).

3.3 Darcy and the Fountains in Dijon

Like Abbé Edmé Mariotte (1620–1684)—cf. Maugin (2014b)—before him, Henry Darcy (1803–1856) was born in Dijon. Nothing predestined this city to breed specialists of hydraulics. This French provincial city in Burgundy is internationally known for gastronomy and its celebrated "mustard". That is a strange destiny when we think that Dijon in fact was the capital of a powerful state—the Duchy of Burgundy (1032–1477)—, extending from the present day Burgundy region—indeed famous for its gastronomy and wines—to the actual Netherlands, often in competition and/or war with the kingdom of France (with capital Paris) in the Middle Ages. Dijon has kept until today the nice and rich outlook of a provincial capital. But it happened that Dijon became worldwide known to all hydraulicians through one of his sons, Henry Darcy. This remarkable individual received the standard high education of civil engineers of the period (Ecole Polytechnique in Paris followed by the Ecole des Ponts et Chaussées). On graduation he became a member of the celebrated Corps of engineers of Ponts et Chaussées and specialized rapidly in hydraulics with an interest in the supply systems of drinkable water to cities. This is what brought him back to his native city when the development of the railways network caused a growth of this city that required an appropriate layout of distribution of water. Although an excellent theoretician by formation and with a good training in applied construction and developmental engineering (railways, canals), the bright Darcy attacked this hygienic problem with a rather strict experimental mind. He published the results of his investigation in a now classic synthesis (Darcy 1856—Fig. 3.1—; complemented by Darcy 1857). That is where he proposed his celebrated "Darcy" law. Like Fourier's law in heat conduction, Fick's law in diffusion theory, and Ohm's law in electricity, Darcy's is a paragon of simple but efficient constitutive equation (in modern terms). In Note *D* of his book of 1856, Darcy describes his (vertical) experimental apparatus, and concisely explains how he was led to the formula that relates the total discharge to the pressure drop in the filtration of water through sand. This represents the most basic law in the flow of porous media. It was to be checked, improved, and physically and mathematically justified for a hundred and fifty years. Note that Darcy himself does not speak about viscosity, but his work provides one of the first expressions of this faculty—to flow more or less easily- in some measurable form related to the flow in some conduit. It is Hagen (1839) who really explicitly introduced the viscosity in Darcy's formula (cf. Nakayama 2007). The subject matter was of course of actuality as shown by the book of Dupuit (1843) on the flow of waters in canals and permeable soils published by the same Parisian editor as Darcy's book. Boussinesq (1904) pursued the theoretical analysis of such underground flows. Darcy's law was also generalized to saturated and unsaturated flows in heterogeneous and anisotropic bodies in a little known technical report by Buckingham (1907).

The *Darcy-Hagen law* can be given in the following modern direct vector form:

$$\mathbf{q} = -\frac{k}{\mu}(\nabla p - \rho \mathbf{g}) \tag{3.4}$$

where $\mathbf{q}$ is the flux or discharge per unit area, p is the pressure, ∇p is the pressure gradient, $\mathbf{g}$ is the gravity vector, k is the permeability, and μ is the dynamic viscosity. The so-called Darcy velocity is defined as

$$v_{Darcy} = \frac{\mathbf{q}}{A} = v_{pore} n, \tag{3.5}$$

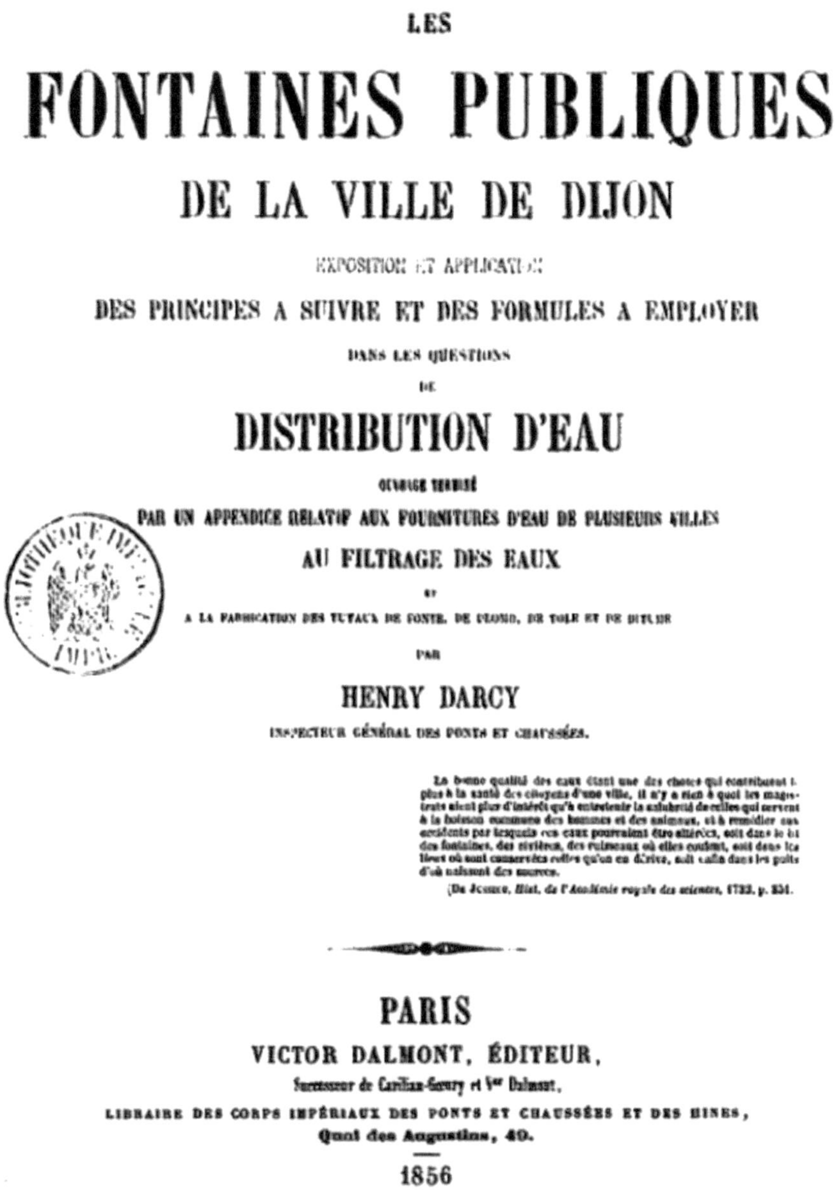

LES

FONTAINES PUBLIQUES

DE LA VILLE DE DIJON

EXPOSITION ET APPLICATION

DES PRINCIPES A SUIVRE ET DES FORMULES A EMPLOYER

DANS LES QUESTIONS

DE

DISTRIBUTION D'EAU

OUVRAGE TERMINÉ

PAR UN APPENDICE RELATIF AUX FOURNITURES D'EAU DE PLUSIEURS VILLES

AU FILTRAGE DES EAUX

ET

A LA FABRICATION DES TUYAUX DE FONTE, DE PLOMB, DE TOLE ET DE BITUME

PAR

HENRY DARCY

INSPECTEUR GÉNÉRAL DES PONTS ET CHAUSSÉES.

La bonne qualité des eaux étant une des choses qui contribuent le plus à la santé des citoyens d'une ville, il n'y a rien à quoi les magistrats aient plus d'intérêt qu'à entretenir la salubrité de celles qui servent à la boisson commune des hommes et des animaux, et à remédier aux accidents par lesquels ces eaux pourraient être altérées, soit dans le lit des fontaines, des rivières, des ruisseaux où elles coulent, soit dans les lieux où sont conservées celles qu'on en dérive, soit enfin dans les puits d'où naissent des sources.

(De Jussieu, *Hist. de l'Académie royale des sciences*, 1733, p. 351.

PARIS

VICTOR DALMONT, ÉDITEUR,

Successeur de Carilian-Gœury et Vve Dalmont,

LIBRAIRE DES CORPS IMPÉRIAUX DES PONTS ET CHAUSSÉES ET DES MINES,

Quai des Augustins, 49.

1856

Fig. 3.1 Darcy's book of 1856

where $\mathbf{q}$ is the filtration flux (discharge per unit area), A is the filtration surface and n is called *porosity* (sometimes noted ϕ) given by

$$n = \frac{volume\ of\ void\ (pores)}{total\ volume}. \tag{3.6}$$

In the original work by Darcy, Eq. (3.4) in a horizontal tube of length L would be given by

$$Q = -\frac{kA}{\mu}\frac{(p_b - p_a)}{L}, \tag{3.7}$$

where Q is the total discharge (volume per unit time), A is the cross-sectional area of the flow, and $(p_b - p_a)$ is the pressure drop. Compared to (3.7), Eq. (3.4) accounts for the "elevation head". The negative sign obviously means that the fluid flows from high pressure to low pressure. Equation (3.5) means that only a fraction of the total formation volume is available for flow. Darcy's equation, in its mere simplicity, provides excellent results for groundwater flowing in aquifers, a case where the validity of Darcy's law for slow viscous flow is verified, i.e., *laminar* flows with a Reynolds number less than about ten. This Reynolds number would here be defined by

$$\mathrm{Re} = \frac{\rho v d}{\mu}, \tag{3.8}$$

where ρ is the density of water, v is the specific discharge (not the pore velocity) and d is a representative grain diameter for the porous medium (e.g., sand).

Darcy is also known for the Darcy-Weisbach equation (Darcy 1857; Weisbach 1855) which provides in hydraulics the head loss in terms of an average velocity, the acceleration of gravity, the diameter and length of a pipe, and a friction factor. This was supposed to improve on empirical equations proposed earlier by Antoine Chézy (1718–1798)—cf. Mouret (1921)—and De Prony (1825). These equations in which both hydraulic radius and slope of the water line provide a mean flow velocity, proved to be of universal usage in the design of canals in the nineteenth century.

3.3.1 Generalizations of Darcy's Law

With Darcy's laws deduced from the Navier-Stokes equation within a certain framework of hypotheses, it can be expected that by relaxing some of these hypotheses a more general form can be obtained for this law. This is in fact the case of the equation proposed by Brinkman (1947) for the case of sufficiently high porosity. In a horizontal pipe, Eq. (3.4) will thus be generalized to

$$\frac{\mu}{k}\mathbf{v} = -\nabla p + \lambda \nabla^2 \mathbf{v}, \tag{3.9}$$

where λ is called *effective viscosity*. Another generalization is due to Forchheimer (1901) when the fluid velocity is large enough, yielding a *nonlinear* generalization of Darcy's law in the form

$$\frac{\mu}{k}\mathbf{v} + a|\mathbf{v}|\mathbf{v} = -\nabla p. \tag{3.10}$$

This equation can be said to be *nonlinear* and to account for *inertial* effects.

In a convective flow where the viscosity varies considerably with temperature, it may be necessary to combine Eqs. (3.9) and (3.10). Higher order nonlinear terms can also be added in the left-hand side of (3.10). All Eqs. (3.4) through (3.10) are valid for incompressible fluids for which $\nabla \cdot \mathbf{v} = 0$. There also exist anisotropic generalizations that involve a second-order tensor of permeability $\mathbf{k}$ instead of the scalar k. Finally, in the same line of thought as what was proposed for heat conduction by C. Cattaneo and P. Vernotte in the 1940s–1950s to give a finite speed to the heat propagation phenomenon, for very short times Eq. (3.4) can be improved to read

$$\tau \frac{\partial \mathbf{v}}{\partial t} + \mathbf{v} = -\frac{k}{\mu}\nabla p, \tag{3.11}$$

where τ is a very small time constant. Such a form will result in a *hyperbolic* groundwater flow equation, perhaps not of much practical use (cf. Cattaneo's modelling applied to porous media in Straughan 2011, pp. 238–240).

3.4 Porous Media and Homogenization Technique

A porous medium can be seen as a heterogeneous body made of two constituents, the matrix and the interconnected voids filled of a liquid or really empty. When the voids are completely filled in the medium is said to be *fluid-saturated*. One may think of representing this global medium by assuming that the constituents are interpenetrating and in fact co-existent in certain relative proportions at each point but without real mass exchange between them. Then an averaging can be envisaged that will provide macroscopic laws of, say, a poro-elastic medium. Such an approach was early used in the mechanics of composites and obviously in mixtures of fluids. We leave aside this type of approach, although we shall return to it in due time (Sect. 3.5) as we prefer to examine first a rather mathematical way to deduce basic laws of porous media, that is, *homogenization*.

In a general manner homogenization consists in replacing a heterogeneous material by a somewhat *equivalent* homogeneous material. To perform such an operation one must identify a representative volume element (RVE) that is most characteristic of the structure of the material at a small scale where some properties and field vary rapidly. In most cases the structure is not well ordered as a porous medium obviously presents a random distribution of pores, themselves of various sizes (for this aspect, see Adler 1992). But, remarkably enough, assuming a rather (in most cases unrealistic—unless man-made) ideal situation with a regular, e.g., periodic, pattern of pores, one can conceive of an easily implemented scheme that is both mathematically sound and practically very efficient as proved by the case of porous media. The RVE then is a basic cell of this convenient periodic arrangement while in the averaging technique the RVE should contain enough statistical information about the heterogeneous medium in order to be truly representative. Anyway, we are facing two spatial length scales, those of the RVE (l) and of the macroscopic scale (L), typically in the ratio $\varepsilon = l/L$, where ε is an infinitesimally small quantity. In physical Euclidean space the macroscopic domain Ω of interest is considered periodic and the rescaled unit cell is $Y = (0,1)^3$ and the two coordinates ($\mathbf{x}$ macroscopically and $\mathbf{y}$—fast variable—in the cell) are given by

$$\mathbf{x} \in \Omega, \quad \mathbf{y} = \frac{\mathbf{x}}{\varepsilon} \in Y. \tag{3.12}$$

The result of this parametrization is that the problem now is embedded in a sequence of similar problems parametrized by a scaling ε. Then homogenization amounts to performing an asymptotic analysis when ε tends to zero. The limit is the solution of the homogenized problem. This mathematical technique is called *asymptotic periodic homogenization* (APH). It was essentially developed in France, Russia and Italy by authors such as Sanchez-Palencia (1980), Bensoussan et al. (1978), Bakhvalov and Panasenko (1989), Oleinik et al. (1991), Dal Maso (1992), Luc Tartar and Allaire (1991). Some technical details are given in the appendix to this contribution.

Application of the APH to porous media was first given by Lévy and Sanchez-Palencia (1975) and Ene and Sanchez-Palencia (1975) in Paris by applying the APH scheme to the steady Stokes equation for an incompressible viscous fluid with vanishing $\mathbf{v}$ on the boundary S of Ω, i.e., in the notation of the appendix:

$$\nabla p_\varepsilon - \varepsilon^2 \mu \nabla^2 \mathbf{v}_\varepsilon = \mathbf{f}, \quad \nabla \cdot \mathbf{v}_\varepsilon = 0, \quad \mathbf{v}_\varepsilon|_S = \mathbf{0}, \tag{3.13}$$

a set of equations which admits a unique solution in the appropriate functional spaces for $\mathbf{v}_\varepsilon$ and p_ε. Proceeding as briefly indicated in the appendix, the authors deduce the Darcy law. Lévy (1983) goes further by studying the Stokes flow through a swarm of fixed (non-connected) particles and justifies thus Brinkman's law in the form [compare to (3.9)]

$$6\pi\mu a c \mathbf{v} = -\nabla p_0 + \mu \nabla^2 \mathbf{v} + \mathbf{f} \tag{3.14}$$

by a technique of matched asymptotic expansions involving a boundary layer around the particles. Notice that this model is a strange one for a porous medium because of the lack of connectivity between particles (the same remark applies to Brinkman's work).

The question of the validity of the various approximations obtained in the homogenization procedure (Lévy, Sanchez-Palencia; Allaire, Auriault et al.), calls for a thorough examination of the respective domains of validity in terms of non-dimensional numbers (scales). For this we refer the reader to Auriault (2009) who underlines the case of so-called "poor separation scales", i.e., when, if we may say so, ε is not so small (Auriault et al. 2005). In this analysis, Brinkman's law is shown to be invalid for connected porous matrices so that this casts some doubt on the real physical meaning of Brinkman's modelling for porous media (more on APH in Ene and Polisevski 1987).

3.5 Porous Media and the Theory of Mixtures

It is easily conceived that a porous medium may be considered a mixture, but a mixture reduced to two constituents or two phases, a solid one (the matrix) and a fluid one likely to flow through the interstices left by the solid with interconnected pores. This is the view adopted by, e.g., Rajagopal and Tao (1995). This is the *continuum theory of mixtures* for which Clifford Truesdell (1957) provided the first thermodynamically based rational theory (also Trusdell 1984). There followed in the 1960s–1970s a multitude of works in the same line, among these, Kelly (1964), Green and Naghdi (1965, 1971), Eringen and Ingram (1965, 1967), Bowen (1967, 1976), Müller (1968), and Atkin and Craine (1976).

The theory of mixtures uses the concept of volume fraction, an example of which is given by Eq. (3.6) in the case of porous media. De Boer (2000, p. 31) attributes to the German hydraulician Reinhard Woltman (1757–1837)—the inventor of the water current meter—the introduction of this concept (Woltman 1794). The general thermodynamic theory of mixtures is a complicated affair that still poses fundamental questions, for example: (i) should a temperature be introduced for each constituent (same question for entropy); (ii) should global thermodynamic laws, in particular the second one, be defined for each constituent or for the global medium; (iii) how the stress condition at a boundary should be formulated since only one global applied traction can be specified while the mixture involves partial stresses? The only constraint on which all authors agree is that by summation over the constituents one obtains the thermo-mechanical equations that govern a standard unique continuum. We shall not deal with this generality as temperature effects are altogether ignored in the rest of this section. We prefer to examine the advances made by a few remarkable contributors during the period 1910–1960.

In the period preceding the full implementation of nonlinear continuum thermomechanics, say in the period 1910–1960, we can identify the breakthroughs made by four specialists, civil engineers or physicists, who brought new constructive ideas in the mechanics of porous media, and more generally to soil mechanics, with many potential applications to the construction of dams, the exploration and exploitation of oil resources, etc. These remarkable scientists are: Paul Fillunger (1883–1937); Karl von Terzaghi (1883–1963), Maurice A. Biot (1905–1985), and Gerhard Heinrich (1902–1983), among whom three (that is, save Biot) were affiliated with the Technical University of Vienna (Wien, Austria).

Both Fillunger and Terzaghi were outstanding professors at TU Wien, but with very different personalities. Fillunger, was the first to place in evidence the so-called *uplift effect* in porous media (Fillunger 1913, 1914). This effect, especially in constructions such as dams, in fact is a manifestation of buoyancy (cf. Archimedes' principle). It is the result of the liquid pressure acting on solid particles. It can be expressed in Fillunger's original form as the force

$$P = kV(\mu - \mu'), \tag{3.15}$$

where k is the vertical gradient of pressure, V is the total volume of a liquid-saturated element and μ and μ' are volume and surface porosity coefficients (not to be mistaken for viscosities). Of course, the sign of the difference between these two quantities is of importance (cf. De Boer 2000, p. 189). But in certain cases the vertical uplift force (3.15) prevails over the gravity force and this may cause serious problems in the strength of structures while justifying its name.

von Terzaghi (1923a, b), often considered the father of soil mechanics as a science, and much more internationally known than Fillunger, proposed a rather intuitive law for the vertical consolidation of porous media in the simple (parabolic) form

$$\frac{k}{a}\frac{\partial^2 w}{\partial z^2} = \frac{\partial w}{\partial t}, \tag{3.16}$$

where $w = p_1 - p$, k is the coefficient coming from Darcy's law, and the pore number ε, the compression number a, and the flux rate q are related through the equations

$$\frac{\partial \varepsilon}{\partial t} = -a\frac{\partial p}{\partial t}, \quad \frac{\partial q}{\partial z} = -\frac{\partial \varepsilon}{\partial t} = -a\frac{\partial w}{\partial t}, \quad q = -k\frac{\partial w}{\partial z}, \tag{3.17}$$

where the last expression is none other than a writing of Darcy's law in Terzaghi's notation. Here p_1 is a constant load. It is Eq. (3.16)—akin to the heat equation or the equation obtained by Fick (1855) in diffusion—that made Terzgahi famous. The above sketched proof is somewhat shaky, and this was to be seriously criticized by Fillunger. But the crucial importance of the intervening of Darcy's law must be emphasized. This will be a recurring fact.

Next, Fillunger (1936) published an approach that would perfectly fit in a modern presentation in the way of the theory of mixtures. Considering a one-dimensional setting for the ease in presentation and the case of incompressible fluid and solid components, Fillunger wrote the local equations of conservation of linear momentum and mass (no exchange of mass between the two constituents with a rigid solid constituent) as the following system in his notation:

$$\begin{gathered}\frac{\partial v_1}{\partial t}+v_1\frac{\partial v_1}{\partial z}=\frac{1}{\rho_1}\left(-Z-\frac{\partial p_1}{\partial z}\right),\\ \frac{\partial v_2}{\partial t}+v_2\frac{\partial v_2}{\partial z}=\frac{1}{\rho_2}\left(Z-\frac{\partial p_2}{\partial z}\right),\\ \frac{\partial \rho_1}{\partial t}+\frac{\partial}{\partial z}(\rho_1 v_1)=0, \frac{\partial \rho_2}{\partial t}+\frac{\partial}{\partial z}(\rho_2 v_2)=0,\end{gathered} \tag{3.18}$$

where subscripts 1 and 2 refer to the fluid and solid constituents, respectively, and Z is an interaction force. If n is the porosity, $p=p_1+p_2$ the total pressure in terms of the partial pressures, and γ_1 and γ_2 are the specific weights, we have

$$p_1=np, \quad p_2=(1-n)p, \tag{3.19}$$

and

$$\frac{1}{\rho_1}=\frac{g}{n\gamma_1}, \ \frac{1}{\rho_2}=\frac{g}{(1-n)\gamma_2}. \tag{3.20}$$

This allows us to transform the last two of Eqs. (3.18) in the form

$$\frac{\partial n}{\partial t}+\frac{\partial}{\partial z}(nv_1)=0, \quad -\frac{\partial n}{\partial t}+\frac{\partial}{\partial z}(1-n)v_2=0. \tag{3.21}$$

Finally, the system is closed on account of Darcy's law simply written as

$$Z=\frac{nv_1}{k}, \tag{3.22}$$

where k is the permeability. Again, it is Darcy's law that is instrumental in the final argument. The above given system can be extended to the case of a deformable solid matrix and to three dimensions of space. It contains the quintessence of the theory of mixtures applied to porous media, and thus is extremely modern. This is much more "mechanical" than Terzaghi's argument that yielded Eq. (3.16). As a matter of fact, the harsh criticism of Terzaghi's work by Fillunger in his pamphlet of 1936 will have for tragic consequence the suicide of Fillunger and his wife, when

Fillunger was accused of unjustified exaggerate criticism and unethical conduct toward his colleague at TU Wien. This tragic story is told in detail by De Boer (1996, 2005) who is an ardent irremediable admirer of Fillunger, in particular, of his work and its anticipated modernity. Heinrich, also an Austrian engineer and professor of mechanics at TH Wien, was a direct disciple of Fillunger and formulated with Desoyer (1961) the proper three-dimensional formulation.

Maurice A. Biot (1905–1985) was a Belgian-American physicist who always brought original solutions to problems of mechanics and/or physics. This may be due to the fact that for most of his life he was not associated with a precise academic institution, but he rather acted as a maverick sometimes affiliated with private oil companies or acting as a consultant, and sometimes giving lectures in prestigious universities (Harvard, Columbia, Brown). To him are due original works in the theory of finite strains in elastic bodies, an incremental theory of deformable solid mechanics, variational principles, irreversible thermodynamics, and, most relevant to the present contribution, a theory of *poro-elasticity* (then called "Biot's theory"). The latter was a real extension of constitutive equations to saturated poro-elastic solids. He paid special attention to the propagation of waves in such media, obtaining thus results that provided useful measurement techniques in geophysical applications (Biot 1956a, b). Biot generally is a follower of Terzaghi (Biot 1941). But, in 1955, he proposed an original theory with a set of two coupled linear momentum equations for the fluid (subscript *F*) and solid (subscript *S*) constituents with respective displacement fields $\mathbf{u}_F$ and $\mathbf{u}_S$ in the following form:

$$\rho^{11}\ddot{\mathbf{u}}_S + \rho^{12}\ddot{\mathbf{u}}_F + b(\dot{\mathbf{u}}_S - \dot{\mathbf{u}}_F) = div\underline{\sigma}^S + \rho^S\mathbf{b}, \tag{3.22a}$$

$$\rho^{12}\ddot{\mathbf{u}}_S + \rho^{22}\ddot{\mathbf{u}}_F + b(\dot{\mathbf{u}}_F - \dot{\mathbf{u}}_S) = \nabla\sigma + \rho^F\mathbf{b}, \tag{3.22b}$$

where the mass (or inertial) coupling quantity $\rho^{12} < 0$, $\underline{\sigma}^S$ is the solid elastic stress, $\mathbf{b}$ is the external force per unit mass of each constituent, and $\sigma = -np$ is the real fluid pressure if n is the porosity. Finally, the interacting constant b is related to Darcy's law by an expression of the kind

$$b = \frac{\mu n^2}{k}. \tag{3.23}$$

The strange coupling inertia terms come from a kinetic energy that is jointly quadratic in the velocities $\dot{\mathbf{u}}_S$ and $\dot{\mathbf{u}}_F$. The coupling involving Darcy's concept is deduced from a dissipation potential quadratic in the relative velocity $(\dot{\mathbf{u}}_S - \dot{\mathbf{u}}_F)$—in the manner of Rayleigh and evoking a friction phenomenon between two phases. Indeed, the set (3.22) really follows from a Lagrangian-Hamiltonian variational with source terms in the resulting Euler-Lagrange equations derived from the dissipation potential. This, at the time (1956), was a formidable achievement and original development in the field of geophysical engineering, all the more that Biot could give a reasonable meaning to the inertial contributions in the left-hand sides of Eqs. (3.22). Again, Darcy's law was a decisive element in the formulation.

Interesting dynamic properties were deduced by Biot (1956a, b) from his model (3.22) that is still considering small strains. But in one of his last papers Biot (1972) formulated a theory of finite deformations in porous solids (still in his own formalism and notation). Many subsequent works will reformulate Biot's theories in the modern nonlinear format of continuum thermo-mechanics, adding in fact little to their physics and main properties. Tolstoy (Editor 1991) provides a handy collection of Biot's most innovative papers in the acoustics, elasticity, and thermodynamics of porous media. In the case of poroelastic solids infused with compressible fluids, new results generalizing those of Biot were obtained by Quilogotti et al. (2004).

An interesting, but seldom cited, paper was proposed by Frenkel (1944) before most of the works by Biot. In modern notation Frenkel's equations of the fluid and solid phases of the porous medium are given by

$$\rho^F \frac{\partial \mathbf{v}_F}{\partial t} = -n\nabla p + \rho^F \mathbf{b} - \frac{\mu}{\kappa}(\mathbf{v}_F - \mathbf{v}_S), \tag{3.24a}$$

$$\rho^S \frac{\partial \mathbf{v}_S}{\partial t} = div\underline{\sigma}^S - (1-n)\nabla p + \rho^S \mathbf{b} + \frac{\mu}{\kappa}(\mathbf{v}_F - \mathbf{v}_S), \tag{3.24b}$$

where we identify the last terms in the right-hand side of these equations as due to Darcy's law.

To conclude this section we note Derski's (1978) model where the mass coupling of Biot between fluid and solid components is apparently neglected but it is assumed that the density of the fluid component can be divided into two parts, one part being the density of the free fluid ρ^{Ff}, moving with the velocity $\mathbf{v}_F$, and the other part being the density ρ^{Fs} of the "trapped" fluid that moves with the velocity of the solid skeleton $\mathbf{v}_S$. It was shown that Derski's equations of momenta are equivalent to those of Biot with the appropriate interpretation of densities in the kinetic energy (see De Boer 2000, p. 305).

3.6 Continuum Thermo-Mechanics and Constitutive Modelling

It is Ray Bowen (1967, 1976, 1982) and Ingo Müller (1968) who provided the basic equations of the theory of mixtures in the format of rational continuum thermo-mechanics after the pioneering work of Truesdell (1957). Other presentations in the same line but emphasizing the case of two-component porous media are those of De Boer (2000), Ehlers (2010), and Wilmanski (1998, 2003). The fruitful role of this Essen group around De Boer must be underlined. This was also the case of O. Coussy (1953–2010) and his co-workers at the Laboratory of Ponts et Chaussées in the east suburb of Paris, where thermal effects and finite strains were taken into account in a definite Truesdellian style (cf. Coussy 1995, 2010). The thermodynamic expression, the implementation of the Clausius-Duhem inequality as a constraint imposed on the constitutive equations, and the deduced constitutive equations can be rather cumbersome (cf. De Boer 2000, Chap. 5). This is also the case of the approach exposed by Rajagopal and Tao (1995) where an overuse of the theory of invariants in the representation of some fields obscures the intuitive physics of the problem (See also Grinfeld 1991, Rajagopal 2007). This drawback is often observed in the continuum mechanics of complex media (e.g., also in electromagnetic continua). But in the present case this complication allows a satisfactory treatment of problems such as the unsteady diffusion of a fluid through a nonlinear elastic solid body or the wave propagation in solids infused with fluids. Such wave phenomena were duly studied by K. Wilmanski and B. Albert in Berlin (e.g., Wilmanski 2005b), and Quiligotti et al. (2004) who all started from a theory with large deformations. Noteworthy also are the original works of Wilmanski (1996, 2005a) where this author considers an equation governing the porosity [cf. Equations (3.21) above], and also the notion of tortuosity in the continuum framework. We finally note that in their recent work, Shriram and Rajagopal (2014) exploit the rational thermodynamics of mixtures and a criterion of “maximal rate of entropy production”, and succeed in reproducing the Darcy, Brinkman and Forchheimer laws—in disguise—and some other generalizations but by agreeing on a number of ad hoc working hypotheses.

3.7 Conclusion

In this contribution we have favoured the historical developments during the nineteenth century and until the years 1960s. It is more than obvious that the experimental results of Darcy—condensed in this famous Note D of his innocuous, but finally magisterial, book on the fountains of his native city—provided the real breakthrough on which all forthcoming models will be based. We cannot ignore the decisive role played by engineers/physicists such as Fillunger, von Terzaghi and Biot of which the works provided, not only a sound basis for the study of geomaterials and construction materials, but also elements for the study of dynamical processes. The Technical University in Vienna paid an important contribution to the field that has now reached a worldwide development with the constitution of an international network (PoroNet), the creation of specialized scientific journals, and the periodic gathering of international symposia. Because of personal interests and experience, we have limited the mechanical behaviour of the solid constituent—

Fig. 3.2 J.L.M. Poiseuille (1797–1869)

when deformable—of the considered porous media to elasticity, favouring thus the theory of *poro-elasticity*. However, it is readily acknowledged that porosity plays a role in the plastic behaviour of some materials, in particular among geomaterials and even in metals with growing cavities. The resulting effect of porosity will be an alteration in the yield criterion, whether of the Mises or Coulomb types. In the case of metals, porosity will intervene in the ductile fracture of these materials. It is no question here to expand this sophisticated theory. Suffice it to note the breakthrough due to Gurson (1977), and improvements on his proposal by other authors, e.g., Cologanu et al. (1997). Finally, to be more or less complete, we note a growing interest in so-called microporomechanics (cf. Dormieux et al. 2006) and also the interest for porous media in applied acoustics (cf. Allard 1993; Bourbié et al. 1987). Note that evoking the theory of mixtures we avoided speaking about truly reacting media that are of utmost interest in problems of propulsion technology, but they are not considered as porous media (for this, see Prud'homme 1988, 2010). Other portraits of main contributors to the field are given in Figs. 3.2, 3.3, 3.4, 3.5, 3.6, 3.7, 3.8, 3.9 and 3.10.

Fig. 3.3 Henry Darcy 1803–1856)

Fig. 3.4 Gotthilf Hagen (1797–1884)

Fig. 3.5 Maurice A. Biot (1905–1985)

Fig. 3.6 Paul Fillunger (1883–1937)

Fig. 3.7 Karl von Terzaghi (1883–1963)

Fig. 3.8 Reint De Boer

Fig. 3.9 Clifford A. Truesdell (1919–2000)

Fig. 3.10 Ray M. Bowen (born 1936). (Copyright: Texas A&M University, 2005)

Appendix: Elements of APH

For a typical diffusion, problem with rapidly oscillating coefficients, we consider the partial differential equation:

$$\nabla \cdot \left(\mathbf{A}\left(\frac{\mathbf{x}}{\varepsilon}\right) \cdot \nabla \mathbf{v}_{\varepsilon} \right) = \mathbf{f}. \tag{A.3.1}$$

This is to be replaced by the homogenized equation

$$\nabla \cdot \left(\mathbf{A}^{\text{hom}} \cdot \nabla \mathbf{v} \right) = \mathbf{f}, \tag{A.3.2}$$

where $\mathbf{A}^{\text{hom}}$ is the homogenized (constant in space) material tensor coefficient.

To arrive at this result and the expression of $\mathbf{A}^{\text{hom}}$, one has to consider the following formal expansion (so-called Ansatz)

$$\mathbf{v}_{\varepsilon}(\mathbf{x}) = \mathbf{v}(\mathbf{x}, \mathbf{y}) = \mathbf{v}_0(\mathbf{x}, \mathbf{y}) + \varepsilon \mathbf{v}_1(\mathbf{x}, \mathbf{y}) + \varepsilon^2 \mathbf{v}_2(\mathbf{x}, \mathbf{y}) + O(\varepsilon^3). \tag{A.3.3}$$

Substitute from this in the equation (A.3.1) yielding thus a hierarchy of problems. The effective coefficients $\mathbf{A}^{\text{hom}}$ are determined by solving a set of problems over the unit cell for the function $\mathbf{v}_1(\mathbf{x}, \mathbf{y})$. It is noted that only the first two in (A.3.3) are justified; to go further one would have to account for boundary-layer terms.

References

Adler P (1992) Porous media: geometry and transports. Butterworth-Heinemann, London

Allaire G (1991) Homogenization of the Navier-Stokes equations in open sets perforated with tiny holes. Arch Rat Mech Anal 113:209–298

Allard JF (1993) Propagation of sound in porous media: modelling sound absorbing materials. Elsevier, UK

Atkin RJ, Craine RE (1976) Continuum theories of mixtures: basic theory and historical development. Q J Mech Appl Math 29:209–244

Auriault JL (2009) On the domain of validity of Brinkman's equation. Transp Porous Media 79:215–223

Auriault JL, Geindreau C, Boutin C (2005) Filtration law of porous media with poor separation of scales. Transp Porous Media 60:89–108

Bakhvalov N, Panasenko G (1989) Homogenization: averaging processes in periodic media. Kluwer, Dordrecht

Bear J (1972) Dynamics of fluids in porous media. Elsevier, New York

Bensoussan A, Lions JL, Papanicolaou G (1978) Asymptotic analysis for periodic structures. North-Holland, Amsterdam

Biot MA (1941) General theory of three-dimensional consolidation. J Appl Phys 12:155–164

Biot MA (1955) Theory of elasticity and consolidation for a porous anisotropic solid. J Appl Phys 26:182–185

Biot MA (1956a) Theory of propagation of elastic waves in a fluid-saturated porous solid, I. Low-frequency range. J Acoust Soc Am 28:168–178

Biot MA (1956b) Theory of propagation of elastic waves in a fluid-saturated porous solid, II. Higher frequency range. J Acoust Soc Amer 28:179–191
Biot MA (1972) Theory of finite deformations in porous solids. Indiana University Mathematics Journal 21:597–620
Bourbié T, Coussy O, Zinszner B (1987) Acoustics of porous media. Editions Technip, Paris
Boussinesq JV (1904) Recherche théorique sur l'écoulement des nappes d'eau infiltrées dans le sol et sur le débit des sources. J Math Pures Appl 10:5–78
Bowen RM (1967) Toward a thermodynamics and mechanics of mixtures. Arch Rat Mech Anal 24:370–403
Bowen RM (1976) Theory of mixtures. In: Eringen AC (ed) Continuum Physics, vol 3. Academic Press, New York, pp 1–127
Bowen RM (1982) Compressible porous media models by use of the theory of mixtures. Int J Eng Sci 20:697–735
Brinkman HC (1947) A calculation of the viscous force exerted by a flowing fluid on a dense swarm of particles. Appl Sci Res 1:27–34
Buckingham E (1907) Generalization of Darcy's law to saturated and unsaturated flows in heterogeneous and anisotropic bodies. Bulletin no. 8, Bureau of Soils, Department of Agriculture, US Government, Washington, DC
Cologanu M, Leblond JB, Perrin G, Devaux J (1997) Recent extensions of Gurson's model for porous ductile metals. In: Suquet P (ed) Continuum micromechanics. Springer, New York, pp 61–130
Coussy O (1995) Mechanics of porous media. Wiley, Chichester
Coussy O (2010) Mechanics and physics of porous solids. Wiley, Chichester
Dal Maso G (1992) An introduction to Γ-convergence. Birkhäuser, Boston, Basel
Darcy H (1856) Les fontaines publiques de la ville de Dijon. Victor Dalmont Editeur, Paris
Darcy H (1857) Recherches expérimentales relatives au mouvement de l'eau dans les tuyaux. Two volumes. Imprimerie Impériale/Malet-Bachelier, Paris
De Boer R (1996) Highlights in the historical development of the porous media theory—toward a consistent macroscopic theory. Appl Mech Rev 49:201–262
De Boer R (2000) Theory of porous media—highlights on historical developments and current status. Springer, Berlin
De Boer R (2005) The engineer and the scandal: a piece of Science history. Springer, Berlin
De Prony G (1825) Résumé de la théorie des formules fondamentales relatives au mouvement de l'eau dans les tuyaux et les canaux. Paris
Derski W (1978) Equations of motion for a fluid-saturated porous solid. Bull Acad Pol Sci, Sér Sci Techn 26:11–16
Dormieux L, Kondo J, Ulm FJ (2006) Microporomechanics. Wiley, UK
Dupuit J (1843) Etudes théoriques et pratiques sur le mouvement des eaux dans les canaux découverts et à travers les terrains perméables. Victor Dalmont, Paris (Second edition, Dunod, Paris, 1865)
Ehlers W (2010) Porous media. Springer, Berlin
Ene H, Polisevski D (1987) Thermal flow in porous media. Springer, Berlin
Ene H, Sanchez-Palancia E (1975) Equations et phénomènes de surface pour l'écoulement dans un modèle de milieux poreux. J Mécanique (Paris) 14:73–108
Eringen AC, Ingram JD (1965) A continuum theory of chemically reacting media I. Int J Eng Sci 3:197–212
Eringen AC, Ingram JD (1967) A continuum theory of chemically reacting media II. Int J Eng Sci 5:289–322
Euler L (1762) Anleitung zur Naturlehre. In: Opera postuma mathematica et physica, Bd. 1-2, St Petersburg, (Reprint, New York, 1969)
Fick A (1855) Über diffusion. Ann Physik und Chemie 94:59–86
Fillunger P (1913) Der Auftrieb in Talsperren. Österreische Wochenschrift für den öffentlichen Baudienst 19(532–556):567–570

Fillunger P (1914) Neuere Grundlagen für statische Berechnung von Talsperren. Zeit des Österreischen Ingenieur und Architekten-Vereines 23:441–447
Fillunger P (1936) Erdbaumechanik. Privately published, Vienna
Forchheimer P (1901) Wasserbewegung durch Boden. Zeit. Ver.deutscher Ingenieure 45:1736–1788
Frenkel I (1944) On the theory of seismic and seismo-electric phenomena in a moist soil. J Phys 8:230–241
Green AE, Naghdi PM (1965) A dynamic theory of interacting continua. Int J Eng Sci 3:231–241
Green AE, Naghdi PM (1971) Entropy inequality for mixtures. Q J Mech Appl Math 24:473–485
Grinfeld MA (1991) Thermodynamic methods in the theory of heterogeneous systems. Longman, Harlow
Gurson AL (1977) Continuum theory of ductile rupture by void nucleation and growth, I. Yield criteria flow rules for porous ductile media. J Eng Mater Technol 99:2–15
Hagen G (1839) Über d. Bewegung d. Wassers in engen zylindr. Röhren. Hdb d Wasserbaukunst 3:1841–1863
Heinrich G, Desoyer K (1961) Theorie dreidimensionaler Setzungsvorgänge in Tonschichten. Ing Arch 30:225–253
Kelly PD (1964) A reacting continuum. Int J Eng Sci 2:149–158
Lévy T (1983) Fluid flow through an array of fixed particles. Int J Eng Sci 21:11–23
Lévy T, Sanchez-Palencia E (1975) On boundary conditions for fluid flow in porous media. Int J Eng Sci 13:923–940
Maugin GA (2014a) Continuum mechanics through the eighteenth and nineteenth Centuries. Springer, Dordrecht
Maugin GA (2014b) Hydraulics: the importance of experiments. Preprint UPMC. Paris (Chapter 2 in this volume)
Mouret G (1921) Antoine Chézy: histoire d'une formule d'hydraulique. Ann Ponts et Chaussées 61:165–269
Müller I (1968) A thermodynamic theory of mixtures of fluids. Arch Rat Mech Anal 18:1–39
Nakayama A (2007) Historical development and current state of porous media theory and Applications. Internal report, Shizuoka University, Hamamatsu, Japan
Oleinik OA, Shamaev AS, Yosifian GA (1991) Mathematical problems in elasticity and homogenization. North-Holland, Amsterdam
Poiseuille JLM (1846) Recherches expérimentales sur le mouvement des liquides dans les tubes de très petit diamètre. Mém Acad Sci 9:433–564
Prud'homme R (1988) Fluides hétérogènes et réactifs: écoulements et transferts. Lecture Notes in Physics, vol 34. Springer, Berlin
Prud'homme R (2010) Flows of reactive fluids. Springer, New York
Quiligotti S, Maugin GA, dell'Isola F (2004) Wave motions in unbounded poroelastic solids infused with compressible fluids. Zeit angew Math Phys 53:75–84
Rajagopal KR (2007) On a hierarchy of approximate models for flows of incompressible fluids through porous solids. Math Models Method Appl Sci 17:215–252
Rajagopal KR, Tao L (1995) Mechanics of mixtures. World Scientific, Singapore
Sanchez-Palencia E (1980) Non-homogeneous media and vibration theory. Springer, Berlin
Shriram S, Rajagopal KR (2014) A thermodynamic basis for the derivation of the Darcy, Forchheimer and Brinkman models of flows through porous media and their generalizations. Int J Non-Linear Mech 58:162–166
Straughan B (2011) Heat waves. Springer, New York
Tolstoy L (eds) (1991) Acoustics, elasticity, and thermodynamics of porous media. Twenty one papers by M.A. Biot. Acoustical Society of America, New York
Truesdell CA (1957) Sulle basi della termomeccanica. Rendi C Accad Naz Lincei 22(33–38):158–166
Truesdell CA (1986) Rational thermodynamics, Second revised and, enlarged edn. Springer, New York

von Terzaghi K (1923a) Die Berechnung der Durchlässigkeitsziffer des Tones aus dem Verlauf der hydrodynamischen Spannungerscheinungen. Sitz Akad Wissens Wien, Math Naturwiss Kl, Abt IIa(No.3/4): 125–138
von Terzaghi K (1923b) Die Berechnung der Durchlässigkeitsziffer der Tones aus dem Verlauf der hydrodynamischen Spannungerscheinungen. Sitz Akad. Wissen. Wien. Math. Naturwiss., Kl. Abt IIa, 132:105–124
von Terzaghi K (1943) Theoretical soil mechanics. Wiley, New York
Weisbach JK (1855) Die Eksperimental-Hydraulik. Engelhardt, Freiberg
Wilmanski K (1996) Porous media at finite strains: the new model with the balance equation for porosity. Arch Mech (PL) 48:591–628
Wilmanski K (1998) Thermomechanics of continua. Springer, Berlin
Wilmanski K (2003) On thermodynamics of nonlinear poroelastic materials. J Elast 71:247–261
Wilmanski K (2005a) Tortuosity and objective relative accelerations in the theory of porous materials. Proc Roy Soc London A461:1533–1561
Wilmanski K (2005b) Critical time for acoustic waves in weakly nonlinear poroelastic materials. Cont Mech Thermodyn 17:171–181
Woltman (1794) Beyträge zur Hydraulischen Architectur. Third and Fourth Parts. John C. Dietrich, Göttingen

Specialized Journals

In addition to standard journals in civil engineering, geomechanics, soil mechanics, continuum mechanics, applied mathematics, and numerical techniques, the following specialized journals focus on various aspects of porous media: Transport in porous media, Mechanics of cohesive-frictional materials, International Journal of multiphase Flows.

Chapter 4
Viscosity, Fast Flows and the Science of Flight

Abstract In line with our previous perusing of the early developments of hydraulics and the foundation works on porous media, the present contribution pays more due attention to the effects of viscosity, to increasing levels of velocity of the fluid flows, and to the inception of the mechanics of vortices. The main characters at work are essentially Helmholtz (his vortex filaments) and Reynolds (his celebrated number). The application side materializes in the groundbreaking ideas of Lanchester and the more mathematical approach of Prandtl that were going to bring a true efficient revolution in the just born science of flight. The role played by non-dimensional numbers, dimensional analysis, and new methods such as good asymptotic expansions is rightly emphasized, leading us to the successes of the mathematics of aeronautics and astronautics in the twentieth century. Again, the offered approach is more discursive and historical than mathematically detailed and rigorous.

4.1 Introduction

One does not need to be a hard-line scientist or a gifted experimentalist in hydraulics to realize that Nature presents to us all kinds of fluid flows, some of them with complicated appearance features. This is classically illustrated by Leonardo da Vinci's remarkable sketches of whirlpools in the Renaissance period and the attractive pictures provided by gifted Japanese illustrators (e.g., Hokusai) on woodprints in the nineteenth century. The rather quiet flow observed by Poiseuille in his celebrated experiments on the flow in pipes at low flow rate of the fluid is not exemplary of this more complex situation which seems to be rather frequent. To go further than Poiseuille (1846) and Hagen (1869) one had to devise more conclusive experiments that would involve "faster" flows. These experiments were performed by Osborne Reynolds (1842–1912) at Owens College (now the University of Manchester). Still there was need to characterize in a rather simple manner the physical circumstances in which the fluid flow occurred as, contrary to what we

G.A. Maugin, *Continuum Mechanics through the Ages - From the Renaissance to the Twentieth Century*, Solid Mechanics and Its Applications 223,
DOI 10.1007/978-3-319-26593-3_4

may have hinted at, velocity is *not* the only quantity to be involved. Geometry and viscosity play also a prevalent part. This was the breakthrough made by Reynolds when he showed that an appropriate combination of velocity, viscosity and a characteristic size of the considered experiment helped him define a particularly important parameter: the *Reynolds number*, the ratio of inertial and viscous forces—a first example of a number without physical dimension—indeed introduced by Reynolds in 1883. This number—together with other numbers such as the Mach number and Prandtl number—was to play a fundamental role in fluid mechanics, in particular with its special value characteristically attached to a transition between *laminar* (conditions of Poiseuille's experiments) and *turbulent* flows. In laminar flows, it can be said that all parts of the fluid move in the same direction, while in turbulent motion one observes vortices in which the direction of motion changes constantly. That is, turbulence visually involves so-called *vortices*. It happens that this notion was mathematically described principally by von Helmholtz. It was to play a fundamental role in studies in turbulence but also in the early developments of aeronautical flight. Immense progress in the theory of flight was due to Ludwig Prandtl (1875–1953). The two giant scientists in the present contribution therefore are Reynolds and Prandtl. The two are in fact related as Prandtl and his co-workers were instrumental in disseminating some of Reynolds' ideas.

4.2 The World of Vortices

This may be considered a prerequisite to the next two sections although the theory of vortices provides a discipline that may practically be considered on its own with a refined beauty and elegance due to the genius of one individual: Herrmann von Helmholtz (1821–1894; Fig. 4.1), one of the most brilliant, versatile and original minds of the nineteenth century. Helmholtz contributed with an equal success to the fields of physiology (in particular physiological optics, the theory of colour vision), energetics, continuum mechanics (hydrodynamics), acoustics (the Helmholtz's equation), applied mathematics (the celebrated Helmholtz decomposition of a vector field), and electricity (electric oscillations), and electrodynamics (a theory in competition with Maxwell's). He was an empirist, but one equipped with an excellent command of mathematics. Above all, he is one of the three scientists who formulated the first law of thermodynamics. This was in 1847. But he also proposed epoch-making contributions in all the above-mentioned fields. For the present concern, however, we underline his very original contribution to the flow of inviscid (i.e., non-viscous) fluids where he singlehandedly but definitely introduced a mechanics of vortices and vortex filaments. This we briefly describe now for the benefit of the non-specialist reader.

First, a *vortex*—as illustrated by a whirlpool, for example, observed in draining a bathtub—is a region in the fluid flow where the flow is mostly spinning in motion about an imaginary axis; the latter can be straight or curved. Vortices provide the major component of turbulent flows. But in the absence of applied forces, viscous

Fig. 4.1 Herrmann von Helmholtz (1821–1894)

friction (if it exists) tends to organize the flow in a collection of "irrotational" (or "free") vortices. In such vortices the fluid velocity is largest near to the imaginary axis; it decreases from this axis in inverse proportion to the distance. Viewed as objects, vortices can move, deform, and interact between fellow vortices. From the point of view of mechanics, they can carry linear and angular momenta, energy and mass. A vortex line is a line of which the tangent is everywhere parallel to the local vorticity vector $\underline{\omega}$. The latter describes the local rotary motion at a point in the fluid. If $\mathbf{v}$ is the velocity field, then in the nabla notation: $\underline{\omega} = \nabla \times \mathbf{v}$. The vortex lines drawn through each point of a closed curve constitute the surface of a vortex tube. A *vortex filament* is the limit of this picture when the cross section of the vortex tube becomes infinitesimally small. Helmholtz expanded the mechanics of such filaments.

Obvious general properties directly follow from the definition of vorticity. Thus,

$$\nabla \cdot \underline{\omega} = \nabla \cdot (\nabla \times \mathbf{v}) \equiv 0 \tag{4.1}$$

$$\int_S \underline{\omega} \cdot d\mathbf{S} = \int_S (\nabla \times \mathbf{v}) \cdot d\mathbf{S} = \int_{L=\partial S} \mathbf{v} \cdot d\mathbf{l} = \Gamma(L), \tag{4.2}$$

where $d\mathbf{S} = \mathbf{n}dS$ is the oriented surface element and Γ is the *circulation* along the closed line L on which the surface S—equipped with unit outward normal $\mathbf{n}$—is leaning. That is,

$$\Gamma(L) := \int_L \mathbf{v} \cdot d\mathbf{l} = \int_S \underline{\omega} \cdot \mathbf{n}\, dS, \tag{4.3}$$

if $L = \partial S$, and Stokes' theorem was used in writing (4.2) or (4.3). These equations are tantamount to saying that vorticity is the circulation per unit area taken around an infinitesimal loop. As to Eq. (4.1), it can be viewed as a simple conservation equation. Sometimes Γ is called the strength of a vortex tube—see below. Indeed, if we consider that the surface S encloses a simply connected continuous volume V, then by the divergence theorem

$$\int_S \underline{\omega} \cdot d\mathbf{S} = \int_V (\nabla \cdot \underline{\omega}) dV \equiv 0. \tag{4.4}$$

Applying this to a vortex filament of end sections S_1 and S_2 and accounting for the fact that the vorticity is, by definition of the filament, tangent to the lateral curved surface of the filament, this leaves as only contributions those at the end surfaces, i.e.,

$$\int_S \underline{\omega} \cdot d\mathbf{S} = \omega_2 S_2 - \omega_1 S_1 = 0. \tag{4.5}$$

We may view this as a continuity equation for the *vortex intensity* (the product of the vorticity and the cross-section area) that is constant along the considered filament. A direct consequence of this is that a vortex filament cannot end inside the fluid region. Indeed, if it were the case, the cross-section area would have to vanish at some point, but the vorticity would have to become infinite in order to respect this continuity equation. Accordingly, a vortex filament must either form a closed vortex ring, or it must terminate at the fluid boundary. Furthermore, if we consider a vortex tube as a bundle of vortex filaments—just as a large electric wire containing a bunch of closely packed small wires—the net intensity of the tube is the sum of the intensities of individual filaments. From this it follows that the intensity of a vortex tube remains constant along the tube. This we can also write as the evolution equation

$$D_C \underline{\omega} = \mathbf{0}, \tag{4.6}$$

where D_C denotes a convected time derivative (in modern terms, a Lie derivative in following the flow motion), so that in an inviscid fluid, the flow remains irrotational if it was so initially. The proof of this requires taking the curl of the equation of motion for an inviscid incompressible fluid in a potential field of external forces, i.e., the Euler equation

$$\rho\left(\frac{\partial \mathbf{v}}{\partial t} + (\mathbf{v} \cdot \nabla)\mathbf{v}\right) = -\nabla p + \rho \mathbf{f}, \quad \mathbf{f} = -\nabla \varphi, \nabla \cdot \mathbf{v} = 0. \tag{4.7}$$

All these results are given by Helmholtz—in his own notation—in his paper of 1858 that is a real pearl and a paragon of excellent scientific literature.[1] In his brilliant introduction Helmholtz states in his own words his main results: *if all the forces that act on the fluid have a potential* [then]*: (1) no water particle that was not originally in rotation is made to rotate (2) the water particles that at any time belong to the same vortex line, however they may be translated, will continue to belong to the same vortex line; (3) the product of the cross section and the velocity of rotation of an infinitesimally thin vortex filament is constant along the entire length of the filament and retains the same value during all displacements of the filament. Hence the vortex filaments must run back into themselves in the interior of the fluid or else must end at the bounding surface of the fluid.* Our own statements just paraphrase Helmholtz's ones. But this author achieved much more in his epoch-making paper. In particular, he thoroughly studied interactions between straight parallel filaments and between circular vortex filaments. Exemplifying his truly aerial view of physics, Helmholtz is not without mentioning at several points the analogy of his vortex filaments with electric currents. Such analogies remain useful until today in magneto-hydrodynamics as illustrated in the works reported by Moffatt (2008) related to knottedness and topology and Kelvin's extensions of Helmholtz's considerations. Moffatt recalls on this occasion that the quantity called the *helicity* (discovered by J.-J. Moreau in 1961) and defined by

$$H := \int_V \underline{\omega} \cdot \mathbf{v}\, dV \tag{4.8}$$

is an invariant of the Euler equations.

Of course Helmholtz's work refers to a case where viscous effects are negligible, but we know while observing real fluids that the strength of vortices will always decay gradually in presence of viscosity. As a consequence some of the above-mentioned remarkable results will be altered, e.g., a viscous diffusive term will appear in the right-hand side of Eq. (4.6).

As indicated by the title of this section the study of vortices opens up horizons of a real world (cf. Saffman 1995), especially with the concept of vortex sheet and its applications to aerodynamics (cf. Anderson 1997). Before returning to this point in Sect. 4.4, we must necessarily deal more inclusively with the influence of viscosity

[1]The reader can check by himself that Eqs. (7), (6), and (5) are none other than Eqs. (1, p. 44), (3, p. 49) and the unnumbered equation in p. 52 in the English translation of the original German text; cf. Helmholtz (1858). In particular Eq. (3, p. 49) is none other than the modern equation.

$$\frac{\partial \underline{\omega}}{\partial t} - (\underline{\omega} \cdot \nabla)\mathbf{v} = \mathbf{0}.$$

For the notion of connectedness Helmholtz refers to recent work by Riemann in the previous volume of the same journal.

in guided flows, where in fact the study was to be perfected after Poiseuille and Hagen, that is, with the experiments of Reynolds.

4.3 Reynolds and the Transition to Turbulence

The general equations that govern (linear isotropic) viscous fluids were proposed by three authors: C.L.M.H. Navier (1785–1836), Adhémar Barré de Saint-Venant (1797–1886), and Gabriel Stokes (1819–1903), respectively in Navier (1822), Saint-Venant (1843) and Stokes (1845). Compared to Eq. (4.7), they read

$$\rho\left(\frac{\partial \mathbf{v}}{\partial t} + (\mathbf{v}\cdot\nabla)\mathbf{v}\right) = -\nabla p + \mu\nabla^2\mathbf{v} + \rho\mathbf{f}, \tag{4.9}$$

and present a formidable challenge for their solution. The new quantity here is the dynamic viscosity μ. The ratio $\eta = \mu/\rho$ is called the kinematic viscosity.

Poiseuille's and Hagen's works were experimental (cf. Maugin 2014b), but their result was also deduced from the Navier-Stokes equation a short time thereafter by Eduard Hagenbach (1833–1910) from Basel in 1860. To go further than the experimental work of these pioneers, one had to carefully study a wider spectrum of physical conditions in which the flow of water in pipes takes place. The ideal person for this endeavour proved to be Osborne Reynolds (1842–1912; Fig. 4.2). Born in Ireland and showing an early interest in mechanical engineering, Reynolds was nonetheless educated in science at Cambridge with a high level of mathematics. Graduating in 1867, he benefited in 1868 from the creation of a professorship of

Fig. 4.2 Osborne Reynolds (1842–1912)

engineering at Owens College in Manchester—where he was to spend an active professional life as an engineering scientist until retirement in 1905. He concentrated on fluid dynamics starting in 1873 after doing some work in general physics. Although educated as a successful wrangler and elected to a scholarship at Queen's College, Reynolds proved to be an excellent experimentalist in Manchester, but an experimentalist with a deep insight in the bases of his new field of interest. Nobody being entirely perfect, he was a poor teacher to students—often wandering among topics with little or no connection—but still he gave excellent oral presentations to specialists (probably something easier than lecturing students as we all know). His research style may have been described as "individualistic"; In particular, he did not do much preliminary reading before attacking a specific problem; this may be one reason for the novelty contained in many of his papers. More on the personality of Reynolds is to be found in Lamb (1912–1913). His creative works span some thirty years. In this long time span we can easily identify three main breakthroughs.

First of all Reynolds' fame remains attached to the introduction of the non-dimensional number of which a critical value characterizes the transition between laminar and turbulent flows. We must remember that the problem of interest in the 1850–1870s was that of the pressure drop observed in the flow of water in pipes; this typically is a hydraulician's concern as was the case with Poiseuille and Hagen. As a matter of fact, it seems that Hagen (1869) was the first to report on the two distinct types of flow but without quantitative criterion and further clarification. The matter was taken over by Reynolds in a series of experiments of which he published the results in a long memoir (Reynolds 1883). It is worth recalling the whole title of this memoir because it says it all: "*An experimental investigation of the circumstances which determine whether the motion of water shall be direct or sinuous, and the law of resistance in parallel channels*". Here "direct" means "laminar" (e.g., in a straight line) and "sinuous" will mean "turbulent". The word "circumstances" implies that the physical conditions (velocity of incoming fluid, diameter of the pipe) will be varied within some range as a good experimentalist should do with a sufficient control of conditions and experimental means for quantification. The experimental apparatus used by Reynolds gave a visual demonstration of the transition. It is illustrated artistically in many books and we cannot refrain from reproducing it once more (Fig. 4.3).

The most striking result obtained by Reynolds is that it is *not* the flow velocity alone that characterizes the flow transition between different regimes. It is a combination in some *nondimensional* number such as

$$\mathrm{Re} = \frac{\rho U D}{\mu} = \frac{U D}{\eta} \tag{4.10}$$

where U stands for the velocity of the entering fluid and D for the diameter of the pipe. Of course, the symbol Re and name of "Reynolds number" were not given by Reynolds himself (he used the symbol K). It seems that it is Sommerfeld (1908) in a lecture at the 4th International Congress of Mathematicians (Rome) who called it the "Reynolds number" but this coinage is sometimes also attributed to Prandtl (1910)

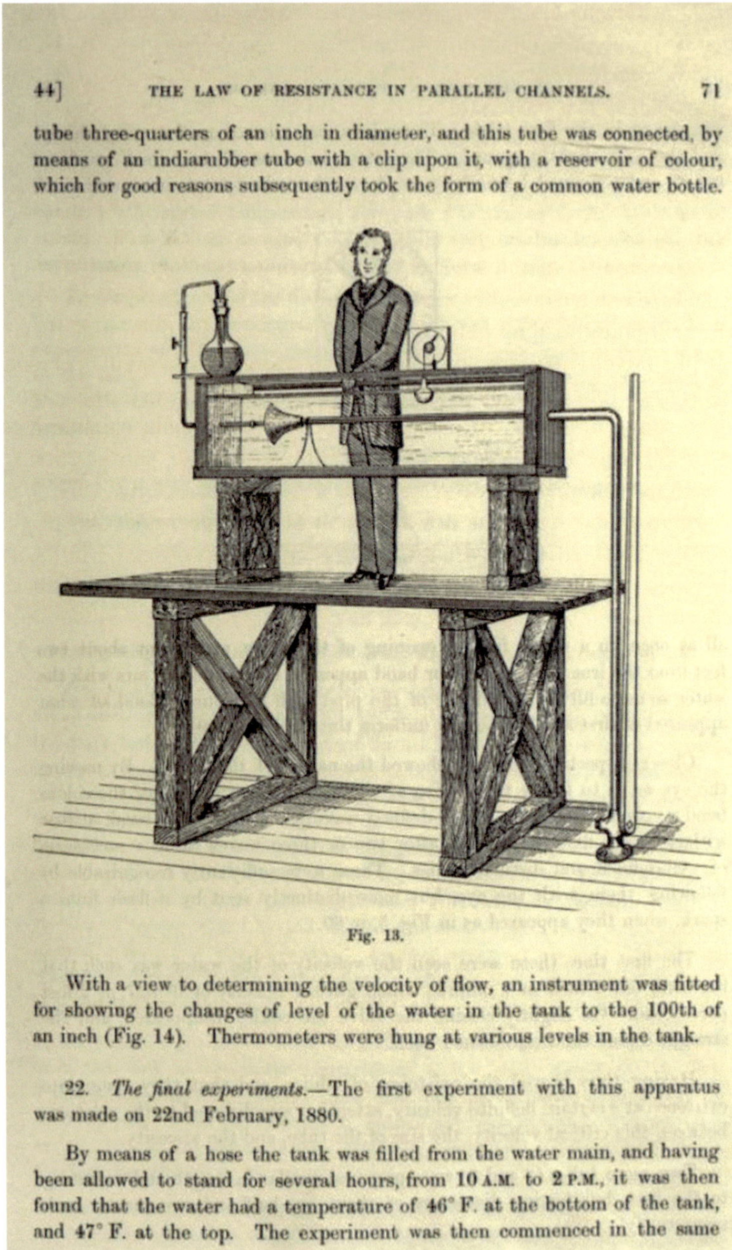

44] THE LAW OF RESISTANCE IN PARALLEL CHANNELS. 71

tube three-quarters of an inch in diameter, and this tube was connected, by means of an indiarubber tube with a clip upon it, with a reservoir of colour, which for good reasons subsequently took the form of a common water bottle.

Fig. 13.

With a view to determining the velocity of flow, an instrument was fitted for showing the changes of level of the water in the tank to the 100th of an inch (Fig. 14). Thermometers were hung at various levels in the tank.

22. *The final experiments.*—The first experiment with this apparatus was made on 22nd February, 1880.

By means of a hose the tank was filled from the water main, and having been allowed to stand for several hours, from 10 A.M. to 2 P.M., it was then found that the water had a temperature of 46° F. at the bottom of the tank, and 47° F. at the top. The experiment was then commenced in the same

Fig. 4.3 Reynolds' experiment from his original publication (1883) as reprinted in Reynolds collected papers

who uses the expression "Reynolds'sche Zahl", or to or Blasius (1912, pp. 640, 642), anyway in the fluid-mechanical influent place of Göttingen. A pleasant and instructive history of the Reynolds number is given by Rott (1990)—also Eckhardt (2009).

Reynolds reported a critical value of the number (4.10) as about 2020, which is not far from contemporary estimates, although it is wiser to avoid giving a precise value and rather to announce that the transition takes place in an interval of values between, say, 1900 and 2000. This attitude is justified by the existence of the phenomenon of *intermittency* (alternation of turbulent and laminar flows) of the flow in the neighbourhood of the critical Re.

But we should not forget that the initial aim pursued both by Poiseuille and Reynold was the study of the drop in pressure over a length L of a circular pipe. Reynolds' result for this quantity can be expressed as (in today's notation)

$$p(L) - p(0) = \frac{1}{2}\rho U^2 \frac{L}{D} f(\mathrm{Re}), \tag{4.11}$$

i.e., essentially a comparison between pressure drop and kinetic energy if we note that the last two factors are without physical dimension. This is called a *similarity law,* so that we can say with Rott (1990, p. 3) that the epoch-making Eqs. (4.10) and (4.11) were "apparently the beginning of a new era in the use of names for nondimensional numbers" (Reynolds, Mach, Prandtl, Froude numbers) and also the inauguration of a whole science of "similarity" (or "similitude" depending on the authors) with due exploitation of such numbers. Here it remains to specify the expression of the function f in Eq. (4.11). A simple example is a power law of the form

$$f(\mathrm{Re}) = c(\mathrm{Re})^{-n}, \tag{4.12}$$

where c is a pure constant. For slow flows one can take $n = 1$ while for very fast flows $n = 0$.

This seems to be compatible with Hagen's (1869) empirical results, but of course Hagen ignored similarity so that the right-hand side of his Eq. (1) would simply be written with a certain power of U. In fact, this last author credits the German hydraulician Reinhard Woltman (1746–1822) for proposing an appropriate exponent of 1.75 for U in Eq. (4.11) above in agreement with $n = 1/4$. A power less than 2 was already observed by the French hydraulician Pierre du Buat (1734–1809) according to Rouse and Ince (1957). More recently, important experiments performed by Saph and Schoder (1903) agreed with the value $n = 1/4$ But A. Flamant (French civil engineer, friend of Barré de Saint-Venant and Boussinesq; cf. Maugin 2014a) claimed that he had given the same formula in 1892 (see his textbook on Hydraulics of 1891/1900 in his second edition: Flamant 1900) indeed in agreement with Saph-Schoder's formula. Reynolds himself gave a value of 1.723 that is close to 1.75.

Anyway, von Kármán (1911) advised the use of Reynolds similarity following his mentor Prandtl, and Moritz Weber (Professor of Naval architecture in Berlin, 1919) put both Reynolds and Froude numbers to use adding a new number of his own (Weber number related to capillarity).

The second remarkable achievement of Reynolds in the field of turbulence was his proposal of a specific approach to the equations of fluid mechanics in the turbulent regime. Probably inspired by recent ideas in the kinetic theory of gases (e.g., by J.C. Maxwell), Reynolds conceived of the fluid motion and allied quantities as being defined in terms of averaged quantities and superimposed fluctuations (Reynolds 1894). This is based on the observation of Reynolds—and Hagen before him- that a turbulent flow is really characterized by a continuous recurrence of turbulent eddies of different size that cause time wise fluctuations in the flow at a given point (cf. Anderson 1997, p. 113). Accordingly, a time average of each flow property would be a steady value. As a consequence, each physical quantity of interest in the turbulent flow is the sum of a time mean, noted $\bar{v}$, and a timewise fluctuating component, say $\tilde{v}$, so that $v = \bar{v} + \tilde{v}$. Applying this time averaging to the Navier-Stokes equation, one is led to a local equation of balance of linear momentum in Reynolds' form (cf. Reynolds 1900–1901):

$$\frac{\partial \bar{v}_i}{\partial t} + \frac{\partial}{\partial x_j}\left(\bar{v}_i\bar{v}_j + \frac{\bar{p}}{\rho}\delta_{ij}\right) = \eta\Delta\bar{v}_i - \frac{\partial}{\partial x_j}\langle \tilde{v}_i\tilde{v}_j \rangle \quad \textit{with} \quad \nabla \cdot \bar{\mathbf{v}} = 0. \tag{4.13}$$

The last contribution in the first of these is the divergence of the so-called *Reynolds' stress*. Then there remains to close the problem by finding a relationship of this tensor with the averaged velocity field. This is an insuperable theoretical problem. However, a simple way out of this enormous difficulty is to assume that the extra term is in fact equivalent to an extra viscosity contribution in the Navier-Stokes equation, i.e., replacing the usual viscosity μ by an effective viscosity $\mu + \mu_T$ where μ_T is an apparent increase due to the fluctuating component in velocity. Such formalism provides the so-called *Reynolds averaged Navier-Stokes equations* for turbulent flow. This has been of constant practical use in engineering with a search for a proper value of μ_T. In more recent times attention was focused on the size and energy of the involved vortices during the development of turbulence, with an interesting cascade of scales where the larger vortices split in smaller ones that in their turn split again, etc. This remarkable advance is due in 1942 to the eminent Russian mathematician Andrei Nikolaevich Kolmogorov (1903–1987) otherwise known for his works in probability theory, dynamical systems and topology. Other attempts to model turbulence by means of constitutive equations in a purely continuum framework are due to Rotta (1951), Lumley (1979) and Favre et al. (1979), but are somewhat outside main trends in the field just as is the nonlocal approach proposed by Speziale and Eringen (Speziale 1981; Speziale and Eringen 1981).

A third important contribution of Reynolds which will have a glorious future in aerodynamics is the so-called *Reynolds analogy* (Reynolds 1874). Reynolds was deeply interested in the coupling between flow and heat conduction since viscosity and thermal effects often concur. This problem concerns the relationship between skin friction and heat transfer. The famous analogy—that exists between heat and mass transfer—is now written in the following non-dimensional form:

$$\frac{C_H}{C_F} = f(\mathrm{Pr}), \tag{4.14}$$

where C_H is the local heat-transfer coefficient, C_F is the local skin-friction coefficient, and the right-hand side is a function of the non-dimensional *Prandtl number* defined by $\mathrm{Pr} = \mu c_P / k$, where c_P is the specific heat at constant pressure and k is the thermal conductivity. This can be interpreted as the ratio of kinematic viscosity to thermal diffusivity. Of course the denomination of "Prandtl number" was given much later and (4.14) is a modern rewriting of Reynolds' result of 1874.

We would not be complete without mentioning two other original works of Reynolds. One is none other than a useful theory of *lubrication* (Reynolds 1886) and the other a theory of *dilatancy* (Reynolds 1885). The first of these provided the foundation of tribology—"the science and technology of interacting surfaces in relative motion and the practices related thereto" (British definition of the Department of Education and Science, H.M.S.O., 1966). The second relates to the property of "granular material to alter its volume in accordance with a change in the arrangement of its grains" (cf. Allen in p. 55 of McDowell and Jackson 1970). Reynolds arrived at this idea while speculating about the existence of a medium that must possess properties akin to those of the all-pervading ether considered as substratum for the propagation of light before the inception of the theory of special relativity. But Reynolds was becoming somewhat mentally perturbed when he proposed such a grain-like model as the "sub-mechanics of the universe" (his words of 1902). This obsolete matter is well dealt with by Allen in McDowell and Jackson (1970). But the creative concepts of Reynolds in true fluid dynamics, and those put forward by Helmholtz and Kelvin were to find one of the best places for a faithful and fruitful implementation at the German university of Göttingen, in the environment of a scientist of highest level, Ludwig Prandtl.

4.4 Prandtl and Boundary-Layer Theory

Here we are particularly interested in the applications of new concepts (vortices, circulation, skin friction) of fluid mechanics to the emerging aerodynamics as of the dawn of the twentieth century. It is not our aim to tell the history of aerodynamics for which there already exist successful books to which we refer the reader (among them the outstanding history by Anderson 1997). We rather try to focus on the heritage of Helmholtz and Reynolds and the breakthroughs by new comers such as Prandtl.

The two forces to be considered in the flight of an object through air are called the *drag* (horizontal force opposing the possible flight) and the *lift* (vertical upward force allowing the object to stay suspended in the air while in motion). These two forces were first considered for an inviscid fluid obeying Euler's equations. We recall that in the absence of viscosity, one is led to D'Alembert's paradox in a potential flow: the drag force vanishes in contradiction with observation, so that this non-realistic result had to be corrected in some manner. In so far as the lift is concerned, considering flat objects at some small incidence (angle made by the object with the velocity direction of the flow) or cambered such as an airfoil (two-dimensional cross section of a wing), application of Bernoulli's theorem could explain the higher pressure on the bottom face compared to the pressure on the top face, and thus globally an upward force or lift. For the global object the lift was computed by the German mathematician Wilhelm Kutta (1867–1944) in 1902. His proof was direct and did not rely on the notion of circulation, but in fact is equivalent to the formula magisterially proposed by the famed Russian professor of mechanics and mathematician, Nikolai Joukowski (or Zhukovsky 1847–1921)—the *Father of Russian aeronautics*—, in the remarkably simple form (cf. Fig. 4.4)

$$l = \rho_\infty V_\infty \Gamma, \tag{4.15}$$

where Γ is the circulation computed as the line integral of the flow velocity taken around any closed curved enclosing the airfoil (cf. Eq. 4.3), and ρ_∞ and V_∞ are the density and the uniform fluid velocity far upstream of the airfoil. Equation (4.15), valid for a unit span of an airfoil of supposedly infinite width, and now known as the *Kutta-Joukowski theorem*, is indeed simple but it hides the caveat of the difficult

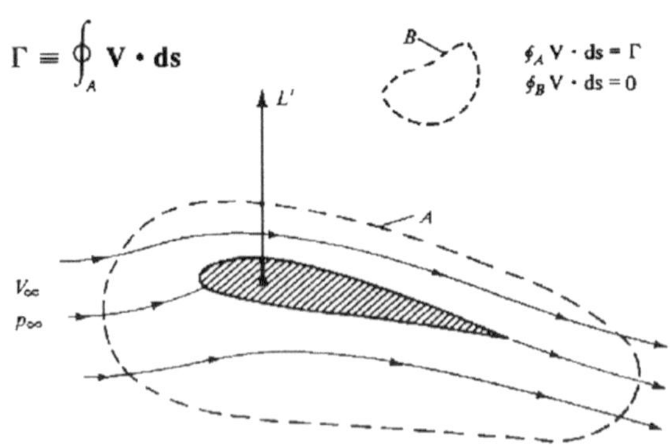

Fig. 4.4 The notion of circulation Γ and the Kutta-Joukowski theorem for lift: $l = \rho_\infty V_\infty \Gamma$

computation of Γ for a specific airfoil at a given angle of incidence. Furthermore, it applies only to low speeds and incompressible two-dimensional flows. It epitomizes the *circulation theory of lift* initiated with ideas of the British engineer/aerodynamician Frederick W. Lanchester (1868–1946) whose line Kutta and Joukowski followed with all computations made as if the flow over an airfoil was the superimposition of a uniform flow without circulation and a "circulatory" flow—a mathematical notion and not an actual rotation of the fluid. This "circulatory" component could be related to the vortex filaments of Helmholtz. Of course, the problem is much more complex for a wing of finite span but vortex lines will play a role at the tip ends of the wing as later proposed by Lanchester and Prandtl. The Kutta-Joukowski theorem predicts a zero drag for two-dimensional inviscid flows. As to three-dimensional, viscous and unsteady flows, the generalization of the Kutta-Joukowski theorem can be formulated only in terms of integral forms.

The situation is altogether different from the lift with the drag since one must necessarily consider the effect of viscosity, and thus the Navier-Stokes equations. The solution of these equations in the whole exterior of the airfoil would be a formidable and practically impossible task. But it is more or less clear that viscosity will play an essential role only in the vicinity of the airfoil with the notion of skin friction, while the rest of the flow can still be considered inviscid and thus amenable by means of Euler's equations. This was the genial idea of Ludwig Prandtl (1875–1953; Fig. 4.5) borrowing the path already indicated by the Italian mathematician Tulio Levi-Civita (cf. Anderson 1997, pp. 252–253). Prandtl was educated at T.H. München (Munich) where he obtained his doctoral degree in 1900 under the supervision of August Föppl, the most reputed mechanical engineer of the period, whose books have educated several generations of German engineers. Prandtl became first a professor of Mechanics at the T.H. Hannover, but then served as a professor of applied mechanics at the University of Göttingen from 1904 to his retirement. He established there a very successful school of aerodynamics and hydrodynamics with both theoretical and experimental components (including large wind tunnels). The Kaiser-Wilhelm Institute (later renamed Max Planck Institute) that he directed for many years achieved world renown. His many students and co-workers spread the Prandtl spirit in many institutions in Germany and outside. Among them, Theodor von Kármán, Theodor Meyer, Heinrich Blasius, Ivan Nikuradzé, Adolph Buseman, Jacob Ackeret, Hermann Schlichting, Oskar Tietjens, Walter Tollmien and others (a total of 85 Ph.D students) became in turn renowned contributors to theoretical or experimental aerodynamics. Although he originally did some work in structural mechanics while in Munich and Hannover and also made a brief incursion in the study of the plastic flow of metals in 1924 (the famous Prandt-Reuss equation), Prandtl deserves above all to be called the "*Father of modern aerodynamics*". His celebrated course on fluid mechanics published as Prandtl (1931) remains a classic of its own, having reached a twelfth edition in 2008 (See also Prandtl and Tietjens 1931). Together with Richard von Mises he was a

Fig. 4.5 Ludwig Prandtl (1975–1953)

founding member of the German Society of Applied Mathematics and Mechanics (GAMM) and an internationally honoured scientist.[2]

The innovative idea of Prandtl is that viscosity plays a prevailing role only in a rather thin layer along the airfoil, where the fluid is subjected to shear forces and the velocity field evolves from zero at the boundary (no-slip condition) to whatever it can be—a maximum—at the matching with the outer flow that may be considered as one of an inviscid fluid (cf. Fig. 4.6). The equations to be solved are those of the Navier-Stokes type within the boundary layer and those of Euler outside with appropriate matching at the outer limit of the layer. With this Prandty created the *science of asymptotics* applied to singular systems of partial differential equations (the singularity comes from the highest-order derivatives). This approach allows obtaining a closed-form solution in both regions of the flow. If thermal effects are taken into account then the fluid temperature at the surface has to be equal to the temperature of the surface. The thickness of the fluid boundary layer can be defined as the distance from the solid body at which the viscous flow velocity is 99 % of the free-stream velocity. The thermal boundary layer is defined in a similar way with temperature replacing velocity. Both layers may be of unequal thickness. It is the *Prandlt number* Pr that measures their ratio. $\mathrm{Pr} > 1$ corresponds to a larger velocity

[2]Of course Prandtl was a very serious man who took all matters at heart but also a bit naïve so that we cannot resist the temptation to report the gossip told by Anderson (1997, p. 259): Having reached the age of 34 he decided he should get married. To that purpose he went to visit his master August Föppl and his wife who had two daughters, asking if he could marry one of them, without further precision. The Föppls made a family decision by selecting the eldest daughter. Prandtl and Gertrude Föppl married, lived happy and themselves had two daughters. The story does not tell if one of them or even the two, married some of Prandtl's students, but this is quite possible if the tradition had to be respected. This almost happened to the present writer.

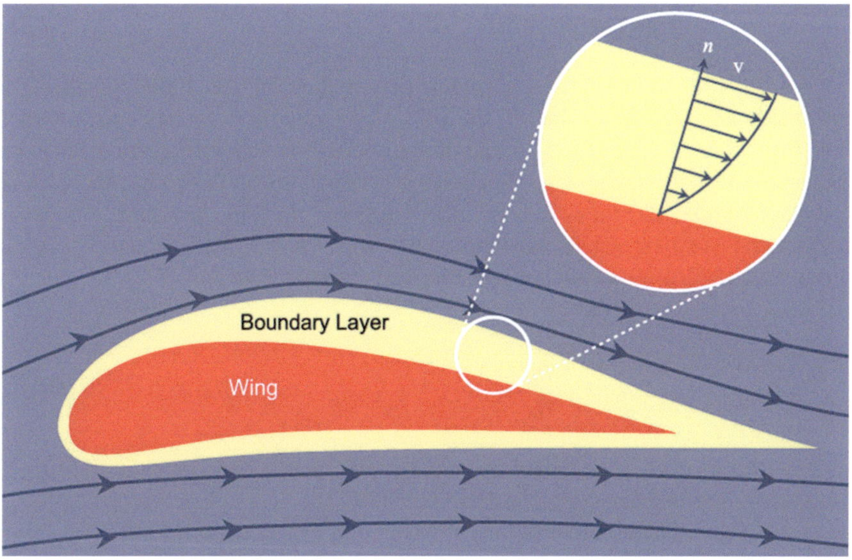

Fig. 4.6 Prandtl's boundary-layer along an airfoil (laminar flow, no separation; NASA document)

boundary layer while for Pr <1 the thermal boundary layer is thicker (case of air at standard conditions). In practice, at high Reynolds numbers the velocity boundary layer inevitably thickens and become less stable as the flow develops along the body. Eventually the flow will become turbulent and the process described by Prandtl and known as the boundary-layer *transition* will take place. Technical means (such as suction of the boundary layer through a porous surface) can be implemented to delay this effect. Laminar flow is preferable in any case.

Mathematically, the groundbreaking idea of the notion of boundary layer proposed by Prandtl in 1904 at the fourth international congress of mathematicians (see Prandtl 1905) used an *order of magnitude* analysis. In the layer the governing equations for a two-dimensional steady incompressible flow are given in direct notation as (with obvious symbols)

$$\mathbf{u}\cdot\nabla\mathbf{u} = -\rho^{-1}\nabla p + \eta\nabla^2\mathbf{u}, \quad \nabla\cdot\mathbf{u} = 0. \tag{4.16}$$

With appropriate nondimensionalization this is rewritten as (keeping the standard notation for $\mathbf{u}$ and p in spite of the scaling)

$$\mathbf{u}\cdot\nabla\mathbf{u} = -\nabla p + \varepsilon\nabla^2\mathbf{u}, \quad \nabla\cdot\mathbf{u} = 0, \tag{4.17}$$

where the parameter ε is none other than the inverse of the Reynolds number, $\varepsilon = (\mathrm{Re})^{-1}$, and thus infinitesimally small for sufficiently high Reynolds number. For $\varepsilon = 0$, system (4.17) reduces to the steady Euler incompressible equations

$$\mathbf{u} \cdot \nabla \mathbf{u} = -\nabla p, \ \nabla \cdot \mathbf{u} = 0. \tag{4.18}$$

Clearly, system (4.17) is a *singular* perturbation of the Euler system. The asymptotic analysis shows that: (i) the vertical velocity (or boundary (wall) normal velocity) is small compared to the horizontal or streamwise component, and (ii) variations in the streamwise direction are generally much lower than those in the boundary normal direction. For an incompressible fluid the pressure is constant through the boundary layer and one is left with the following equations in the appropriate scaling the boundary-layer vertical variable is magnified by the quantity ε^2:

$$(\mathbf{u} \cdot \nabla) u_x = -\frac{\partial p}{\partial x} + (\mathrm{Re})^{-1} \frac{\partial^2 u_x}{\partial y^2}, \ \frac{\partial p}{\partial y} = 0, \ \nabla \cdot \mathbf{u} = 0. \tag{4.19}$$

It is Blasius (1912), a student of Prandtl, who gave a good analytic solution to this system [see technical detail in Germain (1977) or Zeytounian (2002, Chap. 7)].

In the case of a turbulent boundary layer one has to start from the Reynolds equation (4.13). But then, as already noticed, one needs a turbulence model to close the system of equations. The situation is also complicated by the frequent occurrence of a boundary layer separation (from the body), causing a strong coupling between the boundary layer and the outer flow. These technical aspects rest outside the present considerations.

Prandtl's tackling of the boundary layer problem was groundbreaking and a true tour de force in applied mathematics. Many of his disciples brought additional improvements with application to the airfoil theory (in particular, see Schlichting et al. 2004). Simultaneously, Prandtl's method opened up a full horizon for application of asymptotic approximations to partial differential equations, in a discipline that we may call the science of nondimensional numbers and the theory of similarity.

Prandtl contributed also to another aspect of the wing theory by examining, and offering solutions to, the problem posed by the drag of an airfoil of *finite* span. This matter was pondered by Frederick Lanchester (already cited) in the U.K. but without true quantification. Lanchester proposed to account for the finite span by generalizing his theory of circulation. He was advocating the replacement of "the wing model with a model of several filaments aligned along the span and then trailing down from the wing tips as vortex trunks", the whole forming a continuous vortex sheet (Anderson 1997, pp. 283–284). However, it is Prandtl who offered more specific solutions. The first one, quite analogous to Lanchester's but developed independently, consisted in replacing the wing with a single vortex line that would run from one wing tip to the other complemented by vortex filaments escaping from the two tips. This deserved the name of a "horseshoe vortex". In addition to the difficulty of computing the global circulation around the wing, this model yielded a singularity in the induced flow velocity at the tips. The second model by Prandtl in some sense minutely "sliced" this solution along the lifting line (the wing span). Paraphrasing Anderson (1997, p. 284), we can say that it proposed to replace the previous "single horseshoe vortex of finite strength with an infinite number of juxtaposed infinitesimally weak horseshoe vortices. By a limit reasoning this in principle yielded a circulation that would vary continuously along the lifting line, starting form zero at the wing tips to reach a maximum at the midspan of the wing". The vortices that are generated at the tips of the wing and trail downstream from the wing are analogous to mini-tornadoes. There of course remains the difficulty to compute the circulation. In spite of that, this ingenious model justified the existence of an *induced drag*—an idea put forward by Lanchester—added to skin-friction and form drag as a result of the finiteness of the wing span. This smartly devised model is called the *Prandtl lifting-line model* (although British aeronautical engineers adjoin the name of Lanchester to that of Prandtl with some justification). Figure 4.7 shows an artistic rendering of the tip generated trailing vortices. For the amusement of the reader we also provide Fig. 4.8 that shows in a spectacular way the many such vortices generated during their motion in water by two kinds of fish.

In all this approach the most basic notion is the Kutta-Joukowski equation (4.15). Combining this with Prandtl's boundary-layer theory involving the Reynolds number, and the lifting-line model involving Helmholtz's vortices, and the many results from their experiments performed at Göttingen and other places, the students, associates and disciples of Prandtl will take the science of flight to the highest degree of perfection during the twentieth century with a mixture of empirism and good applied mathematics (sometimes devised for this very purpose). Only powerful numerical means will help to find true numbers for the effects still unaccounted for.

Among Prandtl's students we cannot avoid paying special attention, albeit briefly, to Theodor von Kármán (1881–1963; Fig. 4.9). Born Hungarian in "der

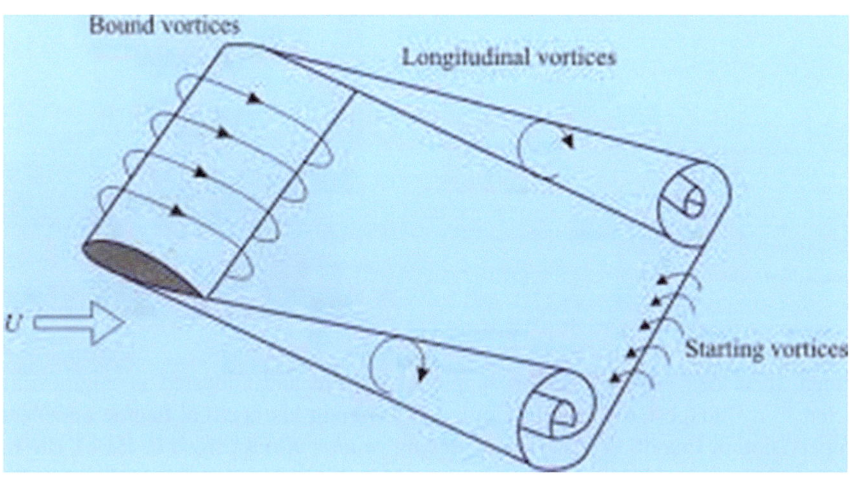

Fig. 4.7 Longitudinal vortices generated at finite-span tips

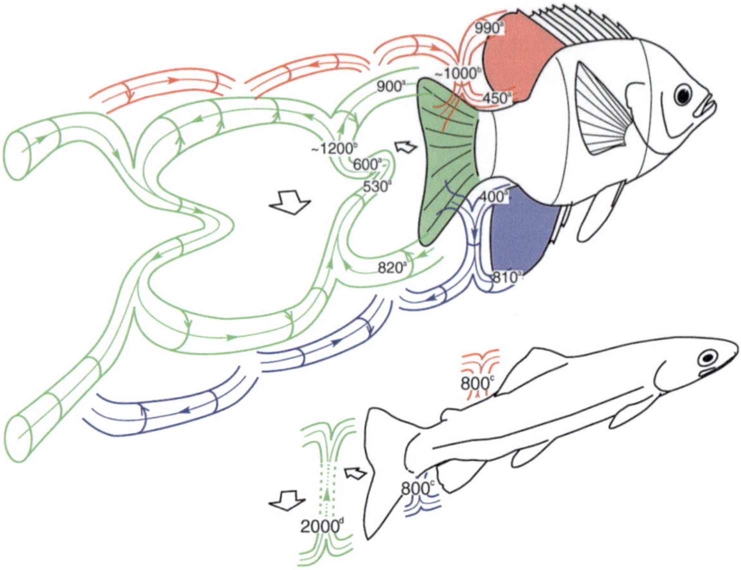

Fig. 4.8 Vortex filament model showing vortices near steadily swimming bluegill sunfish (A; 3D vortex structure for bluegill sunfish swimming at about 1.2 L s^{-1}.) and brook trout (B; Experimentally observed vortices for the dorsal, anal and caudal fins of trout swimming at 0.5 L s^{-1}). Numbers indicate vortex circulation (in mm^2 s^{-1}) (Reproduced from Tytell et al. 2008)

Fig. 4.9 Theodor von Kármán (1881–1963)

guten alten Zeit" of the Austro-Hungarian Empire and educated in engineering science in Budapest, he went to Gottingen to obtain his doctoral degree under the supervision of Prandtl (Thesis: *Untersuchungen über Knickfestigkeit*, 1908). But he also collaborated with the famed physicist Max Born on what was to become the foundation of crystal lattice dynamics (Born and von Kármán 1912), and he also improved on Föppl's theory of large deflections of elastic plates by considering the case of not so thin plates (von Kármán 1910). This is now called the Föppl-Kármán theory of plates.[3]

Von Kármán is acknowledged both as an outstanding scientific contributor to aerodynamics in different regimes, from low speeds to subsonic, supersonic and hypersonic airflows, and astronautics (cf. his scientific autobiography: von Kármán 1961), and a never tired organizer of scientific research in these fields. He spent most of his scientific career at the RWTH Aachen in Germany and then in Pasadena, California, after 1930, with several US government appointments connected with the USA Air Forces, and NASA, and then NATO. In particular, he had joined the Guggenheim Aeronautical Laboratory at the California Institute of Technology (for short, GALCIT) and was, among other things, the founder of the *Jet Propulsion Laboratory* in Pasadena. He made of GALCIT and JPL some of the most creative institutions in aeronautics in the world. He was strongly involved in advising the US military during the Cold War. He was thus the first recipient in 1963 of the National Medal of Science of the USA. He is believed to have founded the *International Union of Theoretical and Applied Mechanics* (IUTAM) having organized its first international conference in Innsbruck in 1922.

A most impressive list given herein below cites some of his scientific achievements. Concerning the subject matter of the present contribution we like to underline his invention of the so-called "von Karman vortex street".[4] This he proposed in 1911 in an analysis of the alternating double row of vortices [see Fig. 4.10; see also Van Dyke's (1982) with splendid photographs] that develops behind a blunt body, i.e., a body having a broad more or less flattened front, in a fluid stream. With a frequency of alternation matching some characteristic frequency of the mechanical system this may have drastic consequences via resonance (so-called vortex shedding).

In concluding this section and just for memory, we recall some quantities, theories or theorems that bear Prandtl's name:

- Prandtl number (compares convection and friction),
- Prandtl's boundary layer,
- Prandtl lifting-line theory,
- Prandtl-Meyer expansion fan in shock-wave theory,
- Prandtl-Reuss flow rule in plasticity,
- Prandtl-Glauert singularity,
- Prandtl-Glauert correction at high-subsonic speeds,
- Prandtl-Busemann supersonic nozzle.

[3]This is a theory in which the elastic potential energy is quadratic in the finite strain measure. The resulting equations surprisingly have a pure geometric meaning.

[4]I personally think that the corresponding French expression "allée des rouleaux (ou tourbillons) de von Kármán" sounds less vulgar and more "chic" (but this is a snob's opinion).

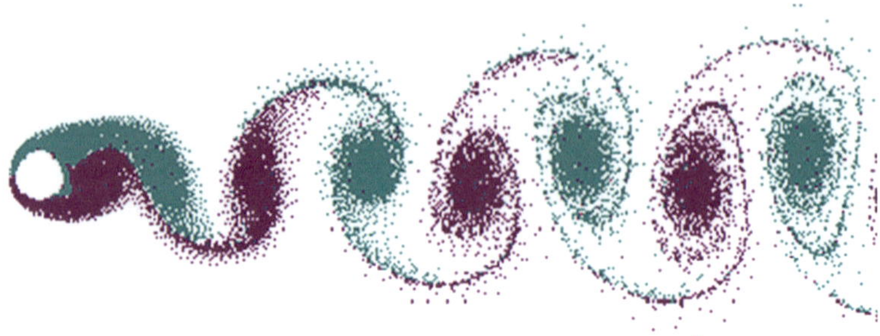

Fig. 4.10 Snapshot of a von Kármán "vortex street" created by the repeated pattern of swirling vortices in the wake of a blunt body

As to von Kármán, the following non-limitative list evokes the formidable range and success of his research themes (both Nikuradzé and Treffz were also students of Prandtl; Max Born was a physicist colleague of Prandtl at Göttingen):

- Föppl-von Kármán plate equations (large deflection of elastic plates),
- Born-Kármán lattice model (pioneering work in the lattice dynamics of crystals),
- Chaplygin-Kármán-Tsien approximation (in potential flow theory),
- Karman-Nikuradsé correlation (in viscous flows),
- Kármán-Treffz transformation (in airfoil theory),
- Prandtl-Kármán law (velocity flow in open channels),
- von Kármán integral equation (in boundary-layer theory),
- von Kármán vortex street (flow past a cylinder),
- vortex shedding (alternating flow in the wake of objects and resulting vibrations under resonance conditions).
- von Kármán ogive (in supersonic aerodynamics).

These two lists provide an idea of the extent of the incommensurable influence of these two individuals in fluid mechanics and aeronautics.

4.5 The Science of Non-dimensional Numbers

Non-dimensional numbers such as the Reynolds (Re) and Prandtl (Pr) numbers evidently play a fundamental role in the appraisal of solutions in fluid mechanics, but more generally in physics. They allow one to emphasize the most relevant effects in prevailing geometric and physical conditions. They usually simplify the general equations, allowing one to retain only the most significant contributions and to compare these contributions in a kind of objective manner, not obscured by obnoxious physical dimensions. The order of magnitude is all what counts and asymptotic techniques are the privileged mathematical techniques to deal with them as clearly illustrated by the paradigmatic treatment of Prandtl's boundary-layer theory. Matched asymptotic expansions are powerful methods in this context (See, e.g., Lagerstrom 1988). This may be called an *Art of modelling* because it requires some deep pondering of the matter by the concerned scientist. This is perfectly illustrated in the rather exhaustive book of Zeytounian (2002) where a large quantity of non-dimensional numbers involved in fluid dynamics are introduced in some rational deductive way yielding well organized hierachies of useful models (Mach, Rayleigh 1892, 1913, Reynolds, Boussinesq, Froude, Strouhal, Weber, etc., numbers). The whole reasoning on models and mock-ups is based on their exploitation as particularly shown by experiments in wind tunnels, although mathematical techniques and direct numerical simulations—helped by *scaling* (a modern term for dimensional analysis)—are taking over these.

Some mathematicians such as Euler, Boussinesq or Poincaré were conscious of the "orders of magnitude" and of the presence of infinitesimally small non-dimensional numbers (more than often a simple ratio of two physical quantities sharing the same physical dimension, e.g., length, velocity or energy) especially in proposing series expansions of solutions in some kind of Ansatz. But the construction of more complex combinations found in modern non-dimensional numbers (although still the ratio of two physical quantities of the same nature, e.g., energies) in fact belongs in the science of *similarity* or *similitude* (depending on the author or the translator) or *dimensional analysis*. This point deserves a brief historical excursion, the detail of which most practitioners are not aware of. We remember that Rott (1990, p. 3) speaking of the Reynolds number says that "this was apparently the beginning of a new era in the use of names for non-dimensional numbers" But without being given specific names non-dimensional groups of various physical quantities were considered before Reynolds. It seems that the French mathematician Bertrand (1878) was the first to ponder this question in a practically unknown short contribution. The matter became more popular with an apparently breakthrough paper by the American engineer Buckingham (1914) whose name was given to a theorem that provided a recipe to build relations between characteristic non-dimensional groups of physical quantities. This he referred to as *physical similarity* between systems. Buckingham cites no references. But it was realized that the French telegraph engineer Vaschy (1892) had formulated the same theorem some twenty years before, so that this theorem on similarity is now universally known as the *Vaschy-Buckingham theorem* or Π-theorem, the "Π" implying the product of quantities. Riabouchinsky (1911) also published on the same subject earlier than Buckingham. One of the most striking applications of this celebrated theorem was the correct deduction of the power of the first American atomic explosion by several scientists, e.g., Taylor (1950a, b) in the UK by analyzing the time sequence of photos of the more or less spherical cloud (to evolve later on in the more spectacular so-called "mushroom") of the explosion, and along the same line, the theory of blast waves by Leonid I. Sedov in Russia in 1942. Concerning the last scientist he had a long interest in non-dimensional analysis, publishing a first book on the subject as Sedov (1944) and reporting on his analysis of the atomic explosion as the innocuous (!) "blast-wave problem" at the Landau seminar in Moscow in 1946.[5] Note that a book on dimensional analysis had already been published earlier in the USA by the physicist/mechanician Bridgman (1931). But this field seems to have been a favourite subject to scientists in the former Soviet Union as G.I. Barenblatt also became a specialist of this field and asymptotic techniques (cf. Barenblatt 1979, 1987). R.Kh. Zeytounian, already cited, was also formed to these techniques in Moscow by Il'ia A. Kibel (1904–1970), a noted specialist of meteorology and weather forecast, who used appropriate sets of equations in fluid mechanics. This, in a nutshell, is our history of the field.

4.6 Summary

It is now clear that after the experimental works of Poiseuille and Hagen, there was need for a pursuit of studies of the effects of viscosity on the flow of fluids in conduits or around objects. The epoch-making work of Reynolds combined with the theory of vortices expanded by Helmholtz provided the tools with which the pioneers of flight science and aeronautics could explain the most relevant phenomena, the lift (cf. Kutta-Joukowski) and drag forces necessary for flight. As exposed above, Prandtl was the most instrumental scientist to contribute to these advances. He created a real school of which both theoretical and experimental results contributed to aeronautics and astronautics by producing a good description of flows at various speeds (compared to the sound speed) and various regimes including coupling between flow properties and thermal effects. The print left was such that all developments in flight science during the twentieth century are due to the extraordinary success of the obtained modelling that made extensive use of dimensional analysis and asymptotic methods. The efficacy of these developments was then complemented by excellent progress in applied mathematics (e.g., applied functional analysis in the direction expanded by Richard Courant, David Hilbert and Jacques-Louis Lions) and enhanced by the development of fantastic powerful numerical means that one could not have imagined at the time of Reynolds and Prandtl.[6]

---

[5]See the snapshots of the atomic explosion in pp. 236–238 in the French edition (1977) of Sedov (1944) [These are in fact reproduced from Taylor (1950a, b)].

[6]This chapter exceptionally bears the print of the initial formation of the author as an aeronautical engineer and simultaneously a university graduate student in fluid mechanics in the 1960s.

References

Anderson JD Jr (1997) A history of aerodynamics. C.U.P, Cambridge, UK

Barenblatt GI (1979) Scaling, self similarity and intermediate asymptotics (translation from the Russian, 1st edn). Consultant Bureau, New York (Reprint, Springer, New York, 2012)

Barenblatt GI (1987) Dimensional analysis (English translation by P. Makerin). Gordon and Breach, New York

Barré de Saint-Venant A (1843) Note à joindre au mémoire sur la dynamique des fluides. C R Acad Sci Paris 17:1240

Bertrand J (1878) Sur l'homogénéité dans les formules de physique. C R Acad Sci Paris 86: 916–920

Blasius H (1912) Das Aehnlichkeitsgesetz bei Reibungsvorgängen. VDI-Z 56:639–643

Born M, von Kármán T (1912) On fluctuations in spatial grids. Zeit Physik 13:297–309 [Also: On the distribution of natural vibrations of spatial lattices. Zeit Phys 14: 65–71, 1913]

Bridgman PW (1931) Dimensional analysis. Yale University Press, Connecticut

Buckingham E (1914) On physically similar systems. Illustrations of the use of dimensional equations. Phys Rev 4:345–376

Eckhardt B (2009) Introduction. Turbulence transition in pipe flow: 125th anniversary of Reynolds paper. Phil Trans Roy Soc Lond A367:449–455

Favre A, Kovasnay LG, Dumas R, Caviglio J (1979) La turbulence en mécanique des Fluides. Gauthier-Villars, Paris

Flamant A (1900) Hydraulique, 2nd edn. Béranger, Paris (First edition, 1891)

Germain P (1977) Méthodes asymptotiques en mécanique des fluides. In: Peube JL (ed) Fluid dynamics, (Lectures at Les Houches, 1973, France). Gordon and Breach, London, pp 1–147

Hagen G (1869) Über die Bewegung des Wassers in cylindrischen nahe horizontalen Leitungen. Math Abh Akad Wiss, Berlin, pp 1–29

Lagerstrom PA (1988) Matched asymptotic expansions. Springer, New York

Lamb H (1912–1913) Osborne Reynolds. Proc Roy Soc Lond 88A:XV–XXI

Lumley JL (1979) Toward a turbulent constitutive equation. J Fluid Mech 41:413–434

Maugin GA (2014a) Continuum mechanics through the eighteenth and nineteenth Centuries. Springer, Dordrecht

Maugin GA (2014b) Hydraulics, Preprint UPMC-Paris 6 (see Chapter 2 in this book)

McDowell DM, Jackson JD (eds) (1970) Osborn Reynolds and engineering science today. Manchester University Press, UK [papers presented at the Osborne Reynolds Centenary Symposium, Manchester, Sept 1968]

Moffatt K (2008) Vortex dynamics: the legacy of Helmholtz and Kelvin. In: Borisov AV et al (ed) Proceedings of IUTAM Symposium on Hamiltonian dynamics. Springer, Heidelberg, pp 1–10

Navier C LMH (1822) Sur les lois du mouvement des fluides. Bull Soc Philomatique 75–79 (also: Mémoire sur les lois du mouvement des fluides. Mémoires Acad Roy Sciences (Paris), 6:389–416, 1823)

Poiseuille JLM (1846) Recherches expérimentales sur le mouvement des liquides dans les tubes de très petit diamètre. Mém Acad Sci 9:433–564

Prandtl L (1905) Über Flüssigkeitbewegung bei sehr kleiner Reibung. In: Proceedins of 3rd international mathematical congress, Heidelberg 1904. Leipzig, pp 484–492

Prandtl L (1910) Eine Beziehung zwischen Wärmeaustausch und Strömungswiderstand der Flüssigkeiten. Phys Zeit 11:1072–1078

Prandt L (1931) Führer durch die Strömungslehre. Vieweg Verlag, Braunschweig (reprint of 12th edition in 2008)

Prandtl L, Tietjens O (1931) Hydro- und Aeromechanik. Julius Springer, Berlin

Rayleigh L (1892) On the question of the stability of the flow of fluids. Phil Mag 34:177–180 (also, in Collected Scientific Papers, 3:575–584, CUP, 1920)

Rayleigh L (1913) Sur la résistance des sphères dans l'air en écoulement. C R Acad Sci Paris 156:109–113 (also in Collected Scientific Papers, 6:136, CUP, 1920)

Reynolds O (1874) Proceedings of Manchester Literary and Philosophical Society, 14, p 7 on, (not seen by the author)

Reynolds O (1883) An experimental investigation of the circumstances which determine whether the motion of water shall be direct or sinuous, and the law of resistance in parallel channels. Phil Trans 174:935–982

Reynolds O (1885) On the dilatancy of media composed of rigid particles in contact. Phil Mag, December issue (see Reynolds collected papers II:203–216)

Reynolds O (1886) On the theory of lubrification and the experimental determination of the viscosity of olive oil. Phil Trans Roy Soc, Part I. (see Collected papers, 1900–1901 II:203–216)

Reynolds O (1894) On the dynamical theory of incompressible viscous fluids and the determination of the criterion. Phil Trans Roy Soc 186:123–164

Reynolds O (1900–1901) Papers on mechanical and physical subjects, 2 vols. Cambridge University Press, UK

Riabouchinsky D (1911). Méthode des variables de dimension zéro et son application en Aérodynamique. L'Aérophile 407–408

Rott N (1990) Note on the history of the Reynolds number. Ann Rev Fluid Mechanics 22:1–11

Rotta J (1951) Statistischer Theorie nichthomogener Turbulenz. Zeit Physik 129:547–572

Rouse H, Ince S (1957) History of hydraulics. Iowa City, State Univ. Iowa (Reprint by Dover, New York, 1963)

Saffman PG (1995) Vortex dynamics. Cambridge University Press, UK

Saph AV, Schoder EW (1903) An experimental study of the resistance to the flow of water in pipes. Trans Amer Soc Civil Eng 51:253–330

Schlichting H, Gersten K, Krause E, Oertel H, Mayes C (2004) Boundary layer theory, 8th edn. Springer, Berlin

Sedov LI (1944) Similarity and dimensional methods in mechanics (original Russian edition), Moscow [English translation of 4th edn, Infosearch Ltd, London, 1959; French translation of the sixth Russian edition of 1967, Editions MIR, Moscow, 1977]

Sommerfeld A (1908) Ein Betrag zur hydrodynamischen Erklärung der turbulenten Flüssigkeitsbewegung, In: Proceedings of the 4th international congress Math., Rome, vol 3, pp 116–124

Speziale CG (1981) On turbulent Reynolds stress closure and modern continuum Mechanics. Int J Non-Linear Mech 16:387–393

Speziale CG, Eringen AC (1981) Nonlocal fluid mechanics description of wall turbulence. Com Meth Applications 7:321–343

Stokes GG (1845) On the theories of the internal friction of fluids in motion, and of the equilibrium and motion of elastic solids. Trans Cambridge Phil Soc 8:287–319

Taylor GI (1950a) The formation of a blast wave by a very intense explosion, I. Theoretical discussion, Proc Roy Soc Lond A vol 201, pp 159–174

Taylor GI (1950b) The formation of a blast wave by a very intense explosion, II. The atomic explosion, ibid, A vol 201, pp 175–186

Tytell ED, Standen EM, Lauder GV (2008) Escaping Flatland: three-dimensional kinematics and hydrodynamics of median fins in fishes. J Exp Biol 211:187–195

Van Dyke M (1982) An album of fluid motion. Parabolic, Stanford

Vaschy A (1892) Sur les lois de similitude en physique. Annales Télégraphiques 19:25–28

von Helmholtz H (1858) Über Integrale der Hydrodynamischen Gleichungen, welche denWirbelbewegungen entspreche. J reine angew Math 55:25–55 [English translation available on the web at: www.21stcenturysciencetech.com/Articles_2009/Helmholtz.pdf; pp 41–68]

von Kármán Th (1910) Festigkeit im Maschinenbau. Encycl Math Wiss 4:311
von Kármán Th (1911) Über die Turbulenareibung verschierdener Flüssigkeiten. Phys Zeit 12:283–284
von Kármán Th (1961) From low-speed aerodynamics to astronautics. Pergamon Press, Oxford, UK
Zeytounian RKH (2002) Asymptotic modelling of fluid flow phenomena. Kluwer, Dordrecht

Chapter 5
Duhem on Hydrodynamics and Elasticity

Abstract Pierre Duhem is best known nowadays for his works on epistemology and the history of science. But in his time he also contributed original and innovative ideas not only to energetics (for which he was a never tired propagandist), but also to a priori more classical domains of continuum mechanics such as hydrodynamics and elasticity. Here we critically examine these last contributions in the light of his lecture notes for a course in Lille in 1890–1891 and his collected works on hydrodynamics (1903–1904) and elasticity (1906). This allows one to put in evidence Duhem's evolution from a young ambitious teacher—with a need to show his remarkable capture of a whole field at a given time—and his more personal contributions at a mature age. Most new results are related to wave phenomena, especially in the formalism proposed by Hadamard for the classification of discontinuity fronts, and generalizing to three dimensions of space the pioneering works of Riemann and Hugoniot on shock waves. Moreover, in a rigorous framework, themes of interest such as thermo-mechanical couplings, stability questions, and appropriate historical remarks clearly recur in this general landscape.

5.1 Introduction

Pierre Duhem (1861–1916)[1] is probably one of the most powerful intellects of his period of activity, say between 1880 and 1920. He is a rare combination of a brilliant and sharp mind and highly cultivated person, a never tired writer, a creative contributor to phenomenological physics, the most forceful champion and propagandist of energetics, a philosopher of science, and the true creator of the history of medieval science. We have examined elsewhere his general view of science and style of

[1]Biographies of Duhem are of two kinds, some are a bit hagiographic (Ariew 2007; Jaki 1984; H. Pierre-Duhem 1936), while others are more balanced or more precise from the scientific viewpoint (Brouzeng 1987; Humbert 1932; Miller 1970; Picard 1921). A lengthy comprehensive analysis of Duhem's science is given by Manville (1927), who had been a close collaborator of Duhem in Bordeaux.

G.A. Maugin, *Continuum Mechanics through the Ages - From the Renaissance to the Twentieth Century*, Solid Mechanics and Its Applications 223,
DOI 10.1007/978-3-319-26593-3_5

exposition by perusing the detail of two of his most well known works, his general treatise on energetics or general thermodynamics (Duhem 1911) and his visionary text on the "Evolution of mechanics" (Duhem 1903b)—(cf. Maugin 2014a, Chap. 10). We leave aside the most interesting epistemological text on the "physical theory" (Duhem 1906a, 1908) and his immense studies on the evolving cosmological views of the world (in ten volumes) and on pre-renaissance dynamics (Leonardo and the Scholastic school of mechanics). Here we want to focus on purely technical texts that relate to more mundane matters such as elasticity and hydrodynamics, the main two branches of continuum mechanics. Of course, we must reckon that these works are not as much cited nowadays as Duhem's work on epistemology and history of science. One reason for this vanishing interest may be that with full integration in the scholar corpus (in thermochemistry, irreversible thermodynamics, wave theory), there is no more need to refer precisely to his personal works except in a few well-known, almost of everyday use, expressions such as the Clausius-Duhem inequality, the Gibbs-Duhem equation, the Duhem-Margules equation—and also the Duhem-Quine hypothesis in the philosophy of Science. But Duhem considered himself first as a physicist. When invited to occupy the chair of History of Sciences at the famous College de France (the highest French institution of education) in Paris, he proudly rejected the offer arguing that he would enter the College only through the main door, that is, by accepting a chair of Mathematical Physics, and thereby clearly willingly decided to stay in Bordeaux. It is possible that each individual is not the best judge of his own achievements (Fig. 5.1).

Fig. 5.1 Pierre Duhem (1861–1916)

The most relevant texts of Duhem for our endeavour consist of his fantastic lecture notes of a general course delivered at the Faculty of Sciences in Lille in the year 1890–1891 (Duhem 1891), and collections in the form of two books of already published papers, one devoted to his *researches in hydrodynamics* (Duhem 1903a, 1904) and one to his *researches in elasticity* (Duhem 1906b). These works were not published in a complete vacuum. On the contrary, it seems that each great scientist in the field of continuum mechanics wanted to leave a durable print with his own lecture notes or specially prepared treatises. Among such works we must note: the pioneering, now classical, treatises of Lamé (1852) and Clebsch (1862) in elasticity with the largely augmented French translation of the latter by Barré de Saint-Venant (1883); the epoch-making treatises by Love (1892) in elasticity and Lamb (1879) in hydrodynamics, the unavoidable courses by Kirchhoff (1876) and Neumann (1885), the general treatise of Thomson (Kelvin) and Tait (1867, 1879–1883), the detailed historical developments of Todhunter and Pearson (1886), and the more contemporary long papers or books in German by Voigt (1887), Hilbert (1906–1907) and Hellinger (1914), or in French by Brillouin (1884–5), Koenigs (1897), Poincaré (1892), and Cosserat (1896, 1909), the treatise on rational mechanics of Appell (first edition, 1900), and the Ecole Polytechnique course of L. Lecornu (published in three volumes, 1914–1918, also 1925).[2]

But Duhem's writings have a special flavour and specific traits that can be identified. First of all, Duhem is a brilliant writer and stylist with an easily flowing sentence. But in addition one easily acknowledges the following recurring original properties and themes:

- a high accuracy in the expansion of arguments always accompanied by detailed steps in proofs and not avoiding criticisms of the lack of accuracy even by his most famous predecessors (e.g., Cauchy). This particularly applies to the proof of mathematical theorems;
- a need for mathematical precision concerning continuity, analyticity, and good behaviour of physical entities, not excluding a loss of these properties (such as in discontinuity waves);
- a marked interest in matters of stability [A.M. Lyapunov (1857–1918) is a contemporary];
- a specific interest in combining mechanics and thermodynamics;
- a great care in giving accurate citations of previous works in both German and English (an attitude not shared by many French authors);
- a clear will at being rather exhaustive at the time of writing (although practically an impossible task),
- and of course an inevitable historical perspective from the part of a practically professional historian of sciences.

[2]We examined in detail the cited works of Hilbert, Hellinger, Poincaré and Appell in previous contributions (Maugin 2014a, b).

5.2 The Lectures of 1890–1891

It is hard to imagine that this text of more than 600 pages in print corresponds exactly to the contents of lectures delivered orally by a "chargé d'un cours complémentaire" (assistant lecturer granted a complementary course) in the general course of Mathematical Physics and Crystallography at the Faculty of Sciences of Lille in the academic year 1890–1891, a series of lectures that could not amount to more than a few hours only and obviously delivered to a very limited audience. Duhem was then starting his academic career, staying in Lille from 1887 to 1893, before joining Rennes (1893–1894) and then Bordeaux (1894 to his death). He must have written these lecture notes with an unbounded enthusiasm, extending much the scope and detail of the oral lectures and providing in fact one of the most complete, detailed, and documented courses in the field at the time. Although produced in autographed form but published as a book in two volumes by the renowned Paris scientific publisher A. Hermann, this fantastic opus was to have a great influence on both physicists and mathematicians because of its axiomatic and deductive rigour. Duhem's friend Jacques Hadamard will often refer it (cf. Hadamard 1903, 1927). Before this real "bookish monster" Duhem had only (if we may say so) produced three books, one on his unsuccessful text for a Doctoral degree in physics (Duhem 1886), one on his successful doctoral thesis in mathematical physics (Duhem 1888), and a text on diamagnetism in 1889. For simplicity we shall refer to these lectures of 1890–1891 in the abbreviation form HEA, while the books on his researches in hydrodynamics and elasticity will be referred too as RH and RE respectively.

5.3 Duhem's General Views on Continuum Mechanics

These are mostly expanded in the first part of HEA (First volume or HEA-1, pp. 1–59, Second volume or HEA-2, pp. 205–275). Like most continental mechanicians of the nineteenth century, in order to formulate the general theorems of mechanics Duhem exploits the principle of virtual power and its generalization by d'Alembert to which, in a characteristic move, he adds the notion of potential and a criterion of stability. This follows Varignon, Lagrange, Gauss, Cauchy, C. Neumann, Clausius for the potential, Lejeune-Dirichlet for stability, and Thomson and Tait (1867) for thermodynamics. The description of the general motion of a continuous body (HEA-1, p. 19 on) closely follows the original works of Cauchy (1823, 1828a, b, c) and Poisson (1829). An essential statement typical of Duhem (HEA-1, p.17) is that Thomson (Kelvin) in 1855 was the first to show that thermodynamics was a logical foundation for elasticity. Infinitesimal deformations of a body are introduced next. This borrows from the innovative work of Cauchy (1823, 1828a). Note that Duhem still uses the expressions "dilatations" and "pressures" where we now use "strains" and stresses. The last word was introduced, we believe, by Rankine. The first and second ellipsoids of dilatations are attributed to G. Lamé and Cauchy, respectively. For "pressures", Duhem considers the interaction between two bodies and again refers to Cauchy but avoids the classical argument of the latter, still introducing six a priori independent quantities. Here also the discussion focuses on the first and second ellipsoids of "pressures" and their principal axes. These considerations are entirely independent of the precise behaviour of the considered body. An interesting remark (HEA-1, p. 44) that now sounds trivial, is that in considering a surface cut in the body and the pressures acting on it, we need not know the exact shape of the surface. This obviously means, in agreement with Cauchy, that only the unit normal plays a definite role.[3]

Duhem will return to the general principles of continuum mechanics in his researches in elasticity (RE)—see below.

5.4 Advances in Hydrodynamics

It s true that usually Pierre Duhem is not remembered as a contributor to fluid mechanics. But this vision is erroneous. As a matter of fact, he was instrumental in developing some critical points that are well illustrated in HEA-1 and Duhem (1903a, 1961). HEA-1 contains no less than 200 pages devoted to classical (non-viscous) fluid mechanics to which are appended almost a hundred pages on the propagation of sound in various spatial forms (cavities, resonators, and tubes according to Helmholtz). This was written at the beginning of Duhem's academic career so that we should not expect much originality. Duhem has to dwell with standard subjects such as: general equilibrium equations, internal pressure in a fluid, Euler (1757) and Lagrange (1788) writings of the motion equations and the equation of continuity (mass conservation), vortex motion after Kirchhoff and Helmholtz (and a lengthy transformation of Lagrange's equations due to Cauchy; HEA-1, pp. 129–140), the hypothesis of flows by slices (plane sections of some thickness) in tubes after Daniel Bernoulli, stationary fluid vibrations in tubes, open and closed tubes, spherical waves with reference to very recent (1889) works by Beltrami and the distinction between progressive and regressive waves, interferences and beats (this may be out of place at this level of the given course), and reflection and refraction of plane waves (following Poisson, Green and Lord Rayleigh's theory of sound), and a detailed treatment of Helmholtz's equation (here also recent works by H. Schwarz, E. Picard and H. Poincaré are referred to). In spite

[3]In modern terms this translates to: "Only the geometry of the cut surface at the *first order* (the unit normal) is involved in the natural stress boundary condition". This is valid only in the standard theory where the surface is regular enough (uniquely defined unit normal). The second-order description of the said surface, say its curvature, is not usually involved. If it is involved then we enter the domain of generalized continuum mechanics (e.g., theory with strain gradients, weak nonlocality; cf. Maugin 2013, Chap. 13)

of this rather standard exposition, we witness the emergence of specific points that will constantly recur in further works.

Among these specific points, we emphasize: (i) the notion of "additional" relation (between pressure, density and temperature), (ii) the interaction between motion and heat propagation (HEA-1, pp. 99–103), the careful distinction between isothermal and adiabatic situations[4] (iii) Lagrange's theorem and the velocity potential with an improved proof due to W. Thomson (Kelvin) and the introduction of the notion of *circulation*—typically, Duhem remarks that Lagrange's proof is not satisfactory and cites the rigorous proof of Cauchy in 1827–, (iv) Green's (1828) original theorem, (v) remarks on friction with reference to recent works by M. Brillouin (HEA-1, p.126), (vi) a detailed study of Kirchhoff theorems related to the Laplace equation and the wave equation, (vii) remarks on the possible lack of analyticity of solutions of the equations of small motions (in relation with a very recent (1889) work of Painlevé on singular lines for analytic functions; cf. HEA-1, p.169), and the study of stability (HEA-1, pp. 80–90) exploiting the second variation of a potential, and thus right in the line of recent works by Lyapunov.

Typically, in his forceful view of thermo-mechanics, Duhem pays special attention to the formula of the speed of sound depending on the thermodynamic assumptions; he devotes no less that nine pages to the relevant discussion illustrated by many experimental results. In the same line, Duhem is especially interested in Hugoniot's method of treating the compatibility between two adjacent motions (HEA-1, pp. 183–198), inaugurating thus the theory of the propagation of discontinuity surfaces following the pioneering work of Riemann with a specialization to the motion in a fluid at rest and the application of Huygens' principle to the moving surface. According to Kampé de Fériet (Duhem 1961, p. VII), Hadamard (1927, p. 644) said that it was Duhem's Lille lectures that revealed Hugoniot's breakthrough work to the scientific world (Hugoniot 1887a, b). The twentieth-century reader may be surprised by the recurring use of expressions such as "small motion in another one" and "the motion of a fluid body". These themes will be further expanded in many papers to be gathered in RH (1903b, 1961). The main problem with HEA is perhaps an unnecessary desire to be exhaustive and up to date. This may hide the main ideas although logics and the rigorous succession of subject matters are well respected. Of course about ten years of hard work by Duhem have elapsed between the writing of HEA on fluids and the final redaction of RH where many parts have been pondered and much perfected. That is why a thorough examination of RH is necessary, with the help of the nice critical introduction of the edition of 1961 by Kampé de Fériet.

"Recherches sur l'hydrodynamique" (RH)

Between 1891 and 1904 Duhem published no less than 24 notes in the *Comptes Rendus* of the Paris Academy and 11 full-length memoirs in other journals on fluid

---

[4]On this occasion Duhem pays lip service to his high-school teacher Jules Moutier (HEA-1, Footnote in p. 107), who impressed him very much and may have been responsible for Duhem's attraction to physics.

mechanics. This explains why the contents of RH are much richer and better pondered than HEA with many advances that will leave a print on the developments of this domain during the twentieth century. Furthermore, this was achieved while Duhem's "codification" (expression of Kampé de Fériet) was achieved so that there will necessary be an emphasis placed on thermo-mechanical couplings. From the very start, Duhem involves normal variables[5] of thermodynamic state and accounts for the foundations of energetics including the possible occurrence of dissipative processes and the existence of a dissipation potential in the manner of Rayleigh. This is much in advance on other treatments with the presence of viscosity coefficients possibly dependent on density and temperature. It is on this occasion that he is the first to really prove (Duhem 1896) the set of well known inequalities (not proved by Stokes 1845):

$$\mu(\rho, T) \geq 0, 3\lambda(\rho, T) + 2\mu(\rho, T) \geq 0. \tag{5.1}$$

He clearly distinguishes between the important classes of fluid media, good conducting (of heat) incompressible viscous fluids, and badly conducting compressible fluids, and physical situations: isothermal and adiabatic evolutions, all of these necessary for a further comprehensive study of the propagation of waves (see below). His approach is fully thermo-mechanical and could be reproduced in modern treatments to the small price of a few adjustments in vocabulary only. What he calls the "additional relation" ("relation supplémentaire") is none other than the fully coupled (to the motion) heat equation (RH, p. 39). First steps in this direction had been taken by Kirchhoff and C. Neumann. He is not without noticing that the resulting general problem would be difficult to solve (in his time). Isothermal and isentropic stabilities are duly examined (RH, pp. 57–60). He does not hesitate, where necessary, to return to a Lagrangian description, and to call for Lagrange and Lejeune-Dirichlet for the stability of floating bodies (thereby much improving on the former analysis of buoyancy by Bouguer, Euler, and Abbé Bossut).

Wave propagation is the subject that seems to meet most of Duhem enthusiasm and creativity for he devotes no less than 150 pages to the subject. That is where he exhibits his deep understanding of the phenomenon of *shock wave* with an excellent capture and exploitation of concepts recently put forward by Riemann, Christoffel, Hugoniot, Hadamard, and also Jouguet.

The presentation of the **theory of shock waves** by Duhem (RH, pp. 75–128) is really magisterial. What distinguishes it from modern treatments is the lack of use of vector and tensor notations and of short-hand notations such as those for a jump and a mean value of a field at the crossing of a discontinuity surface. But this is un-consequential although Duhem directly considers the three-dimensional formulation while most previous authors confined their study to one space dimension (thus a synthetic notation would have helped). What is more original in Duhem's

[5]A *normal* set of thermodynamic parameters is one in which entropy is clearly assigned a specific role in the functional dependence of the internal energy function. This definition became standard and is no longer referred to Duhem.

case is that he considered from the beginning coupled thermo-mechanical processes. In particular, the departure point is a formulation of the principle of virtual power with virtual and isothermal "modifications" even in the presence of field discontinuities[6] (cf. RH, pp. 83–89).

Duhem builds on the anterior works of Riemann (1860) and Jouguet (1901) but considers a general discontinuity surface with numerous self-explanatory illustrations of this moving surface. He includes the possibility of viscosity to obtain interesting expressions involving integration through the non-zero thickness of the wave, and shows the impossibility of the propagation of (true) shock waves in a viscous fluid. In the absence of viscosity he obtains Hugoniot's jump relations but in three-dimensional space [HR, p. 104, Eq. (59); cf. Riemann and Jouguet in one space dimension]. The role of the thermodynamic potential is emphasized in the case of good conductors of heat. The case of contact discontinuities—with no mass transfer across the wave—is also introduced. Hugoniot's (1887a) lemmas are expanded. So is the then recent classification of discontinuity surfaces (introduced in 1900) by J. Hadamard (see Hadamard 1903).

[Here we remind the reader that Hadamard proposed to classify moving discontinuities in terms of the order of the space and time derivatives of a field that suffer a true discontinuity. Thus a discontinuity surface across which the medium velocity and the strain are discontinuous is a discontinuity surface of the first order (a shock wave in usual jargon) while one across which the acceleration (second-order time derivative of the motion) is discontinuous is called a discontinuity surface of the second order or "acceleration wave". Jump conditions at the crossing of such surfaces replace the field equations across the surface while usual (continuous) field equations remain valid—but obviously with different field values—on either sides of the surface. Furthermore, Hadamard distinguished between jump relations resulting from pure kinematics—so-called compatibility conditions—and those resulting from the jump of field equations (balance laws in the jargon of mechanicians)].

Duhem constructs somewhat painstakingly the last type of jump relations. But for the rest he is a faithful disciple of Hadamard (RH, pp. 147). Again he wants to deal with waves of the first order (i.e., shock waves) in viscous fluids; he thus shows that first-order waves cannot persist while some second-order waves can exist if all solution parameters (velocity, temperature, pressure) are of second order according to Hadamard. In some cases the wave is of different order (e.g., second and third) for the different parameters. This depends on the possible compressibility and conducting property of the fluid. Perfect fluids are the object of one chapter (RH, pp.175–198). This obviously relates to Euler's theory of fluids.

What Duhem himself considers his great invention in the domain of discontinuity waves is the discovery of so-called **quasi-waves**. This requires some words of explanation. We indeed observe the propagation of sound in air while this material is (very) weakly viscous, and thus in principle forbids such propagation. This

[6]In the late twentieth century the only comparable approach is by Daher and Maugin (1986) developed independently of Duhem's or Jouguet's ideas.

paradox is solved by this notion of quasi waves. In truth such waves are very thin layers across which fields do not suffer exactly a finite discontinuity but they vary continuously albeit rapidly. These quasi-waves play an essential role in the theory of shock waves, although they are now denominated by another expression, that of **structured shock waves**. Without such a structure due to dissipative processes—viscosity and heat conduction in acoustics -, shock waves would not physically exist at all. The quasi-waves obey all laws discovered by Riemann and Hugoniot, but in addition the passing of such waves through the fluid results in an **increase of entropy** [as shown by Jouguet (1871–1943), a direct disciple of Duhem, in his complements to Duhem's results (in Jouguet 1920a, b),[7]—see the three Notes of Duhem reproduced by Kampé de Fériet at the end of the reprint of RH 1961]. Duhem devotes in RH no less than 43 pages to quasi-waves. All this input in the theory of fluid mechanics is well abstracted by Duhem himself in his notice (1913) written a few years before his untimely death. Truesdell and co-workers will capitalize on this type approach in the 1960s–1970s in order to explain the formation of shock waves from weaker discontinuities such as acceleration waves, especially in deformable solids.[8]

In RH Duhem also discusses the generalization of Lagrange's theorem in the presence of viscosity, the problem of boundary conditions (with a nice historical review with references to Du Buat, Coulomb, Prony, Navier, Poisson, Hagen Stokes, Poiseuille, Boussinesq, Darcy, F.E. Neumann, Couette, and others; RH, pp. 301–317), and friction at boundaries (RH, pp. 321–344). The digression on viscosity coefficients (RH, pp. 347–376) is particularly enlightening telling the whole story including the kinetic-theory viewpoints of Maxwell and Boltzmann on the following constraint introduced by Stokes:

$$3\lambda(\rho, T) + 2\mu(\rho, T) = 0. \tag{5.2}$$

We can regret with Kampé de Fériet the non-use of vector and tensor analysis in the whole volume. Fortunately, this does not damage the quality and clarity of

[7]Jouguet's (1920b) memoir develops at length elements of the theory of shock waves in a rather definite manner including the condition of entropy growth through the wave front. Jouguet gives both Eulerian and Lagrangian forms. He very closely follows Duhem and Hadamard's formalism but he favours adiabatic motion. The shock equations (continuity, momentum, energy) are clearly stated (p. 61 and p. 63). For an infinitesimally weak shock he unambiguously shows that the entropy increase across the shock is of third order in general. The case of vitreous (isotropic) nonlinear elastic solids is paid special attention together with its expansion at third order in the line of the Cosserats and Poincaré. This last point is developed in greater detail in a second part which relates to the superimposition of a small motion on an initial state endowed with nonvanishing stresses. This situation was considered by Poincaré in his lectures on elasticity (Poincaré 1892; edition of 1902) and evoked by Duhem in his RE. He shows how an initial pressure allows to treat the wave problem with effective elasticity coefficients in isotropic bodies (e.g., λ and μ replaced by $\bar{\lambda} = \lambda - p$ and $\bar{\mu} = \mu - p$). We witness here the embryonic expansion of what we now call *acousto-elasticity* to be further developed by Brillouin (1925) and others.

[8]On this subject see particularly Chen's (1976) book devoted to discontinuity waves in the sense of Duhem, Hadamard, Jouguet, etc.

exposition of the author and the inquisitiveness of Duhem's reasoning. At least, everybody in his time could read the work without textual difficulties.[9]

5.5 Advances in Elasticity

With regard to Duhem's production in elasticity we must again distinguish between the contributions to the Lille lecture notes of 1891 and his further more original contributions assembled in a special volume entitled "Recherches sur l'élasticité" (Researches in elasticity = RE = Duhem 1906b).

Indeed, Duhem returns to the general principles of continuum mechanics with a well documented discussion on the bases of the theory of elasticity in Chap. 3 of Part IV (pp. 242–264) of the second volume of HEA. Duhem's contribution to solid mechanics and acoustics in the second volume of these Lille lectures (Duhem 1891 = HEA-2) is composed of three essential parts, the last one on (Helmholtz) audible acoustics—that does not require our perusal here—and the first devoted to one-dimensional and two-dimensional elastic bodies (about 200 pages) and the second devoted to true three-dimensional elasticity (about seventy pages) on which we shall more particularly focus our interest. The first part in fact deals with strings (threads) and membranes. This should be close to the strength of materials useful to the engineer. But true to himself, Duhem rather exposes an exhaustive mathematical description of works achieved in the theoretical field during the nineteenth century, including such problems as those posed by strings stretched over a given curved surface or membranes supported by a given contour, or still membranes in contact with a fluid. Reported recent results are due to E. Picard in France and H. Schwarz in Germany. Many aspects on the vibrations of strings—going all the way back to Jacob Bernoulli and d'Alembert but improved during the nineteenth century —and membranes are also reported with mathematical details. Developments by Euler and Fourier concerning series solutions are given. We doubt that all these details were of direct interest to many students in Lille. For instance, Duhem does not hesitate to recall in details the theory of geodesic lines (Gauss, Jacobi, Bertrand, Darboux) where needed. Occasionally, he expands the theory of capillarity and surface tension. But we note already some interest in stability problems, a trademark—if we may say so—of Duhem' scientific interests. All this is not very original although quite involved and much above the head of the students in Lille.

But Duhem provides here a true and rich investigation which carries the mark of the historian. First, Duhem rightfully notices that the main problem rests with the three-dimensional theory of continua since theories in one and two space dimensions have been proposed before with a natural identification of, e.g., longitudinal

[9]We have noticed elsewhere the meagre enthusiasm exhibited by French mechanicians and physicists to accept these relatively new mathematical formalisms (see our contributions on the Cosserats and Appell in Maugin 2014a).

forces and shear forces (by, among others, Euler). Duhem presents the original developments in three-dimensional elasticity theory from the initial work of C.L.M. H. Navier (1785–1836) that is based on molecular interactions, and the real continuum approach (avoiding superfluous microscopic details) such as by Cauchy (1823, 1828a). Priority to Navier relies on his memoir presented to the Paris Academy of Sciences on May 14, 1821. The reasoning is based on the fact that elastic bodies in their natural state are made of elements that exercise no force on each other. In presence of applied forces, the bodies are being deviated from their natural state, with a force acting between each pair of elements a function of their distance. Equilibrium with the externally applied forces then provides the continuum equilibrium equations with only one elasticity coefficient (as if the Lamé coefficients were such that $\lambda = \mu$). Navier also obtains the natural boundary condition (for stresses in our vocabulary) under the same constraint on coefficients.

This contribution above all shows a certain hubris on the part of Duhem who, as a recently appointed young teacher, wants to prove to the academic world that he has reviewed everything on the subject matter although he has not yet contributed original works to it (as on 1891).

The same holds true of the second part devoted to three-dimensional elasticity. Per force, this has to deal with some conventional matter, but it brings us closer to our own focus. Here Duhem critically remarks—perhaps in some arrogant tone of superiority—that most readers of Navier's original memoir of 1821 read it with little attention (HEA-2, p. 244). He then recalls the beautiful work (several memoirs on double refraction between 1821 and 1827) by A. Fresnel (1788–1827), which involves a different starting working hypothesis: in a natural state it is *repulsive* molecular action which maintains the equilibrium, the resulting molecular actions that act on each material point from all sides summing up to zero. Out of equilibrium Fresnel determines the magnitude and direction of the resultant force on each material point. Duhem gives the corresponding proof in detail. We know from Cauchy himself (1823, p. 9) that he discussed the matter with Fresnel in 1822. This led him to formulate the general equations of continuum mechanics (Cauchy 1823, 1828a) after a short oral communication to the Paris Academy of Sciences on September 30, 1822 (Theory of Cauchy-I). Duhem reminds the reader of the way Cauchy proved his basic equations—equilibrium of a small parallelepiped for the local body equation and the famous tetrahedron argument for the associated boundary condition as reported in many books. This avoided the consideration of molecular forces while introducing elegantly the notion of three-dimensional stress tensor (not called this way by Cauchy) with six independent components at most. The reasoning consists in keeping only terms of the same order of magnitude in the equations of balance of forces and moment of forces applied to the infinitesimal body. Cauchy's proofs were completed by Lamé and Clapeyron (1831). Poisson (1829, 1833) proposed another theory that followed the attempts of Navier and Fresnel based on the notion of molecular interactions. Cauchy, deviating from his original proposal, took the same direction in another memoir of 1828 (Cauchy 1828b). Some of these works yield what is now known to be the wrong number of elasticity coefficients in the case of isotropic bodies.

Anyway, it was Cauchy and Poisson who tried to relate the six "pressures" and the six deformations in the linear theory. Cauchy in fact first guesses a plausible form of relations but this tentative approach provided one elasticity coefficient only for isotropic bodies (see the above recalled equality$\lambda = \mu$). In a subtle reasoning Cauchy then postulates that contrary to his first assumption where the principal "pressures" were only proportional to the corresponding principal "dilatations", each principal pressure must be given by a linear combination of the corresponding principal dilatation and another term proportional to the *cubic* (i.e., volume) dilatation, all this in the understanding of functions that are linear and homogeneous. This is what yields the existence of two different elasticity coefficients, called λ and μ by Lamé in the case of isotropy. In modern times (Truesdell and Toupin 1960, Truesdell and Noll 1965), this method which consists in making direct hypotheses concerning the form of the stress-strain relations is called "Cauchy's theory of linear elasticity" (not involving the existence of an elasticity potential). Cauchy (1830) then tried to generalize this reasoning to the case of anisotropic bodies. He writes thus the most general linear relations between the two sets of (6) pressures and (6) dilatations. This involves at most 36 coefficients. But using modern algebra we can say that this is a symmetric linear application of R^6 onto R^6, so that a trivial argument shows that only 21 coefficients at most are independent in agreement with Lamé (1852). This number was in fact obtained by Cauchy on the basis of a theory involving molecular attraction (CAUCHY-2). Poisson, in his molecular approach obtains a number of 15. The debate lasted until Clebsch, Saint-Venant, and Poincaré (cf. HEA-2, p. 262). However, with George Green taking over Lagrange's method with the existence of an elasticity potential (Green 1839) and its thermodynamic justification by W. Thomson (Kelvin), we know that the final answer indeed is 21 for anisotropic elastic bodies and 2 for the isotropic case.

"Recherches sur l'élasticité" (RE)

When Duhem gathered in 1906 four long papers previously published in the *Annales Scientifiques* of the *Ecole Normale Supérieure* in a single volume of more than 200 pages (Duhem 1906b), some fifteen years had elapsed since his writing of his lecture notes in Lille. In this interval of time, the general landscape of continuum mechanics had drastically evolved as a result of works by Voigt, Koenigs, Poincaré and, above all, the Cosserat brothers with their work of 1896. In this long interval Duhem himself had published a large number of Notes in the *Comptes Rendus* of the Paris Academy of Sciences in parallel with his original contributions to hydrodynamics, especially in the domain of wave propagation (see below). Contrary to his lectures of 1891 where he stated that three-dimensional elasticity was essentially dealing with "small" motions and thus a linear theory while one-dimensional and two-dimensional elasticity theories had to be nonlinear—as we obviously witness with the theory of the elastica—he realized the full potential of a three-dimensional theory that involves finite strains. This change of view must have been kindled by the Cosserats' long paper of 1896 and a better learning of the elasticity theory and its sources. Accordingly, he had to return to a more basic general description calling for both Euler and Lagrange kinematic approaches complemented by the original results of Cauchy, Piola (1836, 1848), Kirchhoff (1850, 1852, 1859, 1876), Boussinesq and others related to strains and the stress tensor. It is because of this contribution that Duhem is probably better known in solid mechanics than in fluid mechanics. A reason for this may be the direct interaction he had with some well known contemporaries such as Poincaré and the Cosserat brothers (see also Hellinger 1914), and the fact that he had become much interested in large deformations, the study of which had a tremendous development some fifty years later on.

The subtitle of the book deserves to be remembered. It reads: "On equilibrium and motion of vitreous media; slightly deformed vitreous media; stability of elastic media; General properties of waves in viscous and nonviscous media". This says it all as a kind of abstract giving the title of four original contributions.

In Duhem's overall view, the need to consider finite deformations originates from the will to envisage a whole spectrum of material behaviours, a spectrum that includes intermediate behaviours between true solids and fluids (Duhem 1901). This in turns involves viscosity and so-called "vitreous" bodies (as opposed to "crystallized"—anisotropic—ones), in the line of ideas put forward by Kirchhoff and Boussinesq, and the German analyst Oskar Emil Meyer (e.g., in Meyer 1874) to whom he will frequently refer. According to Miller (1970), Duhem thus "kept alive a correct finite elasticity inspired by other workers". Duhem's introduction of finite deformations leans heavily on the Cosserats' memoir of 1896, but he is also thoroughly aware of, and praises, the main works by George Greeen, G. Kirchhoff, W. Thomson (Lord Kelvin), W. Voigt (considering isothermal and adiabatic deformations), and J.V. Boussinesq on elements of nonlinear elasticity.

The notation is that used with strains invariants $J_\alpha, \alpha = 1, 2, 3$. The determinant of the deformation $a, b, c \rightarrow \xi, \eta, \varsigma$ is noted D, so that the continuity equations reads $\rho D = \rho_0$. Thermodynamic temperature T is introduced from the start as well as the internal potential Φ (following the initial idea of Duhem 1893). A vitreous medium is supposed to be isotropic and thus such that $\Phi = \Phi(T, J_1, J_2, J_3)$. General equilibrium equations are deduced from the virtual work principle in the presence of external actions, supposing that both internal and external interactions are Newtonian. The full dynamical equations are obtained by adding the virtual work of viscous forces and applying d'Alembert's principle for the inertia forces. The resulting equations in the bulk and at the bounding surface generalize those proposed by Boussinesq. But Duhem, following his previous works in hydrodynamics, pays special attention to the dissipation function and the so-called complementary relation, reflecting the heat conduction (à la Fourier) in what we now identify with the local statement of the first law of thermodynamics (HE, p. 41). This completes the first part of the RE.

The second part (RE, pp. 43–73) first considers the motion of a vitreous medium that slightly deviates (i.e., in "small" motions) from its initial configuration, the latter being not necessarily an equilibrium state under the action of vanishing forces. These yield the Green-Lamé equations as completed by Poincaré in his lectures on elasticity (Poincaré 1892, p. 54). The problem envisaged by Meyer (1874)—who used molecular considerations—corresponds to constant uniform temperature and surface external forces only, but still with viscosity. What drastically simplifies the whole matter [RE, p. 53, Eqs. (30)]. Using a technique due to Clebsch, this can be transformed into equations governing the dilatations and the rotations separately. Because of the presence of viscous terms, these equations are akin to those obtained for wave propagation in the electrodynamics of electricity conductors. Stability imposes that elasticity coefficients and viscosity coefficients satisfy well known inequalities. The resulting waves are of differing orders (in Hadamard's classification) for the components of the velocity and those of the stress components (one order less for these quantities than for the velocity). This is indeed studied in detail in the second chapter of this second part of RE. This study is quite innovative with Duhem concentrating on the case of second order (i.e., acceleration) waves—for the velocity field—in heat conducting vitreous bodies, and first order (i.e., shock) waves for the same bodies, with eventual specialization to the absence of viscosity and bad conductors of heat. This chapter, exploiting Hadamard's lemma (kinematic compatibility) is rather technical because of the visible ambition at generality. This parallels studies proposed before in the researches in hydrodynamics. It seems that Duhem is the first mechanician to study the discontinuities in temperature in thermo-mechanical waves of any order. For instance, a wave of second order with respect to the components of the velocity is at least of second order in the temperature in a good heat conductor (RE, p. 75). Also, in a bad conductor of heat, albeit viscous, waves of order greater or equal to one with respect to the velocity cannot persist in time.

The third long part (HE, pp. 83–157) in this series is exclusively reserved to the question of stability of elastic media. The various points examined first concern sufficient conditions for the initial stability of any—or more specially vitreous—medium. Then establishing necessary conditions for the stability of an elastic medium takes more space and yields longer technical developments (pp. 91–132) requiring great attention to the relative orders of smallness of terms. The discussion involves the second derivative of the global energy yielding complicated lengthy expressions (see, e.g., RE, pp. 108–115). The rigor reached in these developments is that introduced by Lyapunov and Poincaré. But it is not proved that two systems of partial differential equations that differ infinitesimally little from one another can provide integrals that differ infinitesimally little in spite of identical initial conditions (RE, p. 121). This third part concludes with a thorough discussion of the displacement of equilibrium under perturbation of external actions. The case of isothermal perturbations is favored and consequences for the stability of equilibrium are emphasized. Some applied inverse argument is based on Hadamard (1903, p. 253). Isentropic displacement from equilibrium is also envisaged. Finally, application of the obtained results to a very slightly deformed body about a loaded configuration is briefly discussed in comparison with results by W. Voigt (1887). One should mention here the importance of so-called *principal waves* which propagate in the direction of one of the principal axes of the initial deformation (many works in the 1960s–1980s concern this special case).

The last part of RE (pp. 159–213) probably is the one of which Duhem is most proud. It constitutes the apotheosis of Duhem's researches in waves in continuous media since, apart from the writing of his thick opus on energetics and general thermodynamics (Duhem 1911), he will devote the ten-year period from 1906 to his death in 1916 mostly to works of a historical and epistemological nature. Indeed, Duhem's contribution in this last part of his RE shows the real professionalism reached by Duhem in this original field of research. The detail and accuracy in reasoning exhibited in this work are stupendous. If it were not for the lack of use of a more synthetic and condensed notation of the vector kind—Duhem still uses the now antiquated independent components with a special, Latin or Greek, letter for each of them—and the introduction of his own neologisms, the reading of this work by twenty-first century students would be an easy matter. These fifty pages are splendid. Duhem is faithful to himself by giving all definitions and proofs in great detail. Contrary to Green and Poisson who understood waves as plane-wave perturbations of small amplitude, Duhem applies rigourously the rather new notion of *discontinuity fronts* as beautifully formalized by his friend Jacques Hadamard, the two ancestors in this line of research being Riemann and Hugoniot (for one-dimensional motions). His approach is in two steps, one giving the general description of discontinuity fronts in both reference (Lagrangian) and actual configurations with the accompanying transformation formulas (e.g., Piola transformations although not cited this way), and the resulting general properties of such waves in the presence or absence of viscosity. In accordance with Hadamard he clearly distinguishes between kinematic compatibility conditions and dynamics ones—that involve the jump of the momentum equation and of energy-based quantities, temperature effects being only exceptionally neglected. The second step is reserved to an in depth examination of discontinuity waves of various orders in media exhibiting both elasticity and viscosity. Without exposing the reader to the detail of the proposed developments, we simply note the originality and novelty of some most striking results.

First of all, Duhem is the first to study the relationships between waves in isothermal and adiabatic finitely deformed systems without viscosity. Again, the specific interest of Duhem in thermodynamic properties led him to establish conditions of stability and those of the existence of waves (in the sense of Hadamard): no true shock waves can exit in the viscous non-linear thermo-elastic bodies (a result already shown in viscous fluids; see above concerning RH). In these studies he is also led to introducing the notion of "ondes-cloisons" ["partition (-wall) waves", GAM] through which there is no exchange of matter. These waves are similar to contact discontinuities (no discontinuity in the velocity). They separate the volume of the considered body into cells, so that Duhem notes their resemblance with H. Bénard's cells in a fluid where large differences in temperature generate convection currents and the formation of cells. All this is well documented with the usual persuasive style of Duhem. One will have to await works by Truesdell and co-workers[10] in the 1960s–1970s, sometimes in a pure modernized

---

[10]See, in particular, Truesdell (1961) and Coleman et al. (1965). Then a real pandemic fashion spread among theoretical mechanicians who studied acceleration and shock waves with the same type of Hadamard-Duhem approach in all kinds of continuous media. The present writer did not eschew this kind of irresistible feverish phenomenon.

adaptation of Duhem's work, to reach the same level of generality and rigor. But a direct (contemporary) disciple of Duhem in the application of thermodynamics to the propagation of discontinuity fronts is Emile Jouguet (1871–1943) who developed the theory of detonation waves with obvious applications to explosives (cf. Jouguet 1905–6, 1917).

We would not be complete without mentioning that Duhem, in addition for an above demonstrated interest in elasticity, was also interested in behaviours of a more complex thermomechanical type such as plasticity and the phenomena of permanent deformations and hysteresis. Some of these related studies may have been premature due to the lack of reliable experimental measurements and also difficulties in developing the related rather singular (irreversible) thermodynamic description. This was fully exposed by Duhem in some of his works (Duhem 1903b, 1911), that we examined elsewhere (Maugin 2014a, b, Chap. 10).

We may conclude this section with the following remark that seems to have escaped the attention of all writers. We have seen that Duhem considered the notion of "vitreous" elastic solids—as opposed to "crystallized" elastic solids with a more complicated material symmetry in their natural state—as essentially isotropic (in their initial configuration) thermo-elastic solids able to exhibit some viscosity. This he wants because he likes to contemplate a rather continuous spectrum of nonlinear elastic behaviours between such "solids" and fluids. The fact that a fluid is necessarily isotropic became a credo in the Truesdell-Noll overall vision of continuum mechanics. But what about *liquid crystals* that can flow while exhibiting an ordering typical of crystals, and thus appear as nonsensical from Duhem's viewpoint? It happens that in the very last sentences concluding RE (p. 213), Duhem mentions Lehmann's[11] experimental work on "fluid crystals" as a possibility of existence for an easily deformable, although anisotropic, crystalline medium. The idea will be taken over by J.L. Ericksen in the 1960s. But this short spot-on remark by Duhem shows the large extent to which he kept aware of any development in physics; he was decidedly a never-at-rest reader with a gift for easy understanding of all new physical facts.

5.6 Contemporary Reception of Duhem's HEA and His Two "Recherches" Volumes

Paul Painlevé (1863–1933), a brilliant mathematician and successful politician, was probably the first to write a detailed (ten pages long) review of Duhem's HEA (Painlevé 1893). Painlevé became in time professor of Rational Mechanics at Ecole

[11]This is Otto Lehmann (1855–1922), German physicist considered to be the "grandfather of liquid crystals" with his pioneering work published as "Über fliessende Krystalle", Zeit. für Physikalische Chemie, 4, 462–472, 1889. A classification of various liquid crystals was given by Georges Friedel in 1922.

Polytechnique, and was the first professor of Mechanics of Aviation at the Ecole Supérieure de l'Aéronautique (created as soon as in 1909 and to become ENSAé or "Sup Aéro"). We cannot ignore that Painlevé and Duhem were students in mathematics and physics at the Ecole Normale Supérieure (ENS) in Paris at the same time, Painlevé being there from 1883 to 1886, and Duhem from 1882 to 1885 with an additional year in 1885–86 and one year as assistant in 1886–1887 before joining Lille. Like the other French contemporary mathematicians (H. Poincaré, J. Hadamard, E. Picard), all these people were friends and more or less shared the same vision of mathematical physics and phenomenological theories in particular. Thus we may qualify Painlevé's review as benevolent, but not avoiding criticism of the form, his somewhat lengthy proofs, and Duhem's love for generalities, the cause of reading difficulties. But he praises the extreme mathematical rigour of Duhem, the neatness of the solution of some problems (e.g., in the compatibility of two small motions), the quality of the treatment of Hugoniot's method for the wave equation after accounting for previous results of Beltrami, Kirchhoff and Riemann, of Helmholtz's equation accounting for Poincaré's general results, of the two-dimensional wave equation in space on account of results by Schwarz and Picard, and the high pedagogy of Duhem's rendering of audible acoustics (following of course Helmholtz and physiological data). He also highly recommends Duhem's historical comments that we commented above. He also notes that other researchers, such as V. Volterra, pursued some of the studies of Duhem on Helmholtz's equation. Poincaré and the Cosserat brothers (1909) as well as Paul Appell (1921) will often cite the relevant works of Duhem. Jacques Hadamard and Emile Picard will write complimentary eulogies of Duhem (Hadamard 1927; Picard 1921). In Germany, Hellinger (1914) generously cites Duhem (1891) many times as one of the "modern" developers of continuum mechanics. Léon Lecornu, himself author of a known course in mechanics (published in 1914–18; also 1925), wrote a kind of pamphlet (Lecornu 1929) in which he praised Duhem's works on elasticity. This was favourably reviewed by Buhl (1929), a former colleague of one of the Cosserat brothers in Toulouse.

5.7 Conclusion

Duhem's texts critically perused in the present contribution unequivocally place in evidence two characteristic aspects of Duhem's style and approaches. In his HEA of 1991 he shows his enthusiasm doubled by the obvious hubris of the young academic who wants to prove that he has mastered (in a very short time indeed) most of the matter existing in hydrodynamics and elasticity since the eighteenth century onward. This is achieved without much original input from himself except for the already typical Duhemian style and some traces of what will become some of his personally recurring interests, such as the notion of internal potential and questions of stability. The published books of "recherches" (RH and RE) clearly show (i) a much more mature scientist who took more time to ponder previous achievements and (ii) a greater originality in the expansion of ideas concerning stability and waves in both fluids and elastic solids. These were the results that will remain and be re-captured by mechanicians of the continuum in the 1960s–1970s.

References

Appell P (1921) Traité de mécanique rationnelle, Third edition. Gauthier-Villars, Paris [facsimilé reprint by Gabay, Paris, 1991]

Ariew R (2007) Entry « Pierre Duhem » in: The open-access Stanford Encyclopedia of Philosophy (revised, 2011)

Barré de Saint-Venant AJC (1883) Théorie des corps élastiques (French translation from the German with many comments and additions of "Clebsch A., (1862), Theorie der Elastizität fester Körper, Leipzig"; Translation with the help of A. A. Flamant) [Reprint by Johnson Reprint Corp. New York, 1966]

Bouasse H (1920) Théorie de l'élasticité. Résistance des matériaux. Librairie Delagrave, Paris [Second edition, 1947]

Brillouin L (1925) Les tensions de radiation, leur interprétation en mécanique classique et Relativiste. J Phys Radium 6:337–353

Brillouin M (1884–5) Leçons sur l'élasticité et l'acoustique. Librairie Labouche, Toulouse

Brouzeng P (1987) Duhem 1861–1916: Science and Providence. Editions Belin, Paris

Buhl A (1929) Review of Lecornu's book « Théorie mathématique de l'élasticité 1929». L'enseignement mathématique 28:149–150

Cauchy AL (1823) Recherches sur l'équilibre et le mouvement intérieur des corps solides ou fluides, élastiques ou non élastiques. Bull Soc Filomat Paris; January issue, 9–13 [This a kind of long abstract of a work that will materialize in the next listed publication]

Cauchy AL (1828a) Sur les équations qui expriment l'équilibre ou les lois du mouvement intérieur d'un corps solide élastique ou non élastique. In: Exercices de mathématiques, Vol. 2, p. 160–187, Sept. 1828 [This presents in print the ideas originally submitted to the Paris Academy of Sciences on September 30, 1822]

Cauchy AL (1828b) In: Exercices de Mathématiques, Vol. 3, Sept. 1828, p. 188–212, and « De la pression ou tension dans un système de points matériels » , Oct. 1828, p. 213–236

Cauchy AL (1828c) Sur l'équilibre et le mouvement d'un système de points matériels sollicités par des forces d'attraction ou de répulsion mutuelle, In: Exercises Vol.3, or Œuvres complètes de Cauchy, 2nd series, Volume VIII, p. 227

Chen PJ (1976) Selected topics in wave propagation. Noordhoff, Leiden, The Netherlands

Clebsch A (1862) Theorie der Elastizität fester Körper. Teubner, Leipzig

Coleman BD, Gurtin ME, Herrera R, Truesdell C (1965) Wave propagation in dissipative materials (a reprint of five memoirs). Springer Verlag, New York

Cosserat E, Cosserat F (1896) Sur la théorie de l'élasticité. Ann Fac Sci Toulouse, 1ère série, 10/3–5 :I1- I116

Cosserat E, Cosserat F (1909) Théorie des corps déformables., Hermann, Paris, 226 pages) [Reprint by Editions Gabay, Paris, 2008; Reprint by Hermann Archives, Paris, 2009; English translation: N68–15456: Clearinghouse Federal Scientific and Technical Information, Springfield, Virginia NASA, TT F-11561 (February 1968); another translation by D. Delphenich, 2007] (originally published as supplement in p. 953–1173 to: Chwolson OD (1909) Traité de physique (traduit du Russe), Vol. II. Hermann, Paris]

Daher N, Maugin GA (1986) The Method of Virtual Power in Continuum Mechanics: Application to Media Presenting Singular Surfaces and Interfaces. Acta Mech 60:217–240

Duhem P (1886) Le potentiel thermodynamique et ses applications à la mécanique chimique et à la théorie des phénomènes électriques. Hermann, Paris (xii + 248 pages)

Duhem P (1888) De l'aimantation par influence. Suivi de propositions données par la Faculté, Paris (Doctorat ès Sciences Mathématiques). Published by Gauthier-Villars, Paris
Duhem P (1891) Hydrodynamique, élasticité, acoustique, Two volumes. Hermann, Paris (688 pages) [This was immediately favourably reviewed by the Paul Painlevé, 1893]
Duhem P (1893) Le potentiel thermodynamique et la pression hydrostatique. Ann Ecole Norm (Paris) 10:183–230
Duhem P (1896) Théorie thermodynamique de la viscosité, du frottement, et des faux équilibres chimiques. Mémoires de la Société des Sciences Physiques et Naturelles de Bordeaux, 5ème série, 2:1–208
Duhem P (1901) Sur les déformations permanentes et l'hystérésis, Part 6, Mém. Acad. Roy. Belgique, 62, 1–33; Parts 7, ibid, 33–136
Duhem P (1903a) Recherches sur l'hydrodynamique, 1st series, 212 p. Gauthier-Villars, Paris [Collection of papers]
Duhem P (1903b) L'évolution de la mécanique (published in seven parts in: Revue générale des sciences, Paris; as a book: A. Joanin, Paris) [English translation: The evolution of mechanics, Sijthoof and Noordhoff, 1980]
Duhem P (1904) Recherches sur l'hydrodynamique, 2nd series, 153 p., Gauthier-Villars, Paris [collection of papers]
Duhem P (1906a) La théorie physique. Editions Chevalier et Rivière, Paris [English translation: The aim and structure of physical theory, Princeton University Press, New Jersey, 1954; paperback reprint, 1991]
Duhem P (1906b) Recherches sur l'élasticité. Gauthier-Villars, Paris [collection of four papers with a total of 218 pages; Facsimile reprint by Indigo, Toronto, 2010]
Duhem P (1908) SOZEIN TA PHAINOMENA, French edition, Hermann, Paris [English translation: To save the phenomena, Chicago University Press, 1969]
Duhem P (1911) Traité d'énergétique ou de thermodynamique générale. Two Volumes. Gauthier-Villars, Paris (528 + 504 pages)
Duhem P (1913) Notice sur les titres et travaux de Pierre Duhem [self-description of his works by P. Duhem when applying for full membership in the Paris Academy of Sciences), Bordeaux (printed posthumously in Mém Soc Sci Phys et Nat de Bordeaux, 7ème série, p. 71–169, 1927) [Partial English translation by Y. Murciano and L. Schramm as « An account of the scientific titles and works of Pierre Duhem » . Science in context 2:333–348, 1987]
Duhem P (1961) Recherches sur l'hydrodynamique [New edition with a long preface by J. Kampé de Fériet, 396 pages. Volume hors série des Publications Scientifiques et Techniques du Ministère de l'Air, S.D.I.T.A., Paris]
Euler L (1757) Principes généraux de l'état d'équilibre des fluides. Mém Acad Sci Berlin 1:217–273 (Memoir presented to the Academy in 1755)
Green G (1828) An essay on the mathematical analysis of electricity and magnetism, Privately printed, Nottingham [Also in: Mathematical papers of George Green, Ed. N.M. Ferrers, Macmillan, London, 1871; reprinted by Chelsea, New York, 1970]
Green G (1839) On the laws of reflection and refraction of light at the common surface of two non-crystallized media. Trans Cambridge Phil Soc 7:245–269
Hadamard J (1903) Leçons sur la propagation des ondes et les équations de l'hydrodynamique. Hermann, Paris
Hadamard J (1927) L'oeuvre mathématique de Pierre Duhem. In: Special issue of Mém Soc Sci Phys Nat Bordeaux, 7ème série, p. 637–665
Hellinger E (1914) Die allgemein Ansätze der Mechanik der Kontinua. In: Enz. Math.Wiss. Eds F. Klein and K. Wagner, Vol. 4, Part 4, p. 602–694. Springer, Berlin
Hilbert D (1906–1907) Vorlesung über die Mechanik der Kontinua (Lectures in the mechanics of continua), Gehalten in Wintersemester 1906–907, zu Göttingen [mimeographed, 239 pages; with an introduction (in German) by J. Stickforth, TU Braunschweig, 1970]
Hugoniot P (1887a) Mémoire sur la propagation du mouvement dans un fluide indéfini, J Math Pures et Appl 4th series, 3:477–492 [also, ibid, 4:153–167]

Hugoniot P (1887b, 1889). Mémoire sur la propagation du mouvement dans les corps et spécialement dans les gaz parfaits. J Ecole Polytechnique 57:3–97 [ibid, 58:1–125 (1889)]
Humbert P (1932) Pierre Duhem. Paris
Jaki SL (1984) Uneasy genius: The life and works of Pierre Duhem. Martinus Nijhoff, The Hague
Jouguet E (1905–1906) Sur la propagation des réactions chimiques dans les gaz [On the propagation of chemical reactions in gases], Journal de Mathématiques Pures et Appliquées, Series 6, 1:347–425 (1905), continued in 2:5–85 (1906)
Jouguet E (1917) Mécanique des explosifs, étude de dynamique chimique. Doin, Paris
Jouguet E (1920a) Sur la variation d'entropie dans les ondes de choc des solides élastiques. C R Acad Sci Paris 171:789–791
Jouguet E (1920b) Notes sur la théorie de l'élasticité. Ann Fac Sci Toulouse 3ème série, 12:47–92
Kirchhoff G (1850) Über das Gleichgewicht und die Bewegung einer elastischen Scheibe. Journ f d reine und angewandte Math 40:51–88
Kirchhoff G (1852) Über die Gleichungen des Gleichgewichts eines elastischen Körpers bei nicht unendlich kleiner Verschiebungen seiner Teile. Sitzgsber Akad Wiss Wien 9:762–773
Kirchhoff G (1859) Über das Gleichgewicht und die Bewegung eines unendlich dünnen elastische Stabes. Journ f d reine und angewandte Math 56:285–313
Kirchhoff G (1876) Vorlesungen über mathematische Physik: Mechanik (known as "Lectures on Mechanics"). Teubner, Leipzig [available on the site Gallica of the Bibliothèque Nationale de France (BNF)]
Koenigs G (1897) Cours de cinématique. A. Hermann, Paris [with notes by E. and F. Cosserat]
Lagrange JL (1788) Mécanique analytique, First edition.Vve Desaint, Paris; Second edition, Vve Courcier, Paris, 1811 [Facsimile reprint of 4th edition with notes and comments par J. Bertrand and G. Darboux, Gabay, Paris; English translation of the 2nd edition (1811), Eds. A. Boissonnade and V.N. Vagliente, Springer, Dordrecht, 1997]
Lamb H (1879) Hydrodynamics, 1st edition. Cambridge University Press, Cambridge (6th edition, CUP, 1932; Dover reprint, New York, 1945)
Lamé G (1852) Leçons sur la théorie mathématique de l'élasticité des corps solides, Paris, (2nd edition, 1866) [Facsimile reprint, Gabay, Paris, 2006]
Lamé G, Clapeyron E (1831) Mémoire sur l'équilibre intérieur des corps solides Homogènes. Journal de Crelle (J für reine und angewandt Mathematik), Vol. VII [also Mémoires de Savants étrangers, Acad Sci Paris IV:463–562, 1833; available on the website of the BNF, Paris]
Lecornu L (1925) Cours de mécanique à l'Ecole Polytechnique. Gauthier-Villars, Paris
Lecornu L (1929) Théorie mathématique de l'élasticité, Vol. 35 of Mémorial des Sciences Mathématiques. Gauthier-Villars, Paris
Love AEH (1892) A Treatise on the mathematical theory of elasticity, [1944, 4th edition, Dover reprint, New York; originally published in two volumes in 1892–3]
Manville O (1927) La physique de Pierre Duhem. Mém Soc Sci Phys Nat Bordeaux, 7ème Série, 171–636 [The most comprehensive and unsurpassed analysis of Duhem's scientific works by one of his direct disciples]
Maugin GA (2013) Continuum mechanics through the twentieth century: A concise historical perspective. Springer, Dordrecht, The Netherlands
Maugin GA (2014a) Continuum mechanics through the eighteenth and nineteenth centuries: Historical perspectives from John Bernoulli (1727) to Ernst Hellinger (1914). Springer, Dordrecht, The Netherlands
Maugin GA (2014b) Poincaré and Hilbert on continuum mechanics, Preprint UPMC, Paris [Chapter 6 in this volume]
Meyer OE (1874) Zur Theorie der inneren Reibung. J für die reine und angewandt Mathematik, 78:130 on (not seen by the author)
Miller DG (1970) "Duhem, Pierre-Maurice-Marie". In: Dictionary of Scientific Biography, Ed. C. C. Gillispie, Vol.3. Charles Scribner's Sons, New York, p. 225, 233
Neumann FE (1885) Vorlesungen über die Theorie der Elasticität der festen Körper und des Lichtäthers, Leipzig (edited by O.E. Meyer when Neumann was 87 years old)

Painlevé P (1893) Review of "Duhem (P.)—Hydrodynamique, Elasticité, Acoustique", Bull Astronomique Sér. 1, 10:390–400 [available on the site of the Smithsonian/NASA Astrophysics Data System]

Picard E (1921) La vie et l'oeuvre de Pierre Duhem, Eloge lue dans la séance publique annuelle du 12 décembre 1921. C-CXLII, C R Acad Sc paris, LVII

Pierre-Duhem Hélène (1936) Un savant français: Pierre Duhem. Paris

Piola G (1836) Nuova analisi per tutte le questioni della meccanica moleculare. Mem Mat Fis Soc Ital Modena 21(1835):155–321

Piola G (1848) Intorno alle equazioni fondamentali del movimento di corpi qualsivoglioni considerati secondo la naturale loro forma e costituva. Mem Mat Fiz Soc Ital Modena 24(1): 1–186

Poincaré H (1892) Leçons sur la théorie de l'élasticité. Georges Carré, Paris [Facsimile reprint, Gabay, 1990; also University of Michigan, Ann Arbor, 2012]

Poisson SD (1829) Mémoire sur l'équilibre et le mouvement de corps élastiques. Mémoires de l'Institut de France, VIII:357–570 (and X :578)

Poisson SD (1833) Traité de mécanique, vol II. Bachelier, Paris

Stokes GG (1845) On the theories of internal friction of fluids in motion, and the equilibrium and motion of elastic solids. Trans Cambridge Phil Soc 8:287–319

Thomson W (Lord Kelvin), Tait PG (1867) Treatise on natural philosophy, First edition, Oxford [Second edition 1876]

Timoshenko SP (1953) History of the strength of materials. McGraw-Hill, New York [Dover reprint, New York, 1983] [Review of this book by Truesdell reprinted in Truesdell, 1984, p. 251–253]

Todhunter I (1886) A history of the theory of elasticity and the strength of materials from Galileo to the present time, vol. 1. Cambridge University Press, UK (edited and published posthumously in 1886 by Karl Pearson)

Truesdell CA (1961) General and exact theory of waves in finite elastic strain. Arch Rat Mech Anal 8:263–296

Truesdell CA, Noll W (1965) The non-linear field theories of mechanics. In: Flügge S (ed) Handbuch der Physik, Bd.III/3. Springer, Berlin [Referred to for short as NFTM]

Truesdell CA, Toupin RA (1960) The classical field theories. In: Flügge S (ed) Handbuch der Physik, Bd. III/1. Springer, Berlin [Referred to for short as CFT]

Voigt W (1887) Teoretische Studien über Elasticitätverhältinsse der Krystalle, I.II. Abh. K. Ges Wissen Göttingen 34:3–52, 53–100

Chapter 6
Poincaré and Hilbert on Continuum Mechanics

Abstract Henri Poincaré in France and David Hilbert in Germany are the most creative mathematicians of the transition period between the nineteenth and twentieth centuries. Both manifested a real interest in mathematical physics with a marked philosophical bias on the part of Poincaré and an accentuated inclination towards axiomatics on that of Hilbert. Although interested in various branches of continuum mechanics, Poincaré delivered a series of lectures specifically on the theory of elasticity in the early 1890s, while Hilbert gave a semester course on general continuum mechanics in 1906–1907. Each of these was redacted in written form by a pair of auditors. The surviving documents provide us with an opportunity to appraise and contrast the different styles and approaches of the two great mathematicians in a field that does not belong in their usual scientific preoccupations. As usual Poincaré is more intuitive than his competitor. He also demonstrates an immoderate interest in the propagation of light as a specific path of application of elasticity, while simultaneously exhibiting a surprising curiosity for the so-called "Saint-Venant problem". Hilbert's exposition is more in a line that will materialize in further developments such as by Hellinger and then mechanicians in the second part of the twentieth century.

6.1 Introduction

Henri Poincaré (1854–1912) (cf. Gray 2013) in France and David Hilbert (1862–1943) (cf. Reid 1990) in Germany are probably the most famous mathematicians of the "belle époque" (roughly the period 1880–1914). They have in common to belong to a now extinguished species of scientists with a wide interest in many domains of mathematics and their applications to mathematical phenomenological physics (and quantum physics for Hilbert). This remarkable aerial view is unmatched by other mathematicians of the period and thereafter. But this in no way implies some superficiality for they both contributed to the definite proof of

G.A. Maugin, *Continuum Mechanics through the Ages - From the Renaissance to the Twentieth Century*, Solid Mechanics and Its Applications 223,
DOI 10.1007/978-3-319-26593-3_6

fundamental theorems, they proposed famous conjectures, and they opened new horizons that participated in the transition to the developments of twentieth century mathematics and physics with the invention of new tools such as topology ("analysis situs"), asymptotic methods, the dynamics of systems, and, among other fruitful notions, the celebrated Hilbert spaces of quantum mechanical fame. They were both deeply interested in the new physics represented by the electron theory, special and general relativity theories, and radiation and quantum mechanics. Involvement of Poincaré in the last two fields was abridged by an untimely death. Although of an equal creativity, they differed in their practice of science, Poincaré being an essentially lonely researcher—but not detached from the scientific and philosophical discussions of his time—while Hilbert was a true leader who mentored many young associates and directed an organized school that blossomed first in Germany and Europe, and then in the USA (Figs. 6.1, 6.2, 6.3 and 6.4).

Much below these stratospheric heights, we find continuum mechanics and its two main branches, solid and fluid mechanics. The reader may find strange the introduction of such mathematical luminaries (and probably unique geniuses) as Poincaré and Hilbert in such mundane fields of sciences situated at the boarder line between applied mathematics and engineering. The interest of Poincaré and Hilbert

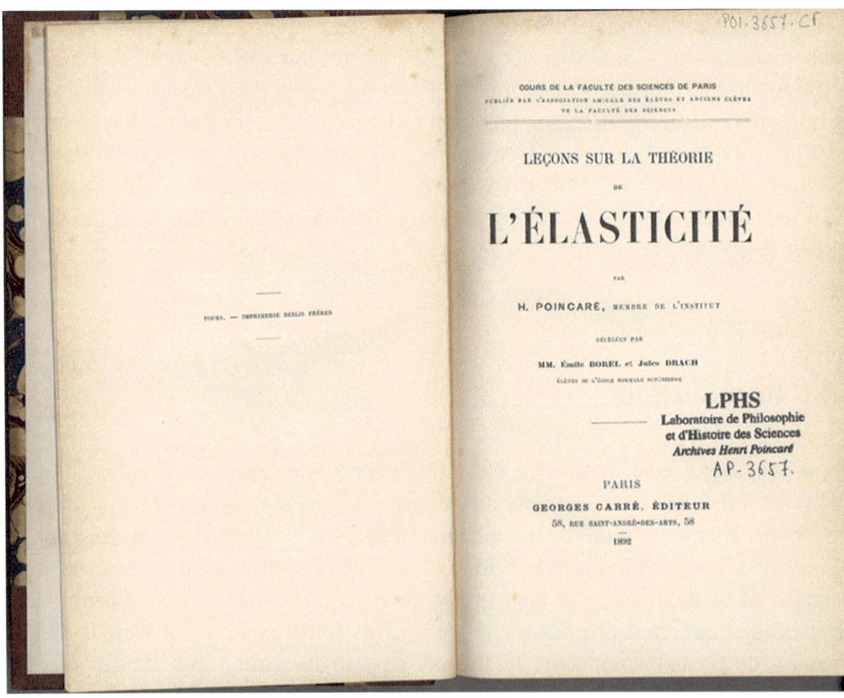

Fig. 6.1 Poincaré's lectures on Elasticity (1892a); from the Archives Henri Poincaré (Laboratoire de Philosophie et d'Histoire des Sciences, Paris)

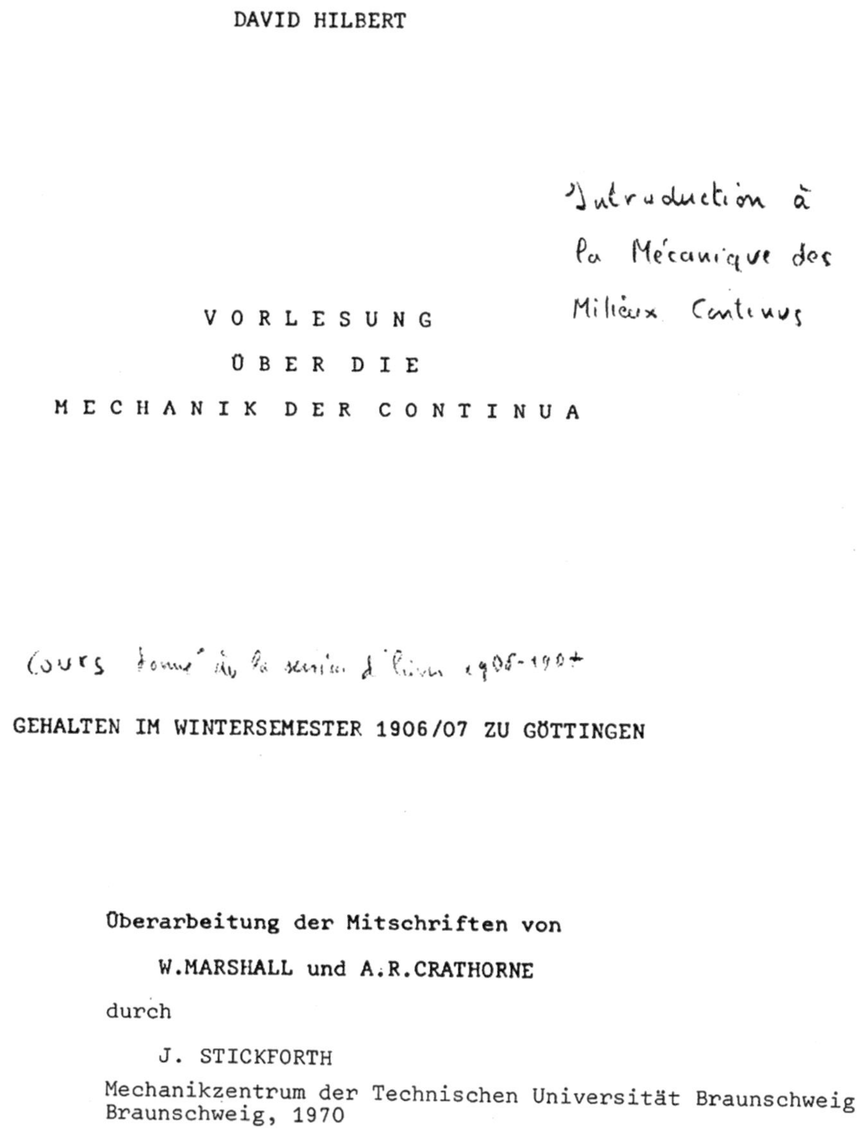

DAVID HILBERT

Introduction à la Mécanique des Milieux Continus

VORLESUNG
ÜBER DIE
MECHANIK DER CONTINUA

GEHALTEN IM WINTERSEMESTER 1906/07 ZU GÖTTINGEN

Überarbeitung der Mitschriften von
W.MARSHALL und A.R.CRATHORNE
durch
J. STICKFORTH
Mechanikzentrum der Technischen Universität Braunschweig
Braunschweig, 1970

Fig. 6.2 Hilbert's lectures on continuum mechanics (1906–1907)—from the Bibliothèque de Mécanique Théorique, IJLRdA- UPMC (Paris 6)

in this field may be due to a teaching obligation (case of Poincaré) or an epistemological interest (case of Hilbert), and above all to the intimate de facto relationship between continuum mechanics and analysis. The notions of "continuum" and "continuity" themselves may have been the main incentives for this inclination, all the more that continuum mechanics at its stage of development in the considered

Fig. 6.3 Henri Poincaré (1854–1912)

Fig. 6.4 David Hilbert (1862–1943)

period presents a favourable framework for pondering the basic principles of mechanics, then the paragon of rational science and phenomenology in some achieved form before the revolution brought in by the "new" physics. So these are favourable objects of thinking for inquisitive minds such a Poincaré and Hilbert after the recent works of the Cosserat brothers and Pierre Duhem in France and Heinrich Hertz and Georg Hamel in Germany and before the expansion and forthcoming differing visions by Henryk Lorentz, Max Planck and Albert Einstein. Note that the axiomatization of mathematical physics—of which continuum mechanics constitutes a most basic building block—was heralded by Hilbert as his problem no.6 among the list he generously offered to the community of mathematicians (see the Appendix below).

6.2 Poincaré on Elasticity

6.2.1 *Preliminary Remark*

Although a superb pure mathematician and a remarkable contributor to the epistemology and philosophy and science, Henri Poincaré also contributed published lectures on field of mechanics and physics for courses (at what we may call undergraduate-graduate level) delivered at the Sorbonne year after year: pure kinematics and fluid mechanics (Poincaré 1886), the mathematical theory of light (Poincaré 1889), electricity and optics (Poincaré 1890, 1901), the theory of vortices (Poincaré 1893) thermodynamics (Poincaré 1892b), electric oscillations (Poincaré 1894), capillarity (Poincaré 1895a), the analytic theory of heat propagation (Poincaré 1895b), the theory of potential (Poincaré 1899a), Maxwell's theory and Hertzian oscillations (Poincaré 1899b), and more to our point, elasticity (Poincaré 1892a, b). These lecture notes were published by the Friendly Association of Students and Alumni from the Faculty of Sciences in Paris Sorbonne. But apart from the first one they were printed by a real publisher, Georges Carré, then transformed into the associated Carré and Naud in Paris. As a whole, they cover a remarkable spectrum embracing all of phenomenological physics as of the end of the nineteenth century. One could think that this conspectus is rather superficial. But none is the case. On the contrary, in each case the author critically reviews what is actually known. He discusses alternate formulations if any, and documents all these with a wealth of material. This bears witness of the tremendous energy of Poincaré—who simultaneously pursued his original research in pure mathematics and celestial mechanics—and his rapid intellectual capture of any scientific domain.

Of course, the final printed lecture notes were edited by a selected group of gifted students, who may not have been numerous to attend the lectures in any case. For the case in point, the lectures on elasticity delivered during a semester of the Academic year 1890–1891 were nicely edited by two gifted students of the celebrated Ecole Normale Supérieure (ENS). One was Emile Borel (1871–1956), a mathematician who became famous for his works on the theory of functions and probabilities, also creator of the Institute of Statistics of the University of Paris and of the centre of mathematics known as Institut Henri Poincaré—and an active politician at the time of the First World War. The other was Jules Drach (1871–1951) with renowned works in the theory of systems of differential equations (in particular, their reducibility) in the line of E. Picard and H. Poincaré. They were classmate at the ENS in Paris and became very close life-long friends. Both were elected to the Paris Academy of Sciences in due time.

6.2.2 The Course on Elasticity

This is not a course on the whole and formal structure of continuum mechanics. On the contrary, the object of study is restricted to linear elasticity and, as we shall uncover, it is only exceptionally intended as an introduction to applications in engineering. Moreover, while the given proofs are fully explicit and sometimes close to a student redaction (this is probably due to the two redactors), still the writing and remarks smell good the spirit of higher geometry as taught in preparatory classes to the competition to admission to Grandes Ecoles (Ecole Polytechnique and ENS), in French: "l'esprit taupin" (the "mole" spirit)—according to the nickname "taupe" (mole) given to these preparatory classes. Finally, the contents of chapters are not formally introduced and there are few comments only.

Chapter 1 is naturally devoted to the kinematics of deformations. This is done at a rather slow path without introducing differential geometry (and the famous ds^2). But orders of terms are well indicated. Displacement components are noted (ξ, η, ς) as in other French works of the period (e.g., Cosserats' works). Emphasis is given to elementary types of deformation (dilatation, shear), the ellipsoid of deformation, principal dilatations (stretches in modern jargon), and the mean rotation (following here Cauchy). The above mentioned higher geometry spirit is illustrated by expressions like (my translation, GAM): "the sphere follows from the ellipsoid by a homographic transformation that conserves the plane at infinity" (p. 12). Rigid-body motion, mean rotation and longitudinal and transverse deformations are introduced. No doubt that this prepares the way for the notion of pure transverse deformation useful in light-wave propagation. What is more surprising is a lengthy discussion about so-called "isotropic polynomials" (rotationally invariant quantities). But this also prepares the way for the introduction of basic second-order scalar invariants in the energy of linear elasticity.

The important Chap. 2 introduces the notion of *elastic forces*. To do so, Poincaré distinguishes between theories based on molecular hypotheses and those based on thermodynamics (in the manner of George Green) where the intimate constitution of matter is disregarded. Nowhere is Cauchy's fundamental notion of stresses (what we call CAUCHY-1 theory, expanded in Cauchy 1828a) mentioned. Poincaré here prefers to start with the *molecular approach* expanded by Poisson (1829, 1831) also advocated by Navier and Cauchy (in the CAUCHY-2 theory, cf. Cauchy 1828b). Here, bodies are made of molecules that are very small compared to their separating distances. They are acted upon by both "internal" and "external" forces. Internal forces are like Newtonian forces between molecules and are ultimately responsible for the macroscopically observed elasticity of the body. They are of the central type, and according to Lamé (1852) vanish at natural equilibrium (in the absence of external forces) with a vanishing attraction. Conservation of energy then implies that internal forces derive from a potential (p. 30), to be called the "force function". External forces may act on all molecules, while at the surfaces they may act on superficial molecules (whatever these be) only. In modern standard continuum mechanics the former are recognized as "body" forces and the latter as applied "tractions". Poincaré cites the example of a gas in a container. Passing to the case of true deformations calls for an exercise in the Taylor expansion of the force functions in terms of the small displacements. The possibility of the existence of mathematical constraints connecting molecules is envisaged and taken care of by the introduction of Lagrange multipliers. Approximation of the expansion of the force function requires the introduction of a small radius of molecular "activity" (interaction), yielding the neglect of far-distance mutual interactions. This is the most disputable part of the procedure, in particular when dealing with molecules placed at the boundary of a body. The case of isotropic bodies indeed involves the previously introduced isotropic polynomials that must be linearly independent. The number of remaining elasticity coefficients in this approach remains a discussed matter as already known to Lamé (1852) and Clebsch (1862).

The equilibrium equations are then introduced in Chap. 3 by means of the principle of virtual velocities. The resulting bulk equations and boundary conditions are finally obtained (pp. 58–59) with identification of what are called "pressures" (in modern jargon, the components of a—Cauchy—*stress tensor*). It is only at this point (p. 62) that Poincaré considers as an alternate approach the energy viewpoint of Green (1839). But he improves on Green's presentation by using a thermodynamic argument that, obviously, Green could not know. Poincaré's argument reminds us of Duhem (without mention of this author) and is worth recalling. With equilibrium already reached the principle of conservation of energy yields the equality between the virtual increase of internal energy and the sum of the virtual work of external forces and the variation of heat gained from the outside. That is, (ignoring the introduction of the mechanical equivalent of heat noted E by Poincaré),

$$\delta U = \delta V + \delta Q. \tag{6.1}$$

With reversible deformation (elasticity limit is not reached), entropy S and absolute temperature T are such that

$$\delta Q = T\delta S. \tag{6.2}$$

Assuming temperature is fixed (to avoid any effect of thermo-mechanical couplings), we have $\delta Q = \delta(TS)$, and thus

$$\delta V + \delta(ST - U) = 0. \tag{6.3}$$

Follwing Helmholtz, the quantity $ST - U$ is called the thermodynamic potential by P. Duhem and available energy by Helmholtz and Gibbs (cf. Maugin 2014, Chap. 7). This will be the potential that govern isothermal elasticity (i.e., minus the Helmholtz free energy). Pressures (stresses) will derive from it through partial derivatives with respect to deformations (strains). Then Poincaré discusses the ellipsoid of pressures as also alternate definitions of pressures from Navier and Lamé, and whether there may exist an initial pressure (as required in the propagation theory of light—theory of double refraction by Fresnel and Cauchy—in the supposedly elastic "ether".[1] It is at this point that we first witness Poincaré's marked interest of elasticity for optical theories. He indeed refers to his own mathematical theory of light (cf. Poincaré 1889, 1890).

Special equilibrium solutions in the absence of external forces are discussed in the short Chap. 4. A particular imposed condition on the two elasticity coefficients of isotropic elasticity, $\lambda + 2\mu = 0$, is paid special attention in relation with Fresnel's theory of light that forbids dilatation, hence longitudinal propagation. This is accentuated in Chap. 5 devoted to small motions of an elastic body. Here dynamic acceleration terms are added by virtue of d'Alembert's principle and allowing for the considerations of periodic oscillations. For a finite body this yields a discrete spectrum of vibrations. This is complemented with a rather thorough study of plane waves in the long Chap. 6 of about fifty pages. There is a deep reason for this emphasized attention. It is that Poincaré, like many, if not all, other physicists of the nineteenth century, remains most interested in the possibility of the propagation of light as a purely transverse elastic wave in an appropriate elastic invisible substratum, the now obsolete "ether" (see Footnote 1). This chapter is very classical in contents introducing first d'Alembert's solution and both longitudinal and transverse wave in the case of full isotropy. Still there are recurring remarks on the case of pure transverse waves. Also, the consideration of reflection (Sect. 59) at a plane with a vacuum on the non material side is certainly inspired by the mathematical theory of light since the vacuum does not allow for the existence of refracted waves. In a more general case, the Snell-Descartes law is obtained (p. 130) when refraction is possible. For a body of finite dimensions, the vibratory state of the body forms a system of stationary waves by constructive interferences. Vibrations of a

[1]Most engineers and theoretical mechanicians ignore that initial developments in the three dimensional theory of elasticity in the early nineteenth were kindled by the need to introduce the notion of an elastic medium that, according to Fresnel's famous discovery, must allow for the propagation of light as purely transverse waves (in contradistinction with longitudinal pressure waves for the propagation of sound in acoustics). Augustin-Louis Cauchy, George Green, and William Thomson (to become Lord Kelvin) are among the proponents of such an approach to light propagation. This interpretation held good practically until the work of Einstein on the electrodynamics of moving bodies and the experimentally proved lack of evidence for the existence of this famous pervasive medium, the "ether". Poincaré and H.A. Lorentz, immense scientific figures, still belong to the previous period in so far as light propagation is concerned.

rectangular parallelepiped, of a rectangle prism (note the vocabulary of optics), and a sphere (a problem partially treated by A. Clebsch) are discussed. A more complicated solution involving both longitudinal and transverse waves is attributed to Marcel Brillouin (p. 152). The chapter is concluded by a discussion of the radiation of elastic living force [i.e., kinetic energy, GAM] in air.

The penultimate Chap. 7 is somewhat surprising as it is devoted to the so-called *Saint-Venant problem* of essential interest in deformable solid mechanics and the strength of materials, subject matters for which Poincaré usually shows not much enthusiasm. We surmise that Poincaré, like many other scientists of the second half of the nineteenth century must have been bedazzled by Saint-Venant ingenuity and dexterity in his solution. In the period 1853–1856 Saint-Venant[2] (see Barré de Saint-Venant 1855) proposed to examine the following problem: consider a long homogeneous isotropic elastic cylindrical body with forces applied only on its bases, the lateral faces being free of load, and determine the resulting field of deformations in the body. This is one of the most famous problems in solid mechanics, what makes Saint-Venant one of the most illustrious contributors to the field in the nineteenth century. Citing our own text (Maugin 2014, p. 82),

> Saint-Venant "approached the problem of combined torsion and flexure of long rods with a very fruitful intuitive idea that we try to describe without the help of formulas: in linear elastostatics it is thought that the difference in effect of two equipollent (i.e., with same resultant) load systems confined to a small part of a boundary becomes negligible away from that boundary. For instance, in the case of the deformation of a prism or a cylinder, we assume that a system of forces is applied only at the two bases. Now, if the transverse dimensions are small compared to the length of the body, the differences in the distribution of forces on the bases produce only local perturbations. As a consequence a practically identical state characterized by longitudinal fibres exerting only tangential actions on their neighbours, or directed along the length, takes place after a small distance from the bases. Exploiting this idea translated into analysis and using the hypothesis of no surface pressure exerted on the lateral faces, Saint-Venant succeeded in justifying from elasticity the accepted laws of flexure and torsion in the strength of materials". This dazzling result met an exceptionally good immediate reception, especially outside France. Thus Clebsch in Germany coined the expression "Saint-Venant problem" for this approach, and later on, with the enthusiasm of a close disciple, Boussinesq called it "*Saint-Venant's principle*". But with the passing of time we think that the use of "principle" was an exaggeration. We must agree that it was more a well thought working hypothesis than anything else, but a very fruitful one indeed. Nowadays, all mathematical considerations of the Saint-Venant's problem are necessarily based on reasoning involving asymptotic analysis and good applied functional analysis that provides a rational measure of the considered approximations".

[2]Adhemar Jean Claude Barré de Saint-Venant (1797- 1886) was an « engineer-scientist » formed as a civil engineer at Ecole Polytechnique and Ecole Nationale des Ponts et Chaussées in Paris. He did fruitful research for almost sixty full years. His last publication in the *Comptes Rendus* of the French Academy of Sciences was dated January 02, 1886—probably presented on December 24, 1885—while he died on January 06, 1886. He had become a member of the Paris Academy of Sciences in 1869, on Poncelet's seat, at a relatively late age of seventy one for a scientist of this magnitude.

But Poincaré's exposition of Saint-Venant's problem is rather original. Let z the axis coordinate of the cylinder. Assume a priori that some stress components vanish inside the cylinder:

$$\sigma_{xx} = \sigma_{yy} = \sigma_{xy} = 0. \tag{6.4}$$

The three components (ξ, η, ς) of the displacement have to be determined from the equilibrium equations on account of the working hypothesis (6.4) and of the side condition of no traction on the lateral face of the cylinder at any point of the cross section at length coordinate z. This leaves only one condition applied to σ_{xz} and σ_{yz}. Poincaré shows that displacements ξ and η satisfy Cauchy's conditions so that $\xi + i\eta$ is a function of the complex variable $x + iy$ and the longitudinal coordinate z:

$$\xi + i\eta = f(u, z). \tag{6.5}$$

Problem: find the form of function *f*, and then determine the third displacement ς. In the end, he will determine the field of forces that must be applied to the bases of the cylinder in order to yield the already obtained elastic solution. We skip the detail of the first part of the argument which boils down to evaluating a function $\Omega(x, y)$ at length coordinate z, which is part of the solution ς, and satisfies the two-dimensional Laplace equation $\Delta\Omega = 0$ in the cross section, with boundary condition of the form

$$\frac{d\Omega}{dn} = V(x, y), \tag{6.6}$$

where the right-hand side is known for a given cross section of the cylinder. It is proved by exploiting Green's theorem that the solution of the problem in all involves twelve constants, but six of them will correspond to a displacement of the cylinder without deformation. The other six corresponds to a set of simple deformations that comprises a traction or compression, one torsion and four kinds of flexure. An argument exploiting a conformal transformation provides a geometric interpretation to the function *f*: the angle made by two curves situated in the cross section is not altered by the deformation. Furthermore, solving for Ω in the case of a circular cross section, the author shows that Saint-Venant's problem can be solved for all sections that can be mapped to a circle by conformal transformation. This is the case of elliptic and rectangular cross sections. Poincaré then establishes the expressions of the six simple solutions of interest. Particularly important solutions are those that correspond to traction, torsion, simple flexure and complex flexure as they yield different deformations of fibres. From these solutions one finds the corresponding forces. This is not the way Saint-Venant conceived "his" problem since he started with the data of forces on the bases of the cylinder. Such external forces are three standard forces along the three axes and three torques of which the axes are parallel to these three forces. What Saint-Venant does is that he replaces in each case the real distribution of applied forces by another much simpler one that yields the same resulting forces and couples. If one has at hand a sufficiently general solution so that forces and couples may take any value, then we can get an idea of the internal deformation of cylinder submitted to any system of forces on the bases. But Poincaré has found the six arbitrary coefficients for such a general solution. He therefore considers that he has solved the practical problem. He deduces from this a classical theorem of the strength of materials, that of the "bending moment" (in French, "moment fléchissant") and also the expression of the so-called "moment of torsion". Saint-Venant's problem solution thus justifies the simplified expressions adopted by engineers in the strength of materials of structures (for instance, when the hypothesis of cross sections remaining plane and perpendicular to the z axis is applied, although Saint-Venant had himself introduced the fruitful idea of warping).

The short last chapter of Poincaré's lectures (Chap. 8) is devoted to what he calls the problem of the "élastique" in which he applies, mutatis mutandis, Saint-Venant's principle for an elastic body of thin lateral dimensions compared to its length. By "élastique", it must be understood the mean fibre (that goes through the centre of gravity) of a long thin structure with diameter much smaller than its length, e.g., a thin rod called "verge" in the French vocabulary of the ending nineteenth century.[3] The reasoning here is purely geometric with some references to previous works by Lagrange, Jacobi and Sophia Kovalevskaya.

6.2.3 *Concluding Comments*

From this rather fast perusal of Poincaré's lectures on elasticity, we take the liberty to draw the following conclusions on both style and contents.

First of all we note the rather dry style of redaction with very few comments, practically no introduction to the different chapters, no expressed motivation, and rather detailed calculations which may help the student but may sometimes be supererogatory. All this may be due to the two redactors—who were still very young at the time of editing the lecture notes—as this is not much in the usual literary style of Poincaré. As already remarked, it is not well understood why Poincaré delivered this semester course as he was not especially concerned with solid mechanics. But his interest in the propagation of light waves in a supposedly elastic medium may be the whole reason supporting this endeavour. It follows from this a clear emphasis on the propagation of harmonic waves and their reflection at a surface, a subject matter very little reported in pedagogical treatises by some of Poincaré's contemporaries, e.g., Appell in his well known treatise on rational mechanics (Appell 1921, First edition 1900; see comments in Maugin 2014,

[3]The expression "verge" is no longer used by French engineers, as it has become customary to designate the male penis; "tige" or "poutre" is preferred to designate the structural element.

pp. 197–198). To appreciate this special inclination of Poincaré we must remember that until the negative experiments of Michelson and Morley (in the period 1881–1887) and subsequent works of Lorentz and Einstein, most physicists still envisaged the possible true existence of this invisible elastic medium, the "ether", incompressible and providing the appropriate substratum for the desired purely transverse waves of which light is made of according to Fresnel. Very soon, many pedagogues and researchers in continuum mechanics (P. Duhem, P. Appell, the Cosserat brothers, E. Jouguet,...) will turn their attention to discontinuity waves (shock waves, acceleration waves), following the pioneering works of G. B. Riemann, P. Hugoniot, and P. Duhem and after the beautiful work of J. Hadamard on the classification of such waves. We surmise that this would have interested Poincaré since these studies involve singular fields and he enjoyed dealing with pathological cases; but Poincaré's lectures were delivered and published some ten years before this sudden passion of the mechanical community.

What is most striking to observers of the author's generation is the absence of mention of Cauchy's beautiful approach to general continuum mechanics (Theory of CAUCHY-1; introduction of Cauchy's lemma, the stress tensor, finite strains). Poincaré prefers to start with the somewhat antiquated molecular approach of Navier and Poisson, although he does mention the energy approach à la Green; but in a general manner, he gives only very few references, which often are not accurate enough. But it is understood that the few students who took his course and got interested could have ready access to most of these previous works by Cauchy, Poisson, Lamé, Clebsch, Saint-Venant et al., on the shelves of the library at the Sorbonne. We finally note that Poincaré, contrary to his contemporaries in France or elsewhere, does not discuss any general principle of continuum mechanics but we would be mistaking by considering this short adventure into elasticity as his only work in continuum mechanics, for he was very much interested in perfect fluids as shown by his first lectures (Poincaré 1886), his many thorough—now considered as classical—studies in equilibrium shapes of fluid masses (cf. Poincaré 1902), and his lecture notes on capillarity (Poincaré 1895a).

6.3 Hilbert on Continuum Mechanics

6.3.1 Preliminary Remark

Like Poincaré, David Hilbert, although basically a pure mathematician, was also interested in mathematical physics, electrodynamics and general relativity in particular. The celebrated course by Courant and Hilbert (Methods of Mathematical

Physics[4]) is very representative of this interest. The school expanded at New York University by Courant, Friedrichs and others who escaped Nazi Germany before World War Two, was instrumental in infusing this spirit of rigorous applied mathematics in the USA, with definite consequences in applied functional analysis, numerical methods, theoretical aerodynamics, shock-wave theory, magneto-hydrodynamics, etc. The series of books in mathematics published by Interscience Publishers in New York was also a spin-off of this immigration.

In the winter semester of the Academic year 1906–1907, Hilbert delivered in Göttingen a series of lectures entitled (in translation) "Lectures on the mechanics of continua" (Hilbert 1906–1907). The corresponding notes were redacted by W. Marshall and A.R. Crathorne (1873–1946).[5] These notes by Marshall and Crathorne were further edited and produced in mimeographed form by J. Stickforth at the Mechanikzentrum der Technischen Universität Braunschweig in Germany in 1970. This is the document on which our perusal is based. Note in passing that Hilbert delivered another session of lectures on the mechanics of systems of points ("discrete masses") during the Winter semester of Academic year 1905–1906 and this kindled an interest in the principles of physics and mechanics according to Otto Blumenthal (1876–1944)[6]—cf. Blumenthal 1922, as cited by Stickforth, p. III.

6.3.2 *Critical Analysis of Hilbert's Lecture Notes*

By way of an introduction to this rare document Stickforth emphasizes the evident strong interest of Hilbert for axiomatics and the fact that physics, in which he includes mechanics, badly requires an axiomatic framework (no.6 in his list of problems proposed in 1900; cf. Hilbert 1901; and the Appendix to the present contribution). This is corroborated by Max Born (1922). But we must also remember that George Hamel (1677–1954)—cf. Hamel (1908)—, about the same time, worked out an axiomatization of mechanics that will have a marked influence, especially in Germany (cf. Hamel 1927, 1949). This followed an earlier attempt by Hertz (1899). The encyclopaedia article of Hellinger (1914) that we analysed in

[4]In the late 1960s, it was customary for Prof. Roman Smoluchowski (educated in Poland and Germany in the 1930s) to give as written exam—for the foreign-language requirement for the Ph.D in exact sciences at Princeton—a translation into English of one paragraph of the introduction of the original German version of this book ("Methoden der mathematischen Physik", Courant and Hilbert 1924). This is what happened to the writer. Note that the book was essentially the work of Courant.

[5]Crathorne obtained his doctoral degree under the supervision of Hilbert in 1907. Then he became assistant professor of mathematics at the University of Illinois. His successful American book "College Algebra" (1919; Henry Holton and Co., NY.), co-authored by H.L. Rietz, went through several revised editions (1924, 1929,...). We do not know what became of W. Marshall, although he may have been affiliated with Purdue University, since his manuscript on Hilbert's course comes from the library of that university.

[6]Blumenthal was Hilbert's first doctoral student and one of his biographers.

some detail in another place (Maugin 2014, Chap. 12) will lean in the same direction. This would later on be taken over by modern authors such as Truesdell and Toupin (1960) and Truesdell and Noll (1965) and other less influential mechanicians.

When we peruse Hilbert's notes we do not expect much deviation from a rather well defined inescapable agenda starting with the definition of a continuum, introducing some required elements of mathematics (here vector analysis), defining the basic kinematics, complemented by the bases of dynamics, and then expanding special illustrating cases (e.g., elasticity, perfect fluids), some problems, and finally exposing some more exotic subjects (here, capillarity, electrodynamics, and thermodynamics). But we also surmise that these will be delivered with a special flavour of a great mathematician, a deep thinker, and an acknowledged master. This is indeed the case. Page numbers refer to the edition by Stickforth in multigraphed form (original chapters are numbered by Roman numerals).

6.3.2.1 On the Introduction (pp. 1–7)

Here Hilbert clearly leans in favour of analytical mechanics, emphasizing the prevailing role of Lagrange and Hamilton's principle for systems of points. This is viewed as the basic principle of the whole of physics with the fundamental underlying idea of minimalization. Of course, this announces the general viewpoint that will be expanded in the forthcoming chapters for the case of deformable media, leaving little room, if any, for dissipative processes.

6.3.2.2 On the Notion of Continuum (Chap. 1, pp. 8–26)

The space-time parametrization of a continuum motion is contrasted between Lagrangian and Eulerian frameworks. It is remarkable that the introduced notation practically is the same as since the late eighteenth century with a typical application

$$(a, b, c) \to (x, y, z), \tag{6.7}$$

between the initial position at time 0 and the actual position at time t. Assuming sufficient regularity, one can also write the inverse transformation

$$(x, y, z) \to (a, b, c), \tag{6.8}$$

still parametrized by time. This notation was practically left unchanged since Lagrange, Piola and others. In modern direct notation these relations read

$$\mathbf{x} = \bar{\mathbf{x}}(\mathbf{X}, t), \quad \mathbf{X} = \bar{\mathbf{X}}(\mathbf{x}, t) = \bar{\mathbf{x}}^{-1}(\mathbf{x}, t), \tag{6.9}$$

from which there follow the definitions of velocity as

$$\mathbf{v} = \left.\frac{\partial \bar{\mathbf{x}}}{\partial t}\right|_X, \quad \mathbf{v} = \bar{\mathbf{v}}(\mathbf{x}, t). \tag{6.10}$$

The print of a mathematician is more cogent in the definitions of a small displacement and its gradient for which the intervening of an infinitesimally small parameter (noted τ by Hilbert) is necessary. This allows for the introduction of the notions of translation, pure deformations and rotations, and what are the related invariants and various quadratic forms and ellipsoids, and what we now call principal stretches ("dilatations" in the vocabulary of Cauchy). Hilbert finally introduces the standard deformation gradient (noted **F** in modern direct notation), i.e.,

$$\mathbf{F} = \left.\frac{\partial \bar{\mathbf{x}}}{\partial \mathbf{X}}\right|_t. \tag{6.11}$$

It is from this quantity and its inverse $\mathbf{F}^{-1}$ that all known finite-strain tensors by Cauchy, Lagrange, Green, Finger and Piola can be defined (cf. Truesdell and Toupin 1960). But Hilbert, contrary to these authors, is not going to exploit them thoroughly in further studies of finite deformations. Evoking the deformation of an ellipsoid, he simply notes the Cayley-Hamilton theorem in a brief discussion of the search for eigenvalues, while introducing a set of standard invariants of a second order tensor (not called this way by Hilbert). No tensor notation is used.

6.3.2.3 Elements of Vector Analysis (Chap. 2, pp. 27–41) Vector Analysis

This is standard in many books of the period as also in further books in the twentieth century. Appell (1921) does the same in Sect. 3.2 of his course on rational mechanics. However, Hilbert is much more "modern" than his competitor in using systematically the notion of vector fields, inner and vector products (with a remark on syzygies in invariant theory, p. 31), an intrinsic notation for the differential operators of gradient (*grad*), divergence (*div*), curl (noted *rot*), and Laplacian (Δ), but the latter is minus our usual Laplacian as a result of the gradient being also minus the usual one (probably inspired by electrostatics). As a result all identities (p. 34) are written in short-hand notation. Also, Hilbert does not make use of he convenient nabla symbol (∇) of Gibbs and Heaviside that emphasizes the vectorial nature of this differential operator. Integral theorems of Gauss and Stokes are recalled in their usual format but passing to the limit of infinitesimally small volume and surface produces definitions of the divergence and curl operator that reminds us of their introduction in electromagnetism (pp. 36–38) [This was also the original definition of the *curl* given by O. Heaviside in his "Electrician" papers]. The chapter ends with an allusion to a classical variational problem of minimization for obtaining the Laplace equation.

6.3.2.4 The Kinematics of Continua (Chap. 3, pp. 41–66)

The material contained in this chapter would be rather classical with its approach to streamlines if it were not for the importance granted to the notion of potential and level surfaces, the transformation of volumes (with the introduction of the Jacobian determinant D of the deformation function—noted $J = \det \mathbf{F}$ in modern notation—, the condition of constant volume (isochoric deformation), the condition of continuity in both Lagrangian and Eulerian forms and, above all, the consideration of possible discontinuity surfaces. This last point borrows from the then recent work of Jacques Hadamard (1865–1963) where the latter introduced the equation satisfied by discontinuities in the field of velocity in the form (in modern notation where square brackets stand for the difference between the limit values on both sides of the surface in approaching uniformly the surface)

$$[\mathbf{v}] \cdot \nabla \phi = 0, \tag{6.12}$$

where $\phi(\mathbf{x}, t) = 0$ is the intrinsic equation of the discontinuity surface. This has led to the notion of "kinematic condition of compatibility" and "Hadamard's lemma". In the Lagrangian description this leads to the conditions (p. 63)

$$\left[\frac{\partial F}{\partial \mathbf{X}}\right] = \lambda \frac{\partial f}{\partial \mathbf{X}}, \quad \left[\frac{\partial F}{\partial t}\right] = -\lambda \frac{\partial f}{\partial t}, \tag{6.13}$$

for a field F continuous across the surface of equation $f(\mathbf{X}, t) = 0$. The attention paid by Hilbert to this very recent work of Hadamard is remarkable,[7] but it is not an exception since a similar attention is paid to this development at the same time by French mathematicians such as Duhem in his research on hydrodynamics (cf. Duhem 1903; Second part, Chap. 2) and also Appell (1909, Chap. 33) in his general course. This will also be considered by Hellinger (1914) and it kindled a feverish activity about shock and acceleration waves in the 1960s–1970s after an initial work by Truesdell (1961) largely influenced by Duhem's works on the subject—see also Coleman et al. (1965). Hilbert concludes this chapter with an evaluation of the ratio of densities between both sides of the discontinuity surface.

6.3.2.5 Bases of the Dynamics of Continua (Chap. 4, pp. 67–121)

In agreement with the above recalled introduction dynamics is constructed by means of Hamilton's principle with the essential introduction of the notions of kinetic energy and Lagrangian function density in analogy with what is already known for systems of points. The included energy will comprise energy of internal

[7]Hadamard's book on waves was published in 1903 although specific proofs were given in previous paper publications.

origin, the *internal* energy, and the potential of externally applied forces. It is the internal energy that characterizes the behaviour of the considered material body with the possible introduction of the deformation gradient and the density, and perhaps (p. 72) the higher-order deformation gradients and the gradient of density (such as in capillarity). Both Lagrangian and Eulerian formulations are given in parallel.

An extended example of behaviour is **elasticity** (Paragraph A, pp. 73–103) where Hilbert first envisages the finite-strain framework. The variation in Hamilton's principle is carefully effected with the introduction of a small parameter (noted σ). In statics, the result of this variation in the Lagrangian framework is what we may now recognize as Piola's equations in the bulk (in modern notation; cf. Hilbert's equations (9) and (10) in p. 78):

$$\frac{\partial}{\partial X_K}\left(\frac{\partial \varepsilon}{\partial F_{iK}}\right)+f_i=0, \tag{6.14}$$

and at the bounding surface of unit normal $\mathbf{N}$ of components N_K,

$$N_K\left(\frac{\partial \varepsilon}{\partial F_{iK}}\right)=\bar{t}_i \tag{6.15}$$

Here

$$T_{Ki}=\frac{\partial \varepsilon}{\partial F_{iK}} \tag{6.16}$$

is none other than the *first Piola-Kirchhoff stress* (in modern jargon) if ε is the internal energy per unit referential volume, and f_i and $\bar{t}_i$ are applied volume and surface forces (parametrized in the Lagrangian system) but still vectors in the actual configuration. Hilbert refers neither to Piola nor to Kirchhoff as if all this was a matter of fact (Caution: his definitions include a minus sign). All equations are given in components (no direct vector or tensor notation). The presentation of elasticity in the Eulerian framework requires some more work. This was already the case in the original works of Piola (1836), what seems to justifies that the Lagrangian viewpoint a priori is the most natural and directly apprehended one (for solids). Of course passing to the Euler-Cauchy equations requires manipulating the Jacobian determinant of the deformation. One will recognize in Eqs. (20) and (21) of Hilbert (p. 89), the Cauchy equations of equilibrium and the associated natural boundary conditions, if one defines the Cauchy stress in terms of the first Piola-Kirchhoff stress by (in modern notation)

$$\underline{\sigma}=J^{-1}\mathbf{FT}\ or\ \sigma_{ij}=J^{-1}F_{iK}T_{Kj}=J^{-1}F_{iK}\frac{\partial \varepsilon}{\partial F_{jK}},\quad J=\det\mathbf{F}. \tag{6.17}$$

The elastic homogeneous *isotropic* body is viewed as one for which the internal energy depends on the strain via its three basic invariants only. Approximations are given by homogeneous quadratic expressions of these invariants producing "normal" forms of the energy density as proposed by Helmholtz (cf. p. 94) with two remaining coefficients of elasticity. Another form is referred to Kirchhoff. The approximation of small strains involving only the displacement gradient is further obtained by a new scaling (p. 96). We are finally left with a Cauchy stress that is linear homogeneous in the displacement-gradient components with two scalar coefficients to be identified with Lamé's coefficients (Eq. (24)). A direct variational deduction of these equations is also evoked as well as limit cases such as incompressibility, and the case proposed much earlier by Lord Kelvin for accounting for the propagation of light as pure transverse waves cases. It is clear that Hilbert is thinking in more mathematical and theoretical physical terms than in engineering ones that he never mentions. His very few citations corroborate this feeling.

The remainder of this chapter (pp. 103–120) is devoted to (perfect) fluids. In a typical mathematician's view, Hilbert defines such fluids as elastic bodies for which the internal energy depends on the strains only through the Jacobian determinant J or equivalently the density ρ in accordance with the continuity equation $\rho_0 = \rho J$ in Lagrange-Piola formulation. As usual for this author this is first done in the Lagrangian format, and thus precisely following Lagrange himself, and then transformation to the Eulerian description is obtained by exploiting Eq. (6.17) above. This obviously yields the notion of pressure p (to be called later on "thermodynamic pressure"). Here, of course, as already noticed by Lagrange himself (cf. Lagrange 1788), Hilbert has to admit that the Eulerian format is more amenable than the Lagrangian one from a solution viewpoint (cf. pp. 105–108). He cannot resist mentioning a variational Hamiltonian formulation. The full Euler equations are then given as Eqs. (31) and (32) in the notes, possibly complemented by an external force derivable from a potential (32)—which combines with the pressure. The associated natural boundary condition involves a purely normal force of pressure in agreement with Euler's original idea. He also deals with the notion of incompressibility by introducing a scalar Lagrange multiplier (to be called later on "mechanical pressure"). Finally, he realizes that the energy function may also depend on an additional parameter noted k, that can parametrize both energy and pressure all the way to the limit condition (pp. 117–119) of incompressibility. He mentions in passing the notion of Legendre transformation of the energy.

6.3.2.6 Special Problems of Hydrodynamics (Chap. 5, pp. 122–171)

This rather long chapter clearly illustrates the accepted saying that the simpler the model of continuum, the more rigorous, complicated and enjoyable are the resulting problems. In the period under scrutiny this is reflected by the intensive use of the technique of complex variables and the solution of problems related to the equilibrium of fluid masses, flows around bodies, and the vorticity field on the path paved by Helmholtz. Hilbert is not the only one to suffer from this temptation as also shown by the lectures of Appell (1921), the works of his disciple Henri Villat (1879–1972) (also in Appell but Edition of 1909), and also works by Poincaré, Riemann, and C.A. Bjerknes and V. Bjerknes (see Maugin 2014, Chap. 11).

As most material given in this chapter has become integrated in usual text-book presentations in the twentieth century, we need not recall the detail of this presentation. Of course, irrotational flows receive the largest share of it, and specialization to two-dimensional flows provides the opportunity to introduce current lines, level curves, and streamlines, the associated representation plane, and the powerful technique of complex variables. Treated exemplary solutions relate to classical flows [flow in a corner, through an orifice (Borda's flow), flow from an outlet (Helmholtz), around a cylinder, or another obstacle (ellipse or perpendicular plate)]. Waves of small amplitude offer an opportunity of some elementary asymptotics.

Vortices also capture Hilbert's attention with due reference to the works of Helmholtz, H. Weber, G. Stokes and a mention of analogy with electric currents and magnetic fields. The latter was in fact duly studied by the Bjerknes', father and son (Norwegian hydrodynamicists and meteorologists), who had interfered with H. Hertz (cf. Maugin 2014, Sect. 11.3): magnetic induction replaces the fluid velocity. On this occasion Hilbert even mentions Kelvin's attempt to represent atoms by vortices (p. 158). Lagrange's theorem for the conservation of circulation is proved. In all, however, most references are to German authors, mostly Helmholtz and Kirchhoff. Nonetheless, in a kind of partial conclusion, Hilbert, albeit quite avaricious of citations in this chapter, alludes to the works of Riemann on partial differential equations in physics (book by Riemann and Weber 1901), the book on waves by Hadamard (1903), the work of Poincaré on equilibrium figures of fluid masses (Poincaré 1902), and works by Gerstner and Rankine on waves, also reported by Kirchhoff (1876).

6.3.2.7 Capillarity (Chap. 6, pp. 172–180)

The original theory of capillarity was developed by Thomas Young (1773–1829) and Pierre Simon de Laplace (1749–1827). It is Laplace who proposed the celebrated equation of capillarity. In both cases this theory was conceived as a Newtonian like theory involving an inverse-square law of interaction between "particles". What Hilbert exposes in a few pages is a continuum theory that relies on the existence of a surface energy. This in fact is not so brutal as the author first considers that the energy across a thin layer of thickness 2d varies like the normal derivative of the density across this layer, i.e., $e = a\partial\rho/\partial n$ where a is a scalar coefficient. By integration this yields

$$E = a \iint_F df \int_{-d}^{+d} \frac{\partial \rho}{\partial n} dl = a \iint_F (\rho_1 - \rho_2) df, \tag{6.18}$$

where F is a surface of element df between the two media in contact. A more mathematical viewpoint consists in parametrizing the surface F in a Gaussian way with squared line element $ds^2 = edu^2 + 2f\,du\,dv + gdv^2$. Then $df = \sqrt{eg - f^2}du\,dv = h\,du\,dv$. Therefore,

$$E = c \iint_{F_O} hdudv, \tag{6.19}$$

assuming that $\rho_1 - \rho_2$ is a constant along the surface F.

The final Laplace equation will be obtained by application of a Hamiltonian principle after appropriate definition of the variations of geometric and energy quantities:

$$p_2 - p_1 = \gamma\left(\frac{1}{R_1} + \frac{1}{R_2}\right), \tag{6.20}$$

where p denotes the pressure, γ is the coefficient of surface tension, and R_1 and R_2 are the principal curvature radii of the interface at the considered point.

To modern eyes this interest for the phenomenon of capillarity within continuum mechanics may seem a bit strange. But it is a subject matter of actuality at the time of Hilbert's lectures as proved by the interest manifested by other great scientists (e.g., Duhem, Poincaré) after a very interesting paper by Korteweg (1901) in the Netherlands. A short time later, the internationally quasi-unknown French scientist Henri Bouasse (1866–1955) will in fact publish a beautiful treatise on capillarity (Bouasse 1924). Studies along the same line as Korteweg and the exposure of Hilbert will know a revival in the 1960s–1980s, especially in France (e.g., Casal 1963) within the framework of what may be called *generalized continuum mechanics*.

6.3.2.8 Electrodynamics (Chap. 7, pp. 181–225)

For a scientist like Hilbert, as also for Duhem and Poincaré, the various branches of phenomenological physics form a whole that needs not necessarily be split in these various branches. That explains while Hilbert devotes a rather long chapter to electrodynamics in a course theoretically concerned with the mechanics of continua, and *not* the physics of continua, unless Hilbert, like Duhem, equals mechanics and physics (but then he should have added a chapter on heat conduction). Nevertheless, it is rather difficult for physicists of the early twentieth century to appreciate at their right value the notes of Hilbert on the subject. We must realize that they are written at a time when Maxwell's theory is rather fresh in Germany in spite of Hertz's work. The first book on the theory for non-specialists was written by the mechanical engineer August Föppl (1854–1924) in 1894, while L. Boltzmann's interesting critical lectures for specialists were printed in 1891. Poincaré's lecture notes and Duhem's survey in France belong in the same period

(Poincaré 1890, 1899a, b, 1902; Duhem 1902). Early in the twentieth century the basic book in Germany is the book by Max Abraham (1875–1922) and Föppl (1904) as a rewritten and much enriched version of the Föppl's (1894) original book. This remained a standard for education in Maxwell's theory of electricity and magnetism during a large part of the twentieth century with revised versions by R. Becker and then F. Sauter. Moreover, we are dealing with a period that is pre-Einstein and pre-Minkoswski works, where it is still thought that light waves (solutions of the dynamical set of Maxwell's equations according to the experimental of Hertz) propagate as pure transverse elastic waves in a hypothetical pervasive elastic incompressible solid, the ill-fated "ether". Hilbert's lecture does not eschew this now obsolete notion, so that in a sense the chapter on electrodynamics here rightfully belongs to the mechanics of continua, albeit of a very special one.

On account of the just made remarks the chapter by Hilbert develops in a very original way. This is our feeling that may not have been shared by contemporaries of Hilbert. Some twelve pages are first devoted to what appears to be a continuous theory of Newtonian gravitation, with an interaction potential between bodies of non-uniform mass densities in the manner of Poisson and Gauss (compare to the classical theory of potential). The corresponding medium deformation will in fact be of the curl type (as developed by other authors such as Kelvin and Larmor in the UK—see Chaps. 5, 9 and 10 in Whittaker 1951) with an elastic energy quadratic in this curl (p. 193). This indeed provides a wave equation of the appropriate form (p. 194) in the absence of dilatation. Then the analogy with the equations deduced from Maxwell's theory is expanded in which wave equations for both electric and magnetic fields are obtained by ad hoc identification (pp. 197–199). Hilbert also refers to the Lorentz force (p. 200) acting on rigid electrons (a totally different theory), to the energy (Poynting) vector $\mathbf{S}$ (p. 202), and he establishes the electromagnetic energy identity (p. 203; in his units but our notations)

$$\frac{\partial U}{\partial t} + div\,\mathbf{S} = -\mathbf{J}.\mathbf{E}, \tag{6.21}$$

where

$$U = \frac{1}{8\pi}\left(\mathbf{E}^2 + \mathbf{H}^2\right), \quad \mathbf{S} = \frac{1}{4\pi}\mathbf{E} \times \mathbf{H}, \tag{6.22}$$

and $\mathbf{J}$ is the current vector. He alludes to the Maxwell stress—or rather a vector associated with it—(p. 206), and even to the global form of Maxwell's equations (in particular, Faraday's and Ampère's laws; p. 212). The final vectorial form of Maxwell's equations (in the absence of displacement current) is given p. 213. Note that Eq. (6.21) above is a pure energy identity deduced from Maxwell's equations alone. It cannot be said to represent a statement of the first law of thermodynamics unless the electromagnetic system is fully isolated from any other physical properties.

It is only at this point that scalar electric potential and vectorial magnetic potentials are introduced (pp. 214–216) while Maxwell originally set forth his system of equations with these—cumbersome—potentials. Even at this later point does Hilbert re-introduce the notions of Eulerian and Lagrangian descriptions, concluding with the retarded potentials introduced by Alfred Liénard (1869–1958) and Emil Wiechert (1861–1928), respectively in 1898 and 1900. Various classes of electromagnetic bodies (dielectrics, conductors, magnets) are mentioned in the last two pages. Abraham (about his electron theory) is the only true contemporary cited in the chapter. We can conclude that Hilbert is more or less aware of the most recent developments in the field at the time of delivery of his lectures.

6.3.2.9 Thermodynamics (Chap. 8, pp. 226–239)

This last chapter of some ten pages is truly written as a mathematician's work. It lacks any physics and any contact with previous thermodynamic theory. It introduces such basic notions as entropy in a purely formal way. As for the rest of the lectures the basic argument is one relevant to a variational Hamiltonian principle with a space parametrization of the Lagrangian type in the form

$$\iiint \varepsilon(\tau,\eta) da\, db\, dc = \min, \tag{6.23}$$

where ε denotes the "internal" energy, τ stands for the specific volume (noted v by Hilbert) and η will be identified as the *entropy* . This applies only in what we now call *thermo-statics,* and also for homogeneous bodies from which there follows under the constraint that total entropy be constant, i.e.,

$$\iiint \eta\, da\, db\, dc = const., \tag{6.24}$$

and the introduction of a scalar Lagrange multiplier λ that

$$\frac{\partial \varepsilon}{\partial \tau} = -p, \quad \frac{\partial \varepsilon}{\partial \eta} = -\lambda = const. = \theta. \tag{6.25}$$

Here p is to be understood as the "thermodynamic" pressure and θ is a thermodynamic parameter called the *temperature*, here a spatially uniform quantity. Hilbert acknowledges the fact that this must be in agreement with Boltzmann's recent kinetic theory. One can define the work (noted A for "Arbeit")

$$A_{12} = \int_{\tau_1}^{\tau_2} \frac{\partial \varepsilon}{\partial \tau}(\tau, \eta(\tau)) d\tau \tag{6.26}$$

and the expanded heat

$$Q_{12} = \int_{\eta_1}^{\eta_2} \frac{\partial \varepsilon}{\partial \eta}(\tau(\eta), \eta) d\eta. \tag{6.27}$$

Then

$$A + Q = 0 \tag{6.28}$$

is none other than an elementary statement of the *first law of thermodynamics* (exchange between work and expanded heat) in thermostatics. Equation (6.27) above allows one to write down the standard (thermostatic) formula

$$d\eta = \frac{dQ}{\theta}. \tag{6.29}$$

This shows that θ will be an *integrating factor* (Caratheodory) if η is to be a true state function.

One can further define the heat capacities at constant volume and pressure by

$$C_V = \left(\frac{dQ}{d\theta}\right)_{V=Const.}, \quad C_p = \left(\frac{dQ}{d\theta}\right)_{p=const.} \tag{6.30}$$

The law of perfect gases and Gay-Lussac law also follow from this analysis by manipulating partial derivatives of functions $\varepsilon, \eta, \theta$. Hilbert concludes by emphasizing the importance of Legendre transformations in considerations of thermodynamics.

All above manipulations are purely formal in a field theory and they lack any real physics. Furthermore, Hilbert, himself a devoted champion of axiomatics, here is much below the remarkable contribution of Constantine Caratheodory (1873–1950) to the axiomatics of thermodynamics (Caratheodory 1909), a study which may have been suggested to him by Max Born and perhaps Hilbert (see also, Maugin 2014, Chap. 9).

6.4 Conclusion

Now we can compare the texts of two great mathematicians, although, in all fairness, the two texts pursue different objectives. On the one hand, Poincaré's text was written some fifteen years before Hilbert's. This time span must be taken into account. It was a condensed form of lectures for students who could have heard the master in person; it was probably intended as a technical support of Poincaré's—now antiquated—theory of the propagation of light. As such it was not very original and did not bear fruits. The strange attention paid to Saint-Venant's problem (or principle) appears more as a tribute to Saint-Venant's ingenuity and right engineering intuition. But overall it still bears the print of a mathematician with spot on geometric remarks and dexterity in analytic manipulations. Hilbert's lectures are more thoughtful and in the German tradition of Helmholtz and Kirchhoff. In that, it prepares the way for the splendid rigorous encyclopaedia article of Ernst Hellinger (1914) that we dissected in a surgical manner in a previous work (Maugin 2014, Chap. 12) and that provided the best and definitive synthesis of the field before the expansion of the field in the second half of the twentieth century. It happens that Hellinger's text, although not translated in any foreign language at the time of its publication (unfortunately during World War One), drew a line to be successfully taken over by other mechanicians (Truesdell et al. in particular).

Appendix

English translation (cf. Bull Amer Math Soc 8/10:437–479, 1902) of the statement of his Problem no.6 by Hilbert as proposed at the Second International Congress of Mathematicians (chaired by H. Poincaré) in Paris, 1900; German original to be found in Hilbert's (Gesammelte Abhandlungen) collected works (available on the website of the University of Göttingen).

Mathematical Treatment of the Axioms of Physics

The investigations on the foundations of geometry suggest the problem: *To treat in the same manner, by means of axioms, those physical sciences in which mathematics plays an important part; in the first rank are the theory of probabilities and mechanics.*

As to the axioms of the theory of probabilities,[8] it seems to me desirable that their logical investigation should be accompanied by a rigorous and satisfactory development of the method of mean values in mathematical physics, and in particular in the kinetic theory of gases.

Important investigations by physicists on the foundations of mechanics are at hand; I refer to the writings of Mach,[9] Hertz,[10] Boltzmann[11] and Volkmann.[12] It is

[8]cf. G. Bohlmann, "Ueber Versicherungsmathematik," from the collection: F. Klein and E. Riecke, *Ueber angewandte Mathematik und Physik,* Teubner, Leipzig, 1900.

[9]E. Mach: *Die Mechanik in ihrer Entwickelnng,* Brockhaus, Leipzig, 4th edition, 1901.

[10]H. Hertz: *Die Prinzipien der Mechanik,* Leipzig, 1894.

[11]L. Boltzmann: *Vorlesungen über die Principe der Mechanik,* Leipzig, 1897.

[12]P. Volkmann: *Einführung in das Studium der theoretischen Physik,* Teubner, Leipzig, 1900.

therefore very desirable that the discussion of the foundations of mechanics be taken up by mathematicians also. Thus Boltzmann's work on the principles of mechanics suggests the problem of developing mathematically the limiting processes, there merely indicated, which lead from the atomistic view to the laws of motion of continua. Conversely one might try to derive the laws of the motion of rigid bodies by a limiting process from a system of axioms depending upon the idea of continuously varying conditions of a material filling all space continuously, these conditions being defined by parameters. For the question as to the equivalence of different systems of axioms is always of great theoretical interest.

If geometry is to serve as a model for the treatment of physical axioms, we shall try first by a small number of axioms to include as large a class as possible of physical phenomena, and then by adjoining new axioms to arrive gradually at the more special theories. At the same time Lie's principle of subdivision can perhaps be derived from profound theory of infinite transformation groups. The mathematician will have also to take account not only of those theories coming near to reality, but also, as in geometry, of all logically possible theories. He must be always alert to obtain a complete survey of all conclusions derivable from the system of axioms assumed.

Further, the mathematician has the duty to test exactly in each instance whether the new axioms are compatible with the previous ones. The physicist, as his theories develop, often finds himself forced by the results of his experiments to make new hypotheses, while he depends, with respect to the compatibility of the new hypotheses with the old axioms, solely upon these experiments or upon a certain physical intuition, a practice which in the rigorously logical building up of a theory is not admissible. The desired proof of the compatibility of all assumptions seems to me also of importance, because the effort to obtain such proof always forces us most effectually to an exact formulation of the axioms.

References

Abraham M, Föppl A (1904) Theorie der Elektrizität: Einfuhrung in die Maxwellsche Theorie der Elektrizität. Teubner, Leipzig [This succeeded Abraham (1894), and was followed in the twentieth century by revised and augmented editions by authors Abraham and R. Becker, and then Becker and F. Sauter]

Appell P (1921) Traité de mécanique rationnelle, Third edition. Gauthier-Villars, Paris [First edition, 1900; facsimilé reprint of the third edition by Gabay, Paris, 1991]

Barré de Saint-Venant A (1855) Mém Acad Sci Savant étrangers XIV:233–560

Blumenthal O (1922) David Hilbert. Die Naturwissenschaften 10:67–72

Boltzmann L (1891, 1893) Vorlesungen über Maxwells Theorie der Elektrizität und des Licht. Two volumes. J.A. Barth, Leipzig

Bouasse H (1924) Capillarité et phénomènes superficiels. Delagrave, Paris

Born M (1922) Hilbert und die Physik. Die Naturwissenschaften 10:88–93

Caratheodory C (1909) Untersuchgen über die Grundlagen der Thermodynamik. Math Ann 67:355–386 [There exists an English translation by J. Kestin in (Kestin, Ed. 1976) p. 206 on]

Casal P (1963) Capillarité interne en mécanique. C R Acad Sci Paris 256:3820–3822

Cauchy AL (1828a) Sur les équations qui expriment l'équilibre ou les lois du mouvement intérieur d'un corps solide élastique ou non élastique. In: Exercices de mathématiques, Vol. 3, pp. 160–187, Sept. 1828 [This presents in print the ideas originally orally presented to the Paris Academy of Sciences on September 30, 1822]

Cauchy AL (1828b) De la pression ou tension dans un système de points matériels. In: Exercices de Mathématiques, vol. 3, Sept. 1828, pp. 188–212, and Oct. 1828, pp. 213–236

Clebsch A (1862) Theorie der Elasticität der festen Körper. Teubner, Leipzig

Coleman BD, Gurtin ME, Herrera I, Truesdell C (1965) Wave propagation in dissipative materials. Springer, New York [collection of reprints of papers]

Courant R; Hilbert D (1924) Methoden der mathematischen Physik, Two volumes (in all, about 1000 pages), First edition. Julius Springer, Berlin [This original German edition is freely accessible on the web at Göttinger Digitalierserungszentrum (GDZ)]; English translation: Methods of Mathematical Physics, vol.1, 1953, vol.2, 1962. Interscience Publishers, New York

Duhem P (1902) Les théories électriques de J. Clerk Maxwell. Essai historique et critique. A. Hermann, Paris

Duhem P (1903) Recherches sur l'hydrodynamique, First series. Gauthier-Villars, Paris [Second series, 1904; reprinted in a single volume with a presentation by J. Kampé de Fériet, Volume hors série of the Publications Scientifiques et Techniques du Ministère de l'Air, SDITA, Paris, 1961]

Föppl A (1894) Einfuhrung in die Maxwellsche Theorie der Elektrizität. Teubner, Leipzig

Gray J (2013) Henri Poincaré: A Scientific Biography. Princeton University Press, Princeton, NJ

Green G (1839) On the laws of reflection and refraction of light at the common surface of two non-crystallized media. Trans Cambridge Phil Soc 7:245–269

Hadamard J (1903) Leçons sur la propagation des ondes et les équations de l'hydrodynamique. Hermann, Paris

Hamel GWK (1908) Über die Grundlagen der Mechanik. Math Annalen 66: 350–397 [This was Hamel's Habilitationsschrift]

Hamel GWK (1927) Die Axiome der Mechanik. In: Handbuch der Physik, Eds. Geiger and Scheel, Bd.5. Springer, Berlin, pp. 1–42

Hamel GWK (1949) Theoretische Mechanik. Springer, Berlin

Hellinger E (1914) Die allgemein Ansätze der Mechanik der Kontinua, In: Enz. Math. Wiss. Eds F. Klein and K. Wagner, vol. 4, Part 4.Springer, Berlin, pp. 602–694

Hertz H (1899) The principles of mechanics (translation from the German edition, Leipzig, 1893 [Reprint, Dover, New York, 1956]

Hilbert D (1901) Archiv der Mathematik und Physik, (3) 1:44–53, 213–237 [Abgedruck in den Nach d K Ges d Wiss zu Göttingen, Math Phys. Kl (1900), H3:253–297]

Hilbert D (1906–1907) Vorlesung über die Mechanik der Kontinua (Lectures in the mechanics of continua). Gehalten in Wintersemester 1906–907, zu Göttingen [Mimeographed, 239 pages; with an introduction (in German) by J. Stickforth, TU Braunschweig, 1970]

Kirchhoff G (1876) Vorlesungen über Mechanik (Lectures on Mechanics). Leipzig

Korteweg DJ (1901) Sur la forme que prennent les équations du mouvement des fluides si l'on tient compte des forces capillaires causées par des variations de densité considérables mais continues et sur la théorie de la capillarité dans l'hypothèse d'une variation de la densité. Arch Néer Sci Exactes et Nat. Sér. II 6:1–24

Lagrange JL (1788) Mécanique analytique, First edition, 1788. Vve Desaint, Paris; Second Edition. Vve Courcier, Paris, 1811 [Facsimile reprint of 4th edition with notes and comments par J. Bertrand and G. Darboux, Gabay, Paris] [English translation of the second edition (1811), Eds. A. Boissonnade and V.N. Vagliente, Springer, Dordrecht, 1997]

Lamé G (1852) Leçons sur la théorie mathématique de l'élasticité des corps solides. Paris, (2nd edition, 1866) [Facsimile reprint, Gabay, Paris, 2006]

Maugin GA (2014) Continuum mechanics through the eighteenth and nineteenth centuries. Springer, Dordrecht

Piola G (1836) Nueva analisi per tutte le questioni della meccanica moleculare. In: Memorie di matematica e di fisica della Società Italiana delle Scienze, Modena, 21:155–321 (received 1835)

Poincaré H (1886) La cinématique pure, mécanismes, potentiel, mécanique des fluides (edited by A. Guillet). Published by the Association Amicale des Elèves de la Faculté des Sciences de Paris

Poincaré H (1889) Leçons sur la théorie mathématique de la lumière (Edited by J. Blondin, M. Lamotte and D. Hurmuzescu). George Carré, Paris

Poincaré H (1890) Electricité et optique (Edited by J. Blondin and B. Bruhnes). Georges Carré, Paris [Translated into German, 1891]

Poincaré H (1892a) Leçons sur la théorie de l'élasticité (edited by E. Borel and J. Drach). Georges Carré, Paris [Facsimile reprint, Gabay, 1990; also University of Michigan, Ann Arbor, 2012; also available as a facsimile reprint by BiblioLife, Charleston, SC, at www.bibliogrande.com]

Poincaré H (1892b) Thermodynamique (Edited by J. Georges Carré, Paris, Blondin)

Poincaré H (1893) Théorie des tourbillons (Edited by M. Lamotte). Georges Carré, Paris [Facsimile reprint, Gabay, Paris, 1990]

Poincaré H (1894) Les oscillations électriques (Edited by C. Georges Carré, Paris, Maurain)

Poincaré H (1895a) Capillarité (Edited by J. Georges Carré, Paris, Blondin)

Poincaré H (1895b) Théorie analytique de la propagation de la chaleur (Edited by L. Rouyer and R. Baire). Carré et Naud, Paris

Poincaré H (1899a) Théorie du potentiel newtonien. Carré et Naud, Paris [Facsimile reprint, Gabay, Paris, 1990]

Poincaré H (1899b) La théorie de Maxwell et les oscillations hertziennes. Carré et Naud, Paris

Poincaré H (1901) Electricité et optique; la lumière et les théories électrodynamiques. Carré et Naud, Paris

Poincaré H (1902) Figures d'équilibre d'une masse fluide. Naud, Paris [Facsimile reprint by Gabay, Paris, 1990]

Poisson SD (1829) Mémoire sur l'équilibre et le mouvement des corps élastiques et des Fluides. Ann Chimie et Physique 42:145–171

Poisson SD (1831) Mémoire sur les équations générales de l'équilibre et du mouvement des corps solides élastiques et des fluides. J. Ecole Polytechnique, Vol. 13. Cahier 20:1–174

Reid C (1996) Hilbert. Springer, New York

Riemann GB, Weber HM (1901). Die partiellen Differentalgleichungen der Physik, Two volumes. Vieweg, Braunschweig [This, in fact, is Riemann's lectures published posthumously by his student Hattendorff, and edited and revised by HM Weber, Professor of Mathematics in Strasbourg; an interesting review is given by AS Ames, Bull Amer Math Soc (1901) 8/2:81–85]

Truesdell CA (1961) General and exact theory of waves in finite elastic strain. Arch Rat Mech Anal 8:263–296

Truesdell CA, Noll W (1965) The Non-linear field theories of Mechanics. In: Flügge S (ed) Handbuch der Physik, Bd.III/3. Springer, Berlin [Referred to for short as NFTM]

Truesdell CA, Toupin RA (1960) The classical field theorie. In: Flügge S (ed) Handbuch der Physik, Bd. III/1. Springer, Berlin [Referred to for short as CFT]

Whittaker E (1951) A History of the Theories of Aether and electricity. Nelson, London [Reprint by Dover, New York, 1989]

Chapter 7
Viscoelasticity of Solids (Old and New)

Abstract Viscoelasticity often is considered less exciting than elasto-plasticity, visco-plasticity and the viscosity and rheology of fluids. Though, it is a branch of continuum mechanics which fostered the expansion of a good general thermo-mechanics of continua (first with Duhem and Eckart) and favoured the creation of "rational continuum mechanics" in the sense of Truesdell, Noll and Coleman. Born with the simple models proposed by Kelvin, Voigt, Maxwell and Boltzmann in the second half of the nineteenth century, it further developed in the hands of Wiechert, Jeffreys (Maxwell's generalized model, standard model, differential model), and then Gross, Biot, Lee, Mandel, Oldroyd, Bland, Ilyushin, Rabotnov, Zener and others, to reach the more abstract view of time functionals of Green and Rivlin, Coleman and Noll, Gurtin and Sternberg, etc., with the application of the notion of fading memory and the introduction of satisfactory functional spaces. Simultaneously, solution techniques such as the application of the Laplace transform, Heaviside operational calculus and the so-called correspondence principle were implemented while viscoelasticity yielded the introduction of a fruitful thermodynamic theory of internal variables of state. It finally provided a nice application frame for the notion of fractional calculus. The approach adopted here is essentially historical, with technicalities kept to a minimum.

7.1 Introduction

As its combined name indicates, *visco-elasticity* is a mechanical behaviour that combines elasticity and viscosity, most often in a linear manner. The following observable facts characterize this behaviour. Two typical related phenomena are *creep* and *relaxation*. The first of these corresponds to an increase of strain in time when stress is held constant, while relaxation is evidenced by a decrease of stress with time when the strain is held constant. Various models render these phenomena more or less correctly (cf. Finley et al. 1976). Moreover, when a cyclic loading is applied a hysteresis manifested by a phase lag is produced. This is related to a

G.A. Maugin, *Continuum Mechanics through the Ages - From the Renaissance to the Twentieth Century*, Solid Mechanics and Its Applications 223,
DOI 10.1007/978-3-319-26593-3_7

dissipation of mechanical energy. A direct consequence of the presence of viscosity, although often small in solids, is the *attenuation* of acoustic waves that propagate through a visco-elastic material. Visco-elastic materials abound and, in principle, all elastic materials can be made to exhibit the effects of a viscosity in particular if we wait long enough, centuries or millenaries—for glassware produced in antique times—so that Heraclitus' old but spot on motto "ta panta rhei" (translated as "everything flows") holds true. But actual examples of visco-elastic materials include some polymers, metals at high temperature, natural bio-materials (soft tissues), and bitumen materials. The range of strains is not necessarily limited to infinitesimal ones, as can be imagined with polymers and bio-materials. The scientific and technical interest in these materials has evolved in time.

Of course a mathematical description of visco-elasticity could not be envisaged before a sufficiency clear understanding of both elasticity and viscosity, hence before the middle of the nineteenth century. A development period in the corresponding modelling extends from the mid nineteenth century to the Second World War (see Sect. 7.2). It needed a firm connection with thermodynamics, but this had to await an accepted thermo-mechanical format (see Sect. 7.3). With the tremendous advances in theoretical rheology and rational continuum mechanics by scientists such as Oldroyd, Green and Rivlin, Truesdell, Coleman and Noll, Gurtin, Sternberg and others, visco-elasticity took a more strictly mathematical form—but substantiated by corresponding experiments in rheology —along with enlightening mathematical results (Sect. 7.5) and the introduction of an efficient description by means of internal variables of state (that typically characterize dissipative effects). This largely improved on old "hereditary" formulations (Boltzmann, Volterra, etc.) after a period of interest in establishing fundamental properties and problem solving by pioneers such as Lee, Mandel, Bland, Ilyushin, etc. (Sect. 7.4). The present contribution is not an abridged course on visco-elasticity (for which there exist many books, while regular graduate courses are delivered in so many places over the World). It rather purports at exposing the way various theories evolved in the self-evolving general background of continuum mechanics over a period of some hundred and fifty years. The approach is historical, with technicalities kept to a minimum.

7.2 Early Developments (1860–1950)

The first simple model of viscoelasticity by William Thomson (Kelvin 1875, but probably devised before) was a direct linear combination of a Hookean elastic behaviour and a Newtonian viscous behaviour, so that in 1D, one can write

$$\sigma = \sigma_e + \sigma_v = Ee + \eta \dot{e} = E\left(1 + \tau \frac{\partial}{\partial t}\right)e, \tag{7.1}$$

where E is Young's modulus, η is a viscosity, and $\tau = \eta/E$ is a characteristic time. Following Zener's (1948) exploitation of illustrative rheological models, behaviour (7.1) is reproduced by an elastic Hookean element (spring) and a viscous element (so-called dashpot or damper) placed in parallel (the stresses in each of these add up while they suffer the same deformation *e*). This model was also considered by W. Voigt in his crystal studies (Voigt 1889, 1895, 1898, 1910) and is therefore referred to as the *Kelvin-Voigt model* of viscoelasticity. It is agreed upon that this model, simple as it is, does not predict relaxation accurately, but it is easily integrated in a thermodynamic (irreversible) format. In contradistinction, the model proposed by Maxwell (1866, 1868) would correspond to an elastic element and a viscous element placed in series (the same stress is transmitted in the two elements while deformations of the two elements add up). Thus one can write

$$e = e_e + e_v, \quad e_e = E^{-1}\sigma, \quad \dot{e}_v = \eta^{-1}\sigma, \tag{7.2}$$

whence the rheological relation

$$\dot{e} = \frac{\sigma}{\eta} + \frac{\dot{\sigma}}{E}. \tag{7.3}$$

This *Maxwell model* is reputed not to predict creep accurately but it applies to soft solids including thermoplastic polymers and numerous metals at a temperature not too close to their melting point. For a long time it was thought that this model could not be integrated in sound irreversible thermodynamics (but see Sect. 7.5 below).

Because of limitations afflicting each of the above described models, a more general model combining both was proposed (cf. Oldroyd 1950) and is usually referred to as the *standard linear model*:

$$\sigma + \tau_e \dot{\sigma} = E_r e + \tau_\sigma E_Y \dot{e}, \tag{7.4}$$

where τ_e and τ_σ are two characteristic times, with $\tau_\sigma > \tau_E$. This corresponds to a Maxwell element in parallel with a Hookean element in the pictorial view of rheological elements. This model, ad hoc as it appears, proves to be accurate in predicting both creep and relaxation responses for many materials.

In his doctoral thesis (1889) at Königsberg Emil Wiechert[1] proposed an astute generalization of Maxwell's model that proved efficient in engineering applications.

[1]Emil Johan Wiechert (1861–1928) is much less known among mechanical circles than Kelvin, Voigt and Maxwell in so far as visco-elasticity is concerned. Wiechert was first a student of W. Voigt (1850–1919) and studied first in Königsberg. He was a disciple of Franz E. Neumann (1798–1895). In 1889 he defended a doctoral thesis "On the elastic after-effect", which was his lasting contribution to viscoelasticity. He turned to geophysics, creating the first Institute of geophysics at Göttingen, where he became one of the foremost geophysicists of his time, in particular with the design of an efficient seismometer in 1903 [Theorie der automatischen Seismographen, Weidmannsche Buchhandlung, Berlin, 1903]. However, he is mostly well known

He proposed to put several Maxwell elements in parallel with a Hookean one in order to incorporate a satisfactory distribution of multiple relaxation times (Wiechert 1889, 1893). Indeed, a real polymer does not relax with a single relaxation time as molecular segments of varying length contribute to the relaxation, shorter segments relaxing faster than long ones. Wiechert's modelling affords an easy representation of the response of *any* linear viscoelastic body. Both standard model (7.4) and Wiechert's work hint at a formal generalization in the form of a *differential equation* such as

$$A\sigma = Be \tag{7.5}$$

with differential (time) operators—orders j and i may be different—defined by

$$A = \sum_i a_i \frac{d^i}{dt^i}, \quad B = \sum_j b_j \frac{d^j}{dt^j}. \tag{7.6}$$

We note the obvious difficulty posed by the question of initial conditions in dynamics. But applying formally Heaviside operational calculus (i.e., brutally replacing d/dt by the scalar p) will conveniently bring the viscoelastic problem to an elastic one (cf. Alfrey 1945, 1948). The general form (7.5) and its various special cases were particularly appreciated by geophysicists, among them, not the least, Harold Jeffreys (1891–1989) in Cambridge, UK.

With the proposal put forward by Boltzmann (1874) we apparently have a totally different view which can be synthesized as follows. It makes physical sense to envisage that the actual value at time t of strain is the result of the whole past history of the load in force (here stress). Taking an initial time very remote in the past, we can then consider that variables σ and e are functionally related in the following symbolic manner:

$$\sigma(t) = F[e(s)]_{s=-\infty}^{s=t}, \quad e(t) = F^{-1}[\sigma(s)]_{s=-\infty}^{s=t}. \tag{7.7}$$

Such a general form was proposed by Green and Rivlin (1957) in more recent times.

Viscoelasticity will be said to be *linear* if the functional F is *linear*. This is the essence in abstract form of the *superposition principle* of Boltzmann (1874)—cf.

(Footnote 1 continued)

in physics for his creative contributions in Maxwell's electrodynamics with the simultaneous introduction (with Liénard from France) of the *retarded Liénard-Wiechert potential*. He had already published a book on electrodynamics [Grundlagen der Electrodynamik, Göttingen, 1899]. His interest in mathematical physics was kept constant during his whole life, and he cultivated a fruitful friendship with many of the German high-class physicists in Berlin and Göttingen. A short time before his death, he published his general views on mechanics and physics [Die Mechanik im Rahmen der allgemeinen Physik, Leipzig, 1925]. Clearly, with his multiple interests and successes, Wiechert belongs in a class of bygone scientists.

Mandel (1958). In less pedantic terms, we can express this principle as follows (cf. Roylance 2001). Let $\sigma_1(t)$ the stress resulting at time t from the application of a small strain increment Δe_1 applied at previous time s_1. Thus one can write

$$\sigma_1(t) = E_{rel}(t - s_1)\Delta e_1, \tag{7.8}$$

where E_{rel} is called a *relaxation modulus*. A similar relation can be written for a stress $\sigma_2(t)$ due to a strain increment Δe_2 applied at previous time s_2, and so on. If the material is linear, then the resulting stresses at time t add up, and as the number of applied strain increments increases so as to approach a continuous distribution, we will have in the limit

$$\sigma(t) = \sum_j E_{rel}(t - s_j)\Delta e_j \Rightarrow \sigma(t) = \int_{-\infty}^{t} E_{rel}(t - s)\frac{de(s)}{ds}ds, \tag{7.9}$$

so that we have a materialization of the abstract form given by the first of (7.7).

If stress rather than strain is the input quantity, then the limit form (7.9) is replaced by a response in strain such as

$$e(t) = \int_{-\infty}^{t} C_{creep}(t - s)\dot{\sigma}(s)ds, \tag{7.10}$$

where C_{creep} may be called the *creep compliance*. Equations (7.9) and (7.10) stand for Boltzmann's proposal as a so-called *hereditary* formulation of viscoelasticity. In practice the coefficients of relaxation and creep do not attribute the same weight to all instants of past time, very past instants influencing much less than recent ones. This is vividly illustrated by the expression of *fading memory* then anthropomorphically attributed to the considered materials. In expressions (7.9) and (7.10) it may be convenient to separate the instantaneous (time-independent) response from the integral (truly time-dependent) function. As a simple example, we note that the Kelvin-Voigt model (7.1) yields by integration a particular expression of (7.10) as

$$e(t) = \eta^{-1} \int_{-\infty}^{t} \exp[-(E/\eta)(t - s)]\sigma(s)ds. \tag{7.11}$$

While Boltzmann's model yields integro-differential equations when carried in the balance of momentum—and then obviously is computationally challenging both in simulation and control design—it practically immediately attracted attention at the end of the nineteenth century and the dawn of the twentieth century because, on the one hand, "it made physical sense", but on the other hand it also introduced a new class of sophisticated mathematical objects, functionals, in their best application setting. Ludwig Boltzmann (1844–1906) was a most admired powerful physicist and his proposal could not be overlooked in spite of the accompanying mathematical difficulties. The Italian mathematician Vito Volterra (1860–1940) was the mathematician most deeply involved in the theory of such functionals with a marked interest in all phenomena exhibiting some hereditary effect, including in a kind of innovative population dynamics—cf. Volterra (1909), and also the mathematical developments together with Joseph Pérès, the godfather of French continuum mechanics in the 1930–1960 period (cf. Volterra and Pérès 1936). Other mechanicians also were seduced by Volterra's approach, including Natanson (1901–1903) and Zaramba (1903) in Poland. Ernst Hellinger (1883–1950), in his brilliant Encyclopaedia article (Hellinger 1914), captured at once the interest for Boltzmann's modelling. Such modelling (cf. Graffi 1928, 1977) was taken over by Dario Graffi (1905–1990)[2] in Bologna where his influence lasted with more works by Mauro Fabrizio—sometimes in collaboration with Angelo Morro from Genova (cf. Fabrizio and Morro 1992). A strong revival of mathematical studies along Boltzmann's line will take place in the 1960s under the influence of Truesdell (see Sect. 7.5 below). To go further than the original developments one has to establish some connection with dissipation and thermodynamics (Sect. 7.3), and to expand appropriate solution techniques (Sect. 7.4).

7.3 Early Thermodynamics (1940–1965)

Although elements of a true thermo-mechanics of continua may be found in various writings of Pierre Duhem in the period 1895–1916, a real apprehending of the matter was reached only with a rather remarkable—but seldom cited—work by Eckart (1948). This American physicist (1902–1973) was originally an acknowledged contributor to quantum mechanics, and later became an as well successful contributor to oceanography and underwater acoustics. In the mean time (1939–1948), he contributed to a then modern approach to continuum mechanics in a series of fours papers with one of the first clear statement of the second law of thermodynamics for anelastic solids (Eckart 1948). In particular, Eckart provided the expression of the elementary (per unit volume) dissipation associated with an anelastic deformation in a solid as

$$\Phi^a = \underline{\sigma} : \dot{\mathbf{e}}^a, \tag{7.12}$$

where $\mathbf{e}^a$ is the anelastic strain, and (7.12) is one contribution part to the more general *Clausius-Duhem inequality* (here written per unit volume in small strains, but correctly stated by Eckart)

[2]The author benefited from a series of invited lectures on hereditary visco-elasticity delivered by D. Graffi in Paris in 1967.

$$-\left(\dot{W}+S\dot{\theta}\right)+\underline{\sigma}:\dot{\mathbf{e}}-(\mathbf{q}/\theta)\nabla\theta\geq 0, \tag{7.13}$$

where W, S, θ and $\mathbf{q}$ denote the free energy per unit volume, the entropy per unit volume, the thermodynamic temperature ($\theta > 0$, $\inf\ \theta = 0$), and the heat flux vector, respectively. Here $\mathbf{e}$ is the total observed strain. Equation (7.13) implies that entropy flux is none other than the ratio of heat flux to temperature. Exploitation of the inequality (7.13) requires the proposal of an expression for the dependence of function W. This is a crucial step in this exploitation, drastically depending on the working hypotheses selected in the considered thermodynamic theory. On accepting the hypotheses of the theory that might include the presence of non-observable and non-controllable *internal variables of state*—that is, following pioneering ideas of Duhem (1911) and Bridgman (1943)—, one can contemplate an a priori functional dependence of the type:

$$W=\overline{W}(\mathbf{e}^e,\alpha,\theta),\quad \mathbf{e}=\mathbf{e}^e+\mathbf{e}^p, \tag{7.14}$$

where $\mathbf{e}^e$ and $\mathbf{e}^p$ are elastic and "plastic" parts of the strain, and α stands for the ordered set of internal variables of state (or hidden parameters according to the coinage of L. Brun and J. Mandel). The laws of state (Gibbs) are given by

$$S=-\frac{\partial\overline{W}}{\partial\theta},\quad \underline{\sigma_e}=\frac{\partial\overline{W}}{\partial\mathbf{e}^e},\quad A=-\frac{\partial\overline{W}}{\partial\alpha}, \tag{7.15}$$

so that (7.13) reduces to the following residual *dissipation inequality*:

$$\Phi=\Phi_{intr}+\Phi_{th}\geq 0, \tag{7.16}$$

where we have set

$$\Phi_{intr}=\underline{\sigma_v}:\dot{\mathbf{e}}^e+\underline{\sigma}:\dot{\mathbf{e}}^p+A\dot{\alpha},\quad \Phi_{th}=-(\mathbf{q}/\theta)\nabla\theta,\quad \underline{\sigma_v}:=\underline{\sigma}-\underline{\sigma_e}, \tag{7.17}$$

where we agree to call *elastic stress* and *visco-elastic stress* the fields $\underline{\sigma_e}$ and $\underline{\sigma_v}$. Note that the latter is in thermodynamic duality with the elastic strain, while the plastic strain is in duality with the total stress.

The simple—but general enough—scheme just exposed concurs with ideas expressed by Meixner (1953), Biot (1954, 1956, 1958),[3] Eringen (1960); in the

[3]The Symposium contribution of Biot (1958)—originally delivered in 1955 as a series of lectures at a Laboratory of Shell Development Co. in Emeryville, USA—deserves a particular attention. **Maurice A. Biot** (1905–1985) was a Belgian-American physicist-geophysicist with doctoral degrees from Louvain and Caltech. Although not exactly a maverick, he was for the least a very original powerful thinker who did not follow a traditional academic career. He worked sometimes at various American universities and for a number of agencies and companies. He is famous for his theory of poro-elasticity (now called Biot's theory), but also on variational principles, the incremental theory of deformable solids, and, not the least, irreversible thermodynamics of continua, the

absence of internal variables, Brun (1967a, b; 1969), Mandel (1966, 1967) and Kluitenberg (1962, 1963). All these authors were influenced by the original work of Staverman and Schwarzl (1952) that considered Onsager's symmetry relations (published in 1931). In particular, the *Kelvin-Voigt model* clearly obeys this scheme. If suffices to put $\mathbf{e}^p = 0$ *and* $\alpha = 0$, and consider an energy W quadratic in $\mathbf{e}^e$ and then $\underline{\sigma_v}$ linear in $\dot{\mathbf{e}}^e = \dot{\mathbf{e}}$ to arrive at a linear combination such as in Eq. (7.1). The representation of the *Maxwell model* is more subtle and requires the consideration of an internal variable of state. As a matter of fact, we can take $\mathbf{e}^p = 0$ *and* $\underline{\sigma_v} = 0$, and select α as a second order symmetric tensor $\mathbf{e}^v$ that we call viscoelastic strain, and then assume that W depends upon $\mathbf{e}$ only through the difference strain $\hat{\mathbf{e}} = \mathbf{e} - \mathbf{e}^v$. With a quadratic expression of W in terms of this $\hat{\mathbf{e}}$ and the fact that we can consider $\dot{\mathbf{e}}^v$ linear in the stress to satisfy the residual dissipation inequality, this yields Maxwell's model indeed. The selection of internal variable of state here is based on the convincing argument proposed by Mandel (1973), an author who shows that the internal variable should be the deformation in the dash-pot or damper in 1D—as illustrated in Maugin (1999), pp. 85–86; Fig. 4.1—, hence the present choice in 3D.

(Footnote 3 continued)

subject matter of Biot (1958) with an emphasis on the modelling of thermoelasticity and viscoelasticity. With this work we are offered a thoroughly documented and rather complete overall conspectus of the field as of the middle 1950s, but with a flavour of things to come in the next thirty years. Biot pays special attention to the initial well-founded definitions of thermostatics generalized to linear irreversible processes in the spirit of Onsager and the Dutch-Belgian school around Prigogine and de Groot. Particularly salient is the introduction of "hidden coordinates" (what we called "internal variables of thermodynamic state) and a dissipation function D, and a systematic exploitation of Heaviside operational calculus. The two basis equations (deduced from a variational formulation improved in the manner of Rayleigh) are (in our notation in Maugin 1999, Chap. 4)

$$\frac{\delta L}{\delta \mathbf{u}} = \mathbf{f}^{\mathbf{d}}, \quad L = T - W$$

and

$$\frac{\delta W}{\delta \alpha} + \frac{\partial D}{\partial \dot{\alpha}} = \mathbf{0}$$

where we introduced Euler-Lagrange functional derivatives, T is the kinetic energy, and W is the free energy, while $\mathbf{f}^{\mathbf{d}}$ is a prescribed bulk force. Note that neither kinetic energy nor prescribed force appear in the equation that govern the internal variable α, in full agreement with the definition of such "internal" variables. Then, by combining the above two equations, the theorem of energy reads:

$$\tfrac{d}{dt}(T + W) = \mathbf{f}^{\mathbf{d}} \cdot \dot{\mathbf{u}} - 2D$$

where D is non-negative, and indeed represents the irreversible conversion of mechanical power into heat.

Appropriate choice of observable and internal variables of state would allow the thermodynamic formulation—and satisfaction of the second law—for a large variety of linear viscoelastic models.

The thermodynamics of the Boltzmann type of viscoelasticity will be more closely examined in Sect. 7.5.

7.4 Engineering Viscoelasticity (1940–1960), Dynamic Studies

Viscoelasticity necessarily involves the time variable. This dictates a special attention to *dynamical* problems and to experiments that favour the evidence of this variable. This in turn explains the points of special focus that attracted the attention of both mathematicians and mechanical engineers in the rough period 1940–1960, once a certain agreement was reached on the basic formulation of the governing equations. Here we must emphasize the contributions of Oldroyd in the UK, Lee in the USA and Mandel in Paris, while other lines were expanded in the former Soviet Union (e.g., by Ilyushin, Rabotnov and others). The synthesizing works of Gross (1953), Bland (1960), Flügge (1967), Mandel (1966, Appendix XXI to Volume II), and Freudenthal and Geiringer (1958) provide a good overview of the results reached in this period for analytical solutions. As to experiments, elementary techniques of vibrations (e.g., applying a small oscillatory stress and recording the resulting strain) allow one to measure the storage modulus, the loss modulus and the phase lag. But more precise techniques such as broadband viscoelastic spectroscopy and resonant ultrasound spectroscopy will be exploited in a fruitful manner later on only (cf. Lakes 1998).

Concerning theoretical dynamic problems, one must acknowledge the importance of initial data, often in the form of step (Heaviside) functions. No wonder, therefore, that given the expressions of accepted linear viscoelastic constitutive equations, the *Laplace transform* was to play a considerable role. Erastus H. Lee (1916–2006) and Jean Mandel (1907–1992) here are fundamental contributors.[4] Recall that the Laplace transform $\hat{f}$ of a function f is defined by

[4]It is interesting to compare the scientific life and achievements of Lee and Mandel. **Erastus H. LEE** (1916–2006) was an English-American mechanical engineer, educated at Cambridge University, UK, and with a Ph.D. obtained at Stanford (1940) with S. Timoshenko. He spent WWII in the UK but moved definitively to the USA in 1948. He taught at Brown University (1948–1962) and then Stanford (1962–1982), and finally moved to Rensselaer Polytechnic for ten years. He has contributed with success to the plasticity of metals, viscoelasticity, and plastic wave propagation. He is often attributed the multiplicative decomposition of the finite total deformation gradient in elasto-plasticity. As to **Jean MANDEL** (1907–1982), he perfectly illustrates the French elite system of engineering education. He belonged to the "Corps des Mines" having studied at the *Ecole Polytechnique* and the *Ecole Nationale Supérieure des mines de Paris.* He became a long-time professor of mechanics in both of these institutions He is the founder of a true school of research in solid and soil mechanics and was most influential in introducing a revived

$$\hat{f}(p) = Lf = \int_0^\infty \exp(-pt)f(t)dt. \tag{7.18}$$

Some French authors (Mandel, Germain) preferred to consider the *Carson transform*, that is none other than the Laplace transform multiplied by p, i.e.,

$$\tilde{f}(p) = p\hat{f}(p), \tag{7.19}$$

one advantage being that the Carson transform of a constant still is the same constant. Laplace and Carson transforms replace operations of time derivative and integration by algebraic operations, hence the convenience of the approach. In particular, one notes that

$$L\frac{df}{dt} = p\left[\hat{f}(p) - f(0_+)\right], \tag{7.20}$$

and

$$L(f \otimes g) = \hat{f}\hat{g}, \tag{7.21}$$

where the symbol ⊗ indicates the convolution product such as in Eqs. (7.9), (7.10) and (7.11). Equations (7.20) and (7.21) contain the strong potential of application to all types of linear viscoelastic constitutive equations. This particularly applies to a differential model such as (7.5 and 7.6) for which one obtains

$$\hat{\sigma}(p) = \hat{r}(p)\hat{e}(p), \tag{7.22}$$

(Footnote 4 continued)

view of thermomechanics in France in the 1960s–1970s. His creative works deal with soil mechanics, solid mechanics (plasticity, viscoelasticity), nonlinear wave propagation, and thermodynamics. He had interests in both theoretical and experimental approaches. He founded the *Laboratoire de Mécanique des Solides* (LMS) at Ecole Polytechnique in 1961 and the French Group of Rheology in 1964. He had no connection with the French University system which he often considered with some scorn.

I personally think that Mandel's book of 1966 that more or less expands his course at Ecole Polytechnique, was the best book on continuum mechanics published in France after WWII in the second half of the twentieth century. It is more adapted to graduate studies than to undergraduate ones. It is rather complete, especially with detailed appendices of which those on plasticity and viscoelasticity are remarkably informative. This book is marked by the real involvement of Mandel in research in soil and solid mechanics, and is full of deep thoughts on the matter. It is deeper than Germain's book (at undergraduate level) published in 1962—a time at which Germain had not yet contributed to thermomechanics and solid mechanics although he had started to give related specialized courses in Paris in 1963—but the latter was much better publicized due to Germain's definite influence among French universities.

with

$$\hat{r}(p) := \frac{\sum_j b_j p^j}{\sum_i a_i p^i}. \tag{7.23}$$

The mathematical difficulty may be in the return from a given $\hat{f}$ or $\tilde{f}$ to a function of real time; in which case one may have recourse to the famous Mellin-Bromwich formula that involves an integration in the complex plane for the complex variable p. In recent times this is circumvented by the exploitation of a numerical package such as Maple™. In the case of (7.22)–(7.23) the real physical response is obtained by first decomposing the rational fraction (7.23) into simple elements. It did not escape some observers (e.g., Alfrey 1948), that a direct application of *Heaviside operational calculus* (replacing d/dt by p) without alluding to the Laplace transform was operationally equivalent although more shaky from a mathematical viewpoint.

The problem is much more complicated when one has to solve a real boundary-value problem, but it may be possible to reduce the viscoelastic problem to an associated elastic one by taking transforms. One then looks for the solution of this associated elastic problem, and next performs a Laplace inversion to return to the time plane. This technique of viscoelastic stress analysis is called the *correspondence principle*, and was particularly developed by Lee (1955, 1962) and Mandel (1955, 1966, Vol. II, Appendix XXI). Uniqueness of the viscoelastic solution follows from that of the associated elastic one.

Corresponding progress was obtained in the former Soviet Union by mechanicians such as Rabotnov (1980) and Ilyushin (1971; Collected works: 2003–2009, Vol. 3). A brief modern review is given in Banks et al. (2011).

7.5 Mathematical Visco-Elasticity (1960–1975)

True mathematical considerations go back to the small opus of Gross (1947, 1953). But much more was to come with the rebirth of continuum mechanics in a stricter mathematical format kindled by Clifford A. Truesdell's (1915–2000) untiring efforts to frame continuum mechanics in a so-called *rational* framework. After the remarkable paper of Noll (1955) that described a kind of continuous spectrum of material behaviours from pure fluidity (Euler) to pure elasticity (Cauchy, Green, etc.), this was devised by applying a rather strict credo in the formulation of sufficiently general constitutive equations, capturing from the start the spirit of Boltzmann's hereditary functional approach abstracted in Eq. (7.7) and enforcing the thermodynamic constraint imposed by the Clausius-Duhem inequality (7.13) while following a definite roster of recommendations. This is sometimes referred to as the Truesdell-Coleman-Noll approach to the thermo-mechanics of continua (exemplified by Truesdell and Noll 1965). Among the "principles" applied in this credo an important one is that of *fading memory*: states of the body at a distant past influence much less the actual state than recent ones do. This is mathematically accounted for by introducing an influence or alleviator function and appropriate functional spaces. This was exposed with talent by Coleman and Noll (1961) in the framework of *linear viscoelasticity*, but published in a rather odd place for a paper on continuum mechanics.[5] What these authors specified was a precise mathematical functional framework with Hilbert spaces as well adapted. We do not need to enter the detail provided by Coleman and Noll as the latter authors offer an informative introduction that says it all mostly in words. They propose to consider a general class of linear viscoelastic materials with a governing constitutive response in the form (in our notation)

$$\sigma(t) = \Omega\{e(t)\} + \Phi(0)\{e(t)\} + \int_0^t \dot{\Phi}(s)\{e(t-s)\}ds, \tag{7.24}$$

with

$$\dot{\Phi}(s) = (d/ds)\Phi(s), \quad \lim_{s\to\infty} \Phi(s) = 0. \tag{7.25}$$

Here $\Phi(s)$ for each s and Ω are linear transformations of the space of symmetric tensors into itself. Function Φ is called the stress relaxation function. For isotropic materials it is completely determined by two scalars, the stress relaxation functions for shear and dilatation, while Ω is determined by the two Lamé coefficients. If one considers a deformation history such that the material is kept in its natural reference unstrained $(e = 0)$ configuration for all times $t<0$, and has strain e^* for all times $t \geq 0$, then Eq. (7.24) yields the following stress response:

$$\begin{aligned} \sigma(t) &= 0 \text{ for } t<0, \\ \sigma(t) &= \Omega\{e^*\} + \Phi(t)\{e^*\} \quad \text{for } t \geq 0. \end{aligned} \tag{7.26}$$

Thus,

$$\lim_{s\to\infty} \sigma(t) = \Omega\{e^*\}. \tag{7.27}$$

This renders the phenomenon of stress relaxation. A proper adaptation of the formulation must be made when the reference configuration is not a natural one (see Sect. 4 in Coleman and Noll 1961). Coleman and Noll discussed the evidence that dependence of the stress should naturally be a smooth one, and this justifies the linearity of (7.24) since "in small neighbourhoods smooth dependence is approximately linear". But in a general rational approach one should start with a generally nonlinear theory based on general physical (self-evident) principles such as the then recently expanded theory of "simple materials" of Noll, i.e., those materials for which the stress depends only on the history of the first spatial gradient of the

[5] I believe this is the only paper on continuum mechanics ever published in the *Review of Modern Physics*.

displacement. To develop a precise mathematical background one must specify a *topology* of functions that characterizes the history of the deformation, hence introducing a *norm* and making precise the notion of closeness of two of these histories. The introduced norm equips the space of such histories as a Hilbert space, but it should also emphasize the fading memory nature of the materials. Constitutive functionals in simple materials are required to be Fréchet differentiable in the Hilbert space of histories. With this mathematical apparatus at hand one first deduces a theory of finite linear viscoelasticity with the classical infinitesimal theory as special case. This is carefully dealt with by Coleman and Noll working first with finite strains. Dealing with fading memory, they introduce a norm $\|G(s)\|$ of a history $G(s)$ by

$$\|G(s)\|^2 = \int_0^\infty |G(s)|^2 h(s)^2 ds, \quad |G(s)|^2 := \operatorname{trace}\left(GG^T\right) \tag{7.28}$$

for a second order tensor G and an influence function h that characterizes the rate at which the "memory" fades. This function with positive real values satisfies a set of limit and decay conditions, an extreme case being the one for which $h(s) = \exp(-\beta s),\ \beta > 0$. The collection of all histories with finite norm (7.28) forms a Hilbert space H whose members do not grow too fast as $s \to \infty$. The assumed Fréchet differentiability of the general functionals allows one to deduce the looked for Boltzmann's hereditary equations (linear functionals) as the first variation or Fréchet differential at the zero history. Hilbert-Riesz's theorem then provides a representation of every continuous linear functional as an inner product, hence the looked for result. Requirement that the functional admits higher-order Fréchet differentials at the zero history would then yield an approximation of the functional by a polynomial functional of a certain degree with a negligible error, exemplified by the putative case of second-order viscoelasticity. A representation by multiple integrals exists under a hypothesis of complete continuity. Rivlin and co-workers had developed such theories (cf. Green and Rivlin 1957; Spencer and Rivlin 1960; see also Pipkin and Rogers 1968; Rabotnov 1980).

The mathematical justification of Boltzmann's modelling of viscoelasticity given by Coleman and Noll was extremely attractive with its enforced rigour allied to common sense. It obviously kindled a large number of highly technical works along the same line. Among these, we note the works of Gurtin and Sternberg (1962), Gurtin (1963), Coleman (1964), Dill (1975), Day (1972), and Christensen (1971). This also permeated the presentation of continuum mechanics by other groups of researchers in pure mechanics (e.g., Eringen 1967) but also in the study of electromagnetic continua (e.g., Eringen and Maugin 1990, Vol. II; Fabrizio and Morro 2003). Of particular importance are the lengthy work of Gurtin and Sternberg (1962) with the proof of many theorems, the work of Coleman (1964) on the compatibility of the functional approach with thermodynamics (i.e., the Clausius-Duhem inequality (7.13)), and that of Gurtin (1963) where this author could generalize *variational principles* to viscoelasticity, exploiting in some way the already mentioned *correspondence principle*. The encyclopaedia article by Leitman and Fisher (1973) closely follows Gurtin since these two scientists were mentored by this author.

One obvious drawback of the above recalled functional formulation is that in theory one should know the whole past history of deformation in order to evaluate the actual stress response. This is not only inconvenient; it is practically impossible. That is why the proposal to replace the functional in time by the evolution of a few parameters called *internal variables of state* by Coleman and Gurtin (1967) was a real breakthrough. As we know, the idea of such variables that are idenfiable and possibly measurable but cannot be directly controlled by means of external associated forces goes back to Duhem (1911) and Bridgman (1943). They are called "hidden parameters" by Brun (1965, 1967a, b) and Mandel (1966, 1967). This was duly exploited by Sidoroff (1975) in the viscoelasticity of solids—in finite deformations, together with the notion of multiplicative decomposition—with one or several "intermediate configurations"—of the deformation gradient already popular in elastoplasticity.[6] This type of approach was exploited to construct the finite-strain generalization of Maxwell's model of viscoelasticity by Gorodzov and Leonov (1968) and Buevich (1968)—see also Drozdov (1996). The influence of the model expanded by Biot (1954) was instrumental in the development of this approach. A recent review on the subject is by Wineman (2009).

Remember for all practical purpose that internal variables of state relate exclusively to internal dissipative processes and are basically governed by the second law of thermodynamics—cf. Eq. (7.17)—of which the exploitation suggests admissible time evolutions. The initial integro-differential system of equations of the Boltzmann-Coleman-Noll hereditary media is thus replaced by a set of time evolution equations, a situation much more pleasant in computation if the number of internal variables is kept small, what is generally the case with an astute choice of these variables.

In conclusion of this section, we cannot avoid emphasizing the very mathematical nature of the functional approach which, while allowing the proof of powerful theorems, may be of little practical importance, save for its reasonable approximation in the theory of internal variables of state. The powerful thermodynamics with internal variables of state was a happy spinoff of the theory.

7.6 Recent Developments

To prescribed (integer) orders of time derivatives in equations such as (7.4) and (7.6), there corresponds a well defined spectrum of frequencies in a vibrational response. But to a recorded spectrum of characteristic frequencies for an unknown material, there is hardly a chance that this will correspond to a modelling involving

---

[6] I perused the historical development of elasto-plasticity in another essay (cf. Maugin 2015).

integer order of time derivatives. This is probably from where the idea of considering *fractional* time derivatives emerged in viscoelasticity. The theory of fractional derivatives finds its origin in works by Abel,[7] but the most commonly associated names are those of F.C. Liouville (in 1832)[8] and G.B. Riemann, with other

---

[7]The expression "fractional derivative" seems to have been used for the first time by Leibniz in a letter to L'Hôpital in 1695.

[8]For the sake of information we remind the reader that Liouville's definition of the fractional derivative D^α of order $0<\alpha<1$ is given by

$$D^\alpha f(x) = \frac{1}{\Gamma(1-\alpha)}\frac{d}{dx}\int_0^x \frac{f(t)}{(x-t)^\alpha}dt \tag{a}$$

Simple examples of application of this definition are given by

$$\begin{aligned} D^\alpha(x^n) &= \frac{\Gamma(n+1)}{\Gamma(n+1-\alpha)}x^{n-\alpha};\\ D^\alpha(\sin(x)) &= \sin\left(x+\frac{\alpha\pi}{2}\right);\\ D^\alpha(\exp(ax) &= a^\alpha \exp(ax). \end{aligned} \tag{b}$$

The composition property of this operation where Γ is Euler's function $[\Gamma(n+1) = n!]$ allows one to use formulas such as

$$D^{3/2}f(x) = D^{1/2}\frac{df(x)}{dx}. \tag{c}$$

The fractional integral—noted J—enjoys similar group properties. With (Liouville)

$$(J^\alpha f)(x) = \frac{1}{\Gamma(\alpha)}\int_0^x (x-t)^{\alpha-1}f(t)dt, \tag{d}$$

so that we have the following property

$$(J^\alpha)(J^\beta f)(x) = (J^\beta)(J^\alpha f)(x) = (J^{\alpha+\beta}f)(x) = \frac{1}{\Gamma(\alpha+\beta)}\int_0^x (x-t)^{\alpha+\beta-1}f(t)dt. \tag{e}$$

Interesting properties in the present context are those related to the Laplace transform of the fractional integral. It is noted that

$$L\{Jf\}(x) = \frac{1}{s}\{Lf\}(s), \quad L\{J^\alpha f\}(x) = \frac{1}{s^\alpha}(L\{f\})(s), \tag{f}$$

and

$$J^\alpha f = L^{-1}\{s^{-\alpha}L\{f\}(s)\}. \tag{g}$$

A corresponding concise presentation is given by Mainardi (1996).

numerous other definitions given by many more or less reputed mathematicians over a period of almost 150 years. Now fractional calculus is well developed in both introductory and treatise forms and very well documented [e.g., books by Mainardi (2010) and Tarasov (2011) with physical applications]. Both Francesco Mainardi (born 1942) in Bologna and Alberto Carpinteri (born 1952) at Torino have been especially productive in the field of applications of fractional calculus to anelasticity, and more particularly viscoelasticity. Along this line we note the early paper of Caputo and Mainardi (1971), the lecture notes edited by Carpinteri and Mainardi (1997), a conference contribution by Mainardi (1994)—where the author generalizes the standard linear model to fractional derivatives—, applications to waves by the same author (Mainardi 1995), the original work where Caputo (1964) introduced a new definition of the fractional derivative—which avoids the definition of fractional-order initial conditions—, and the spot-on work of Koeller (1984). A historical perspective is given in Mainardi (2012). This line of works is in full development at the time of writing and shows interesting connections with the mechanics of fractal structures.

7.7 Conclusion

The branch of continuum mechanics referred to as the *viscoelasticity of solids* developed from simple one-dimensional models combining the primitive notions of elasticity and viscosity in the second part of the nineteenth century to much more sophisticated proposals. With further developments in nonlinear elasticity, the idea of heredity by Boltzmann, and a true understanding of the notion of time functionals, it reached a mathematical level that allowed the proof of fundamental theorems in the expert hands of Coleman, Noll, Gurtin and others. The fruitful idea of fading memory permitted the inclusion of viscoelasticity in the grand scheme devised by Truesdell and his direct disciples. The intrinsic beauty of the functional approach à la Coleman-Noll and the aesthetic satisfaction that it brings were difficult to resist to as witnessed by the very writer in the 1960s–1970s. No doubt that the appearance of polymeric substances and the consideration of new objects of study (e.g., deformable biological tissues) and allied technological questions

(Footnote 8 continued)

An example dealing with the acoustic wave equation for complex media (e.g., biological tissue) typically involves an attenuation that obeys a frequency power law. This is illustrated by a *causal* wave equation that incorporates fractional time derivatives such as given by

$$\nabla^2 u - \frac{1}{c_0^2}\frac{\partial^2 u}{\partial t^2} + \tau_\sigma^\alpha \frac{\partial^\alpha}{\partial t^\alpha}\left(\nabla^2 u\right) - \frac{\tau_e^\beta}{c_0^2}\frac{\partial^{\beta+2} u}{\partial t^{\beta+2}} = 0. \tag{h}$$

Here τ_σ and τ_e are two characteristic time, so that (h) follows from a generalization of the standard linear viscoelastic model with fractional orders α and β.

(e.g., influence of temperature) fostered the expansion of the theory in the second half of the twentieth century, while some smart researchers helped the solution of problems by exploiting the Laplace transform and a welcomed principle of equivalence. But we are inevitably led to envisage a comparison between the assets and successes of viscoelasticity and plasticity. Retrospectively, we personally think that while Boltzmann's initial idea brought continuum mechanics in a full Newtonian predeterministic view of all continuum mechanics (the past influences the present, not only a specific past such as in elasticity, but also through the *whole* past experience of the material), this has permeated both viscoelasticity and plasticity, the two main manifestations of irreversible evolution in the mechanics of deformable solids. Although many contributors shared an equal interest in the two sciences—this is illustrated by the cases of Lee and Mandel—, it seems that in the end plasticity (see Footnote 5), with the challenge posed by its singularity and its more dramatic expressiveness, attracted much more attention than the rather "smooth" viscoelasticity. It favoured richer developments than viscoelasticity in provoking more the thought of both experimentalists and theoreticians. But this may be a too personal apprehending.

References

Alfrey T (1945) Methods of representing the properties of viscoelastic materials. Q Appl Math 3 (2):143–150

Alfrey T (1948) Cited by Mandel (1966) p. 789 (On the use of Heaviside operational calculus)

Banks HT, Hu S, Kenz ZR (2011) A brief review of elasticity and viscoelasticity. Adv Appl Math Mech 3(1):1–51

Biot MA (1954) Theory of stress-strain relations in anisotropic viscoelasticity and relaxation phenomena. J Appl Phys 25:1385–1391

Biot MA (1956) Theory of deformation of a porous viscoelastic anisotropic solid. J Appl Phys 27:459–467

Biot MA (1958) Linear thermodynamics and the mechanics of solids. In: Proceedings Third US National Congress of Applied Mechanics (Providence, RI). ASME, New York, pp 1–18

Bland DR (1960) The theory of linear viscoelasticity. Pergamon Press, Oxford

Boltzmann L (1874) Zur Theorie der elastischen Nachwirkung. Sitz K u K Akad Wien 70:275–306

Bridgman PW (1943) The nature of thermodynamics. Harvard University Press, Cambridge

Brun L (1967a) Sur l'unicité en viscoélasticité dynamique. C R Acad Sci Paris 264:135–139

Brun L (1967b) Thermodynamique et viscoélasticité. Cahiers du Groupe Français de Rhéologie 1:191–201

Brun L (1969) Méthodes énergétiques dans les systèmes évolutifs linéaires, I. J Mécanique (Paris), 8:125–166; II. J Mécanique (Paris) 8:167–192

Buevich YuA (1968) The kinematics of viscoelastic media with finite deformations. Prikl Mat Mekh (English translation) 32:209–221

Caputo M (1964) Linear model of dissipation where q is almost frequency independent. Geophys J Roy Astron Soc 67:529–539

Caputo M, Mainardi F (1971) Linear models of dissipation in anelastic solids. La Rivista del Nuovo Cimento (1971–1977) 1(2):161–198

Carpinteri A, Mainardi F (eds) (1997) Fractals and fractional calculus in continuum mechanics (CISM lecture notes, Udine). Springer, Vienna
Coleman BD (1964) On thermodynamics of materials with memory. Arch Rat Mech Anal 17:1–46
Coleman BD, Gurtin ME (1967) Thermodynamics with internal state variables. J Chem Phys 47:597–613
Coleman BD, Noll W (1961) Foundations of linear viscoelasticity. Rev Mod Phys 33:239–249
Christensen RM (1971) Theory of viscoelasticity: an introduction. Academic Press, New York (Second edition, 1982)
Day WA (1972) The thermodynamics of simple materials with fading memory. Springer, Berlin
Dill EH (1975) Simple materials with memory. In: Eringen AC (ed) Continuum physics, vol 2. Academic Press, New York, pp 284–403
Drozdov AD (1996) Finite elasticity and viscoelasticity. World Scientific, Singapore
Duhem P (1911) Traité d'énergétique ou de thermodynamique générale, Two volumes. Gauthier-Villars, Paris
Eckart C (1948) The thermodynamics of irreversible processes IV. The theory of elasticity and anelasticity. Phys Rev 73:373–382
Eringen AC (1960) Irreversible thermodynamics and continuum mechanics. Phys Rev 117:1174–1183
Eringen AC (1967) Mechanics of continua. J. Wiley, New York
Eringen AC, Maugin GA (1990) Electrodynamics of continua, vol II. Springer, New York
Fabrizio M, Morro A (1992) Mathematical problems in linear visco-elasticity. SIAM Publications, Philadephia
Fabrizio M, Morro A (2003) Electromagnetism in continuous media. Oxford University Press, Oxford
Findley WN, Lai JS, Onaran K (1976) Creep and relaxation of nonlinear viscoelastic materials and an introduction to linear viscoelasticity. North-Holland, Amsterdam (Dover reprint, New York, 1989)
Flügge W (1967) Viscoelasticity. Blaisdell, Waltham
Freudenthal AM, Geiringer H (1958) The mathematical theories of the inelastic continuum. In: Flügge S (ed) Handbuch der Physik Bd. VI. Springer, Berlin, pp 229–243
Gorodzov VA, Leonov AI (1968) On the kinematics, non-equilibrium thermodynamics and rheological relationships in the nonlinear theory of viscoelasticity. Prikl Mat Mekh (English translation) 32:62–84
Graffi D (1928) Sui problemi dell' ereditarietà lineare. Nuovo Cimento 5:53–71
Graffi D (1977) Material with memory. C.I.M.E, Lecture notes, Bressanone
Green AE, Rivlin RS (1957) The mechanics of non-linear materials with memory I. Arch Rat Mech Anal 1:1–21
Gross B (1947) On creep and relaxation. J Appl Phys 18:212–221
Gross B (1953) Mathematical structure of the theories of viscoelasticity. Hermann, Paris
Gurtin ME (1963) Variational principles in the linear theory of viscoelasticity. Arch Rat Mech Anal 12:179–191
Gurtin ME, Sternberg E (1962) On the linear theory of viscoelasticity. Arch Rat Mech Anal 11 (4):291–355
Hellinger E (1914) Die allgemein Ansätze der Mechanik der Kontinua. In: Klein F, Wagner K (eds) Enz Math Wiss, vol 4, part 4. Springer, Berlin, pp 602–694
Ilyushin AA (1971) Mekhanika sploshnoy sredy (Continuum Mechanics). Moscow State University Press, Moscow; Second Edition, 1978; Third edition; 1990
Ilyushin AA (2003–2009; in Russian) Selected collected Works: vol 1 (1935–1945), Fizmatlit: Moscow (2003); vol 2 (1946–1966), Fizmatlit: Moscow (2004); vol 3: Theory of thermoviscoelasticity, Fizmatlit: Moscow (2007); vol 4: Modelling of dynamic processes in solids and engineering applications, Fizmatlit: Moscow (2009)
Kelvin L (1875) In: Math Phys Papers. vol 3, p 27
Kluitenberg GA (1962) On rheology and thermodynamics of irreversible processes. Physica 28:1173–1183

Kluitenberg GA (1963) On the thermodynamics of viscosity and plasticity. Physica 29:633–652
Koeller RC (1984) Application of fractional calculus to the theory of viscoelasticity. Trans ASME J Appl Mech 51(2):299–307
Lakes RS (1998) Viscoelastic solids. CRC, Boca Raton
Lee EH (1955) Stress analysis in viscoelastic bodies. Q Appl Math 3(2):183–190
Lee EH (1962) Viscoelasticity. In: Flügge W (ed) Handbook of engineering mechanics, Chapter 53. McGraw-Hill, New York
Leitman MJ, Fisher GMC (1973) The linear theory of viscoelasticity. In: Flügge S (ed) Handbuch der Physik, Bd. VIa/3. Springer, Berlin, pp 1–123
Mainardi F (1994) Fractional relaxation in anelastic solids. J Alloys Compd 211–212:534–558 [Proceedings of the 10th international conference on internal friction and ultrasonic attenuation in solids, Roma, 1993]
Mainardi F (1995) Fractional diffusive waves in viscoelastic solids. In: Wegner JL, Norwood FR (eds) Nonlinear waves in solids (Proceedings IUTAM Symposium, Victoria, Canada), ASME reprint no. AMR137, ASME-AMR, Fairfield, NJ, USA, pp 93–97
Mainardi F (1996) Fractional relaxation—oscillations and fractional diffusion—wave phenomena. Chaos Solitons Fractals 7(9):1461–1477
Mainardi F (2010) Fractional calculus and waves in linear viscoelasticity. Imperial College Press, London
Mainardi F (2012) An historical perspective on fractional calculus in linear viscoelasticity. Fractional Calc Appl Anal 15(4):712–717
Mandel J (1955) Sur les corps viscoélastiques linéaires à comportement linéaire. C R Acad Sci Paris 241:1910–1911
Mandel J (1958) Les corps viscoélastiques boltzmanniens. Rheol Acta 1, no 203
Mandel J (1966) Cours de mécanique des milieux continus, two volumes. Gauthier-Villars, Paris
Mandel J (1967) Application de la thermodynamique aux systèmes à paramètres cachés. Cahiers du Groupe Français de Rhéologie 1(4):181–190
Mandel J (1973) Thermodynamics and plasticity. In: Foundations of continuum thermodynamics, Domingos JJD, Ninas MRN, Whitelaw JH, (eds) J. Wiley, New York, pp 283–304
Maugin GA (1999) The thermomechanics of nonlinear irreversible behaviors: An introduction. World Scientific, Singapore
Maugin GA (2015) Complements: plasticity over 150 years (1864–2014). Preprint, IJLRdA-UPMC-Paris 6, Paris [Chapter 8 in this book]
Maxwell JC (1866) On the viscosity or internal friction of air and other gases. Phil Trans R Soc London A156:249–268
Maxwell JC (1868) On the dynamic theory of gases. Phil Mag (4) 35:129–145 [Also, Collected Scientific Works, vol 2, p 623, Cambridge, UK, 1890]
Meixner J (1953) Die thermodynamische Theorie der Relaxationserscheimungen und ihr Zusammenhang mit Nachwirkungstheoreie. Kolloid Zeitschrift 134(1):2–16
Natanson L (1901) Über die Gesetse derinnneren Reibung. Z Phys Chem 38:690–704 [also, Bull Acad Sci Cracovie (1901) pp 95, 161; (1902) pp 488, 494; (1903), pp 268, 283]
Noll W (1955) On the continuity of the solid and fluid states. J Rat Mech Anal 4:13–81
Oldroyd JG (1950) On the formulation of rheological equations of state. Proc Roy Soc Lond A200:523–541
Pipkin AC, Rogers TG (1968) A non-linear integral representation for viscoelastic Behaviour. J Mech Phys Solids 16:59–72
Rabotnov YuN (1980) Elements of hereditary solid mechanics (translation from the Russian). MIR Publishers, Moscow
Roylance D (2001) Engineering viscoelasticity. MIT Lecture notes with exercises, 37 pages, Department of Materials Science and Engineering (Dated 24 Oct 2001; available on the w.w. web)
Sidoroff F (1975) Variables internes en viscoélasticité I. J Mécanique 13:679–713; II, ibid, 14:545–566

Spencer AJM, Rivlin RS (1960) Further results in the theory of matrix polynomials. Arch Rat Mech Anal 4(3):214–230

Staverman AJ, Schwarzl F (1952) Non-equilibrium thermodynamics of linear viscoelastic behaviour. Proc Koning Nederl Akad Van Wettenschappen 55(5):474–490

Tarasov VE (2011) Fractional dynamics: application of fractional calculus to dynamics of particles, fields and media. Springer, Berlin

Timoshenko SP (1953) History of the Strength of Materials. McGraw-Hill, New York [Dover reprint, New York, 1983]

Truesdell CA, Noll W (1965) The Non-linear field theories of Mechanics, Handbuch der Physik, Bd.III/3, Ed. S. Flügge. Springer, Berlin. [Referred to for short as NFTM]

Voigt W (1889) Über die innere Reibung der festen Körper, insbesondere der Krystalle. Göttinger Abh Ges Wiss 36, no 1. [Also, Ann d Physik, (2): 47,671]

Voigt W (1895–96) Kompendium der theoretischen Physik, Bd. I: Mechanik und nichtstarre Körper, Wärmelehre, Bd. II: Elektrizität und Magnetismus. Leipzig

Voigt W (1898) Die fundamentalenphysikalischen Eigenschaften der Krystalle in elementarer Darstellung. Veit & Co., Leipzig

Voigt W (1910) Lehrbuch der Kristallphysik. Teubner, Leipzig

Volterra V (1909) Sulle equazioni integro-differenziali dell'elasticità. Rend Acad Naz Lincei (5a) 18:295–301

Volterra V, Pérès J (1936) Théorie générale des fonctionnelles, vol I. Gauthier-Villars, Paris

Wiechert E (1889) Über elasticshe Nachwirkung. Dissertation, Königsberg Universität, Germany

Wiechert E (1893) Gesetze der elasticshen Nachwirkung für constante Temperatur. Ann der Physik 286:335–348 and 546–570

Wineman A (2009) Nonlinear viscoelastic solids—a review. Math Mech Solids 14:300–366

Zaramba S (1903) Bull Inst Acad Sci Cracovie, 85:380, 594, 614 [also, Mém Sci Math (Paris), No. 82, (1937)]

Zener C (1948) Elasticity and anelasticity of metals. Chicago University Press, Chicago

Chapter 8
Plasticity Over 150 Years (1864–2014)

Abstract Plasticity theory in continuum mechanics has evolved from Coulomb's and Tresca's experiments to the exploitation of powerful thermo-mechanical concepts (e.g., second law of thermodynamics) and sophisticated mathematical tools (e.g., convex analysis). The present contribution peruses this development over more than a hundred and fifty years with special attention paid to original but forgotten works of Duhem, the introduction of various yield criteria, the formulation of incremental and normality laws, the influence of thermodynamic concepts (in particular the dissipation inequality) and the resulting variational formulations and inequalities, the fortunate complementarity between, and confluence of, "mathematical" and "physical" plasticity theories, some doomed attempts such as hypoelasticity, the birth of a rational approach to plasticity in finite deformations, and further progress in anisotropic studies, numerical plasticity, homogenization, visco-plasticity, coupling with other physical properties, and the recent gradient plasticity. This is achieved in a historical comprehensive vision with a minimum of technicalities, but always paying tribute to the most influential and well articulated contributors. Both primary and secondary sources are exploited.

8.1 By Way of Introduction

To the question what is "a plastic", most people answer that it is this ubiquitous kind of material that invaded our environment in the 1950s–1960s. These people more or less know that this product is artificial, easily moulded, and is most commonly derived from petrochemicals. On the other hand, consulting *Google* and

G.A. Maugin, *Continuum Mechanics through the Ages - From the Renaissance to the Twentieth Century*, Solid Mechanics and Its Applications 223,
DOI 10.1007/978-3-319-26593-3_8

entering "plasticity" we are nowadays mostly addressed to "brain plasticity" and the like.[1] But this is not what we are interested in because the common word *plastic* should not be confused with the technical adjective *plastic*, which is applied to any material which undergoes a permanent change of shape (e.g., a plastic deformation) when strained beyond a certain point. Some plastics (e.g., Bakelite) will break before deforming and therefore are not plastic in the technical sense. As to the plasticity of the brain it is a mere convenient image to illustrate changes in neural pathways and synapses due to changes in behaviour, environment, neural processes, thinking, emotions, etc. Originally, the word "plastic" derives from the Greek πλαστικός (*plastikos*) meaning "capable of being shaped or molded", from πλαστός (*plastos*) meaning "molded". This explains that. But in physics and materials science, **plasticity** describes the deformation of a material undergoing non-reversible changes of shape in response to applied forces. For example, a solid piece of metal being bent or pounded into a new shape displays plasticity as permanent changes occur within the material itself. This occurs at a transition called *yield*. This is a frequent phenomenon observed in many materials, particularly metals, soils, rocks, foams, etc. Plastic deformation is a property of ductile and malleable solids. "Brittle" materials, such as cast iron, cannot be plastically deformed. Malleability is easily understood in the case of precious metals such as gold—which in fact has no elastic behaviour—which in fact needs the addition of other harder materials to preserve their shape in the form of jewellery. There is also the question of what is "ductility". More precisely, **ductility** is the ability of a solid material to deform under *tensile* stress, for instance, this is often characterized by the material's ability to be stretched into a wire. As to **malleability**, a rather similar property, it refers to a material's ability to deform but under *compressive* stress such as in the forming of a thin sheet by hammering or rolling. Both mechanical properties of ductility and malleability are aspects of **plasticity**, meaning the extent to which a solid material can be plastically deformed without fracture. But note that ductility and malleability are not always coextensive; for instance, gold has both high ductility and malleability, while lead has low ductility but high malleability. Yet, more than often, the word *ductility* is used to embrace both types of plasticity, and is often simply used as opposite to *brittleness*. In the present contribution **plasticity** is the name of a science that refers either to a *mathematical* description of what occurs during plastic deformation in terms of stresses, strains, and loads or to *physical* explanations of plastic flow in terms of atoms, crystals, grains, and motions of structural defects (dislocations) within crystals, so that we make a distinction between mathematical plasticity and physical plasticity. This field of science has gained a tremendous attraction with the occurrence of plastic deformation in many metal-forming processes (rolling, pressing, and forging) and in geologic processes

[1]While participating in an international symposium on plasticity in Baltimore in 1995, another person in the elevator noticed the words printed on my symposium bag and asked me if I was working on brain's plasticity. Certainly an educated person to ask such a question, he was most surprised by my answer mentioning my concern—and the concern of many present at this huge symposium—with the plasticity of metals.

(rock folding and rock flow within the earth under extremely high pressures and at elevated temperatures).

Of all times craftsmen (blacksmiths, goldsmiths) working with metals have noticed and exploited the remarkable properties of ductility and malleability. But a scientific approach to the phenomenon of plasticity has been hindered by its inherent complexity in its physical interpretation and its mathematical modelling. Some of the first recorded measurements go back only to Coulomb in the late eighteenth century and Gerstner in the first part of the nineteenth century (cf. Osakada 2008), but we consider that the real science of plasticity started with Tresca, Saint-Venant and Lévy in the 1860s–1870s, and found a more favourable ground in the twentieth century with a true explosion in research in the second part of that century as we shall see hereafter. Remarkably enough, Timoshenko (1953), a well informed and accurate specialist of continuum mechanics and its applications, and the founder of "applied mechanics" in the USA, says very little on plasticity in his landmark book on the history of the strength of materials. Plasticity is mentioned only in three pages (pp. 242, 395, 396) in this famous opus; But its is true that the first book in English on plasticity appeared only in 1930 as the translation of a German book by Nadai, and most of the new results in mathematical plasticity were published in the 1940s–1950s while physical plasticity was expanded only in the 1930s on, when Timoshenko (1878–1972)—mostly interested in elasticity, structures, vibrations and stability in his most active part of his scientific career—was already in his seventies. Herein after, we try to correct this lack of information by perusing the history of plasticity over two and a half centuries, say between Coulomb and the end of the twentieth century.

8.2 Timid Experimental Steps and First Mathematical Modelling

Plasticity is a subtle phenomenon so that its first quantified observation had to await the conception of true experiences with appropriate recording of data. Along this line it seems that the first true recording goes back to the work of **Charles A. Coulomb** (1736–1806) on the torsion of iron wires (Coulomb 1784) when this gifted experimentalist measured the recovery angle after twisting. This angle increases less and less with the increased number of rotations, with a kind of saturation when, we imagine, plastic deformation that started from the surface of the specimen has reached the centre of the wire with progress of work hardening. The corresponding shearing yield stress estimated from Coulomb's measurements is close to the actual known value for iron (cf. Osakada 2008, p. 23). As noticed by Bell (1973, p. 31), the stress-strain curves recorded in 1824—and published in Gerstner (1831)—by the Bohemian professor of mechanics and hydraulics **F.J. Gerstner** (1756–1832) at the Polytechnicum in Prague—famous for his studies of water waves—for piano wires, clearly indicates the presence of a residual strain after unloading, a patent sign of a plasticity effect. To find much more on these measurements of shearing stresses, one must uncover the tremendous work of **Henri E. Tresca** (1814–1885).

The French Revolution that started in 1789 created many new teaching and research institutions, among them the *Ecole Polytechnique* and the *Ecole Normale Supérieure* to form scientifically educated engineers and competent teachers, respectively. Much less known outside France is the *Conservatoire National des Arts et Métiers* (for short *CNAM*) also founded in Paris in 1794. Somewhat similar to the *Royal Institution* in London, this new educational establishment was to offer public lectures to all, but in time it also became in part a museum of science and technology (Musée des Arts et Métiers) and also entertained some research in engineering science and industrial economy as exemplified by Tresca who started to tackle engineering projects there from 1852. Indeed, in the 1860s–1870s Tresca conducted there accurate experiments on various forms of metal forming and was thus led to studying the phenomenon of extrusion relating the extrusion force to the shear stress (he was using lead as a sample material). As a result of his observation that shear stress dropped in a certain range of loading he concluded that metal flow occurs under a constant *maximum shear stress* $\tau_M = k$. Tresca submitted many notes on the subject starting in 1864 but culminating in his long memoir of 1872. Many of Tresca's figures are reproduced and commented upon in Bell (1973, pp. 427–449). But in that period the leading French mechanician was **Adhémar J.C. Barré de Saint-Venant** (1797–1886)—often shortened to *Saint-Venant*. Of course even Cauchy (1828) was conscious of the general validity of the basic concepts (stresses, strain tensor) that he introduced as indicated by the general title of his groundbreaking memoir—including the "anelastic" behaviour as a possibility. But Cauchy was mostly concerned by elastic solids (and fluids) and practically had no experimental data at his disposal—and perhaps no motivation—to deal more closely with anelastic materials. About fifty years later, it did not take long to Saint-Venant to cast Tresca's results in a sound mathematical frame. To justify the proposed modelling Saint-Venant abstracted from these results three important working hypotheses: first, it is noticed that no change in volume (so called *isochoric* deformation in the modern jargon) is observed during plastic deformation; second, the directions of the principal stresses coincide with those of the principal strains (this assumes an *isotropic* response); third, the maximum shearing (or tangential) stress at a point is equal to a specific constant as noted in Tresca's formula $\tau_M = k$. This can be rewritten in more precise mathematical terms as

$$Sup_{\alpha,\beta}\left|\sigma_\alpha - \sigma_\beta\right| = 2k, \alpha, \beta = 1, 2, 3, \tag{8.1}$$

where the Greek indices label the principal stresses. Introducing then the tangential stresses, this can also be expressed by the following set of three inequalities:

$$2|\tau_1| \equiv |\sigma_2 - \sigma_3| \leq k, etc, \tag{8.2}$$

by circular permutation. In an astute plane representation this is represented by a hexagon (see Maugin 1992a, p. 18). The interior domain (a convex domain with angular corners) is the domain of elasticity. The definition of this elastic limit by pieces of intersecting straight lines offers some difficulties in analytic treatment of problems. Nonetheless, Barré de Saint-Venant (1871a, b, c), a gifted analyst, was able to give the solution of exemplary problems such as: the torsion of a circular shaft, the plane deformation of a hollow circular cylinder under the action of an internal pressure, etc., problems that we still give students to solve without the help of a computer. About the same time, **Maurice Lévy** (1838–1910), a disciple of Saint-Venant, proposed to replace the second working hypothesis of his master by assuming that the directions of *increments* of principal strains coincide with those of the principal stresses (Lévy 1871). This looks like the first attempt at formulating an *incremental* flow rule. But Lévy did not differentiate plastic strain from the total strain—and thus including elastic strain even in the plastic region—so that the validity of Lévy's proposal is indeed limited to the case where elastic strain can be fully discarded and thus corresponds to the approximation of *rigid-plastic* behaviour. This is a rather highly singular behaviour since nothing happens to the strain, not even an elastic one, in so far as the plasticity threshold is not reached and then we have an uncontrolled plastic flow occurring along a plateau in stress.

Other fundamental observations concerning plastic behaviour were made by **Johann Bauschinger** (1833–1893) and **Otto Mohr** (1835–1913), both in Germany. The first author (Bauschinger 1886) has shown that yield stress in compression after plastic tensile deformation is significantly lower than the initial yield in tension. This is called the *Bauschinger effect*. As to Mohr, he introduced an astute graphical representation of the stress state at a point: this state is represented by a circle in a rectangular coordinate system with coordinates as normal and shear stress components. This helped him to devise a yield criterion where the envelope of circles obtained for tension, compression and torsion represents the fracture limit. This is *Mohr's stress circle* and Mohr's (intrinsic curve) criterion (Mohr 1900). The same year (reported by Bell 1973, p. 485), Guest (1900) from University College in London, published results on the strength of ductile materials under combined (tension and torsion) stress states. Like Tresca, he obtained that yielding occurs when a certain maximum shear strain is reached, but while Guess clearly distinguished between yielding of a ductile material and brittle fracture (two effects often summarized under the unique name of failure in past works), his work did not really improved on Tresca's.

All the mentioned works exploit what may be called a *maximum shear stress criterion* with its inherent mathematically awkward form. But **Richard von Mises** (1883–1953) proposed a new criterion (von Mises 1913) that was much easier to apply and in fact used the notion of *maximum-distortion-energy theory* of yielding. It replaces the hexagon of Tresca-Saint-Venant by a circumscribed circle (obviously a convex domain; see Maugin 1992a, Fig. 1.18) of radius k and equation

$$(\sigma_1 - \sigma_2)^2 + (\sigma_2 - \sigma_3)^2 + (\sigma_3 - \sigma_1)^2 = 2k, \tag{8.3}$$

where the maximum, medium, and minimum principal stresses are such that $\sigma_1 \geq \sigma_2 \geq \sigma_3$. It happens that **Maksymilian T. Huber** (1872–1950) had proposed a rather similar criterion in 1903–1904 in a paper published in Polish in Lvov or Lwow (now Lviv in Ukraine) and published in English translation only in 2004 (cf. Huber 1904, 2004). This was identified (Hencky 1923, 1924) as a criterion of strain-energy function, hence the same as von Mises', by Henrich Hencky (1885–1951). Mises' criterion had been introduced as a mathematical convenience but it was also physically founded. In practice, the two Tresca-Saint-Venant and Huber-Mises criteria give equal resistance to simple traction and to simple compression, but different resistances in pure shear. In the case of metals, the experimental results are contained between these two criteria, although they are generally closer to the Huber-Mises criterion as shown in well-publicized experiments conducted by Taylor and Quinney (1931). To conclude this point we note that other criteria were proposed before to measure the energy of deformation, including by the famed William J.M. Rankine (1820–1872) in Scotland as "maximum stress energy" and by Eugenio Beltrami (1835–1900) in Italy as "maximum-strain-energy theory". In the future other criteria will be proposed for anisotropic media, the plasticity of soils, the case of porous media, etc. But all this refers to a rather static view of plasticity where the time parameter does not intervene while elasto-plasticity experiments and problems are rather conceived as scenarios of loading and unloading at a more or less rapid pace. This dynamic or incremental view makes use of the thermo-*dynamic* framework. Still we lastly emphasize that the introduction of yield criteria in the deformation theory of solids, essentially by Tresca, Saint-Venant, Huber and Mises was as great an achievement as the introduction of Reynolds' number in fluid mechanics as it marked a transition to an altogether different mechanical response in continuum mechanics. They were both designed at the same critical period of transition in science. But this was not so readily integrated in a standard cursus of rational mechanics as proved by the lack of reference to plasticity and turbulence in Appell's classical treatise (Appell 1921; first edition in 1900). Probably that these two fields were not "rational enough", before "mathematical plasticity" was formulated (see below).

8.3 Enter Evolution and Thermodynamics

8.3.1 Duhem's Pioneering Works

Of course all engineers and scientists at the dawn of the twentieth century were conscious of the possible existence of some relationship of the plastic phenomenon with dissipation. The most militant energetist of the period, **Pierre Duhem** (1861–1916), had a special interest in all dissipative irreversible physical behaviours in phenomenological physics, including in a possible theory of irreversible processes manifested in the form of *permanent deformations* in solids. He wrote eloquently about so-called *nonsensical branches of mechanics*—that is, those branches like friction, false equilibria, permanent alterations, and hysteresis—which at the time were not yet satisfactorily covered by his general thermodynamics—in his solidly argued essays on the "evolution of mechanics" (Duhem 1903).[2] There he built essentially on pioneering works of Gibbs and Helmholtz. But he was missing our present knowledge (that is still by force limited) in differential geometry, functional analysis, dynamical systems, and convex analysis. Many of the questions he raised and the horizons he opened will find an answer and a broadening in these fields of mathematics, as amply proved by successful developments in the rational mechanics of continua in the second part of the twentieth century. He also tried to have experiments conducted by a team of doctoral students [MM. Marchis, Saurel (from the USA), Pélabon, Lenoble] to work on the possible identification and measurements of the various typical irreversible properties but the success was limited due to the lack of precise experimental techniques at the time. This ambitious programme was in advance on its time. This is illustrated by the thesis of Lenoble (1900) on deformable (i.e., plastic) metals. However, from a theoretical point of view, Duhem had ideas which later on proved to be correct. For instance, with F denoting the free energy and A the external action, he formulates the equilibrium condition for a normal state variable α as (with Duhem's sign convention for dissipation)

$$A = \frac{\partial F}{\partial \alpha}. \tag{8.4}$$

But Sect. 2 of Duhem's relevant chapter in Duhem (1903) rings a more familiar bell to our ears, for it deals with *permanent "alterations" and hysteresis effects* that may be closer to our own plasticity, visco-plasticity and creep concepts. Here Duhem first gives a general idea of what permanent alterations are. He emphasizes the role of infinitesimally slow evolutions, adapting in accordance temperature and external actions. Again, one must go beyond Gibbs' statics and Helmholtz's dynamics. That is, one must elaborate on the generalization of Eq. (8.4) or its incremental form that seems to be more appropriate. More precisely, Duhem discusses the possible generalizations of the following incremental form that follows from equilibrium (8.4):

$$dA = d\frac{\partial F}{\partial \alpha}. \tag{8.5}$$

Here we remind the reader that $\partial F/\partial \alpha$ is computed at constant temperature while noting that systems exhibiting permanent alterations (e.g., residual strains) are quite different from those exhibiting viscosity [but in modern thermo-mechanics we also

---

[2] I gave elsewhere (Maugin 2014) a partial English translation of the most relevant parts of these essays.

envisage a mix of the two effects as in visco-plasticity]. The permanent alterations envisaged by Duhem are exhibited when an unloading (decrease in the "cause") following a loading, does not bring the system back to its initial virgin state. We must thus distinguish between the two possibilities of the incremental law that will be adopted depending on whether we are increasing or decreasing the "cause". For sufficiently slow evolutions, Duhem proposes to generalize (8.5) by [cf. Duhem's Eqs. (7) and (7b)]

$$dA = d\frac{\partial F}{\partial \alpha} + hsign(d\alpha)|d\alpha|, \tag{8.6}$$

where the quantity h may still depend on the state of the system and also on the external action A. Equation (8.6) may seem to be both enigmatic and ad hoc to most readers. It is however in direct line with mathematical works on hysteresis of the 1970s–1990s (see Bouc 1971; Maugin 1992b). To find more elaboration by Duhem on (8.6) and applications one should consult the original works of Duhem published in 1901 in a rather odd place (a physical-chemical publication in Belgium). Here Duhem is satisfied with the cases of deformations, residual magnetization, magnetic hysteresis, and analogous properties for electric polarisation in dielectrics. Duhem emphasizes the interest of his considerations in metallurgical treatments (tempering, annealing, etc.). But much more is also to be found in the exhaustive and clear analysis of Manville (1927). The latter author remarks that it is difficult to summarize the works of Duhem and his co-workers on permanent alterations and the somewhat 400 pages of various memoirs (in particular the seven published in Belgium in 1896, 1897 and 1901) in a short text. But some of the facts and properties recalled by Manville shed light on some aspects that will be of great interest for further comparison with modern developments. One important remark is that from the law of displacement of the equilibrium (a *stability* condition about an equilibrium: cause and effect vary in the same sense), the perturbing work in passing from (A, α) to $(A + dA, \alpha + d\alpha)$ must be non negative [Eq. (9), p. 310, in Manville 1927]

$$dA \cdot d\alpha \geq 0. \tag{8.7}$$

This has for immediate consequence that the slopes of the curves in the (A, α) plane must always be in the same sign whether on increasing or decreasing A. Consequently, the end points of a hysteresis cycle in the plane (A, α) are always sharp and cannot be rounded in the absence of viscosity. Another pertinent remark concerns the extension of Clausius' theorem for a non-reversible closed cycle. The quantity $\int dS = \int dQ/\theta$ over a cycle ought to be non-negative (S = entropy, Q = heat, θ = thermodynamic temperature). With constant temperature, this reduces to the condition $\int dQ \geq 0$. Equivalence between heat and work yields that $dQ = A d\alpha$ for an isothermal cycle, and thus

$$\int_{closed\ cycle} A \cdot d\alpha \geq 0. \tag{8.8}$$

This clearly dictates the sense in which hysteresis cycles are necessarily followed in the plane (A, α). But in practice it was soon discovered that many hysteresis cycles (especially in the mechanics of deformations) are slightly rounded at their ends. The explanation of this phenomenon in contradiction with (8.7) is to be found in the likely presence of small viscous effects. Duhem had several graduate students working out the experimental facets of this research on permanent alterations [Theses of Marchis and Pélabon in 1898, and of Saurel and Lenoble in 1900; see in particular Lenoble (1900)]. Results were not very conclusive from the quantitative point of view. Only a general qualitative agreement was found (cf. Manville 1927, p. 350). Sophisticated experimental techniques and accurate measurement possibilities were not yet available at the time, so that these studies were probably too ambitious and in some sense doomed. Here are Duhem's own words: "It seems that no theory of permanent alterations can obtain from experiments more than qualitative and somewhat vague confirmations" (as cited by Manville 1927, p. 599). What must be retained here is the expression (8.7) that can be rewritten with time rates as

$$\dot{A} \cdot \dot{\alpha} \geq 0, \tag{8.9}$$

while the corresponding dissipation naturally reads

$$\Phi_\alpha = A \cdot \dot{\alpha}. \tag{8.10}$$

In plasticity, A will be the stress and α the anelastic deformation $\mathbf{e}^p$. As a matter of fact, it is **Carl H. Eckart** (1902–1973) who in Eckart (1948) wrote down the anelastic dissipation in small strains as

$$\Phi_{anel} = \sigma_{ij}\dot{e}^p_{ji}, \tag{8.11}$$

opening thus the way to a whole new development in irreversible thermo-*dynamics* of continua (cf. de Groot and Mazur 1962, but these authors show no interest in solid like behaviour). Duhem's and Eckart's original works are seldom directly cited in the beginning of this twenty-first century.

8.3.2 Incremental Laws

Now we can return to the notion of scenario and incremental formulation for the basic laws of plasticity as either the increment $d(.)$ or the time derivative—noted in Newton's way with a superimposed dot—will appear of necessity. For this we need first a qualitative description as follows, considering only the case of isotropic bodies in small stains. Here we closely follow A. M. Freudenthal and H. Geiringer in their lengthy *Handbuch* article of 1958—which gives a state of the Art in the field as in the mid 1950s. The state of an isotropic body is said to be *perfectly plastic* if all admissible states of stress fulfil an isotropic condition, called the *yield condition* $f(\sigma_{ij}) = 0$—in stress space. This function *f*, because of isotropy, is assumed to be expressible in terms of the basic invariants of the stress tensor of Cartesian tensor components σ_{ij}. "This definition implies that at the yield limit the rate of plastic flow associated with any stress increment $d\sigma_{ij}$ would increase so rapidly that states of stresses for which $f(\sigma_{ij} + d\sigma_{ij}) > 0$ cannot be sustained, while any stress increment for which the point remains on the yield surface produces no plastic strain work". Accordingly, "the condition of perfect plasticity is $df(\sigma_{ij}) = 0$ at $f(\sigma_{ij}) = 0$". In the stress space represented by a Cartesian coordinate system with the directions of the principal stresses as axes, the yield condition $f(\sigma_{ij}) = 0$ describes either a single continuously differentiable surface enclosing the origin (e.g., the Huber-Mises case) or a number of such surfaces, intersecting along sharp edges and forming a polyhedron (e.g., the Tresca-SaintVenant case). Points inside the yield surface (for which $f < 0$) are in a pre-plastic state or may be rigid, elastic or visco-elastic. Points outside the yield surface cannot be reached by definition of perfect plasticity. Experimental evidence (remember the first working hypothesis of Saint-Venant) justifies the assumption of *incompressibility* of plastic flow, so that only two stress invariants are involved in function *f*. It is this incompressibility property that allows the representation of the yield stress as a curve in a plane state space described by the three principal stresses at 120° from each other (hexagonal and circle representations of the Tresca and Huber-Mises yield curves, respectively). Because a stress increment $d\sigma_{ij}$ at $f = 0$ produces no strain work, the increment of the rate of plastic strain work along any stress path within the yield surface is such that

$$d\sigma_{ij}\dot{e}^{p}_{ij} = 0. \tag{8.12}$$

This is similar to Duhem's orthogonality relation

$$dA \cdot \dot{\alpha} = 0, \tag{8.13}$$

at the yield. Because of incompressibility, only the deviatoric part of $\dot{e}^{p}_{ij}$ is involved in (8.12). Finally, as a consequence of (8.12), the direction of the plastic strain-increment or strain rate must be *normal* to the yield surface at the point of intersection with the stress-increment vector, provided that this direction (i.e., $\partial f/\partial\sigma_{ij}$) be uniquely defined (case of Huber-Mises yield). In the presence of corners where this normal is not uniquely defined, some caution must be taken (case of Tresca yield). This was circumvented in modern analysis with the notion of cone of outward unit normals to the yield surface at discontinuities of this surface (cf. Maugin 1992a).

In perfect plasticity the yield surface $f(\sigma_{ij}) = 0$ is a *fixed* material property. But it was observed that this yield surface can move, deform and expand with an increase in loading. This is the phenomenon of "hardening". Indeed, a *work-hardening* or *strain-hardening* body is defined by the assumption of a strictly positive value of the rate of plastic strain work, i.e.,

$$d\sigma_{ij}\dot{e}^p_{ij} > 0. \tag{8.14}$$

In more vivid words, once entered in the plastic regime, an extra loading increases the yield limit as if it were defined by the point at which final loading was achieved. The concept of yield function can therefore be replaced by that of "loading function" that is "associated with the consecutive stages of increasing plastic deformation" (Freudenthal's words). Usually a single scalar parameter β is introduced to characterize this effect. It can be the plastic-strain work defined by

$$\beta = W_p := \int \sigma_{ij} de^p_{ij} = \int_0^t \Phi_{anel} dt, \tag{8.15}$$

or an invariant cumulated plastic strain defined by (Odqvist 1963, 1966)

$$\beta = \bar{e}^p := \int_0^t \left(\frac{2}{3}\dot{e}^p_{ij}\dot{e}^p_{ij}\right)^{1/2} dt. \tag{8.16}$$

In both cases, the variable β accounts for the time-integrated past behaviour of the material up to the actual point of loading, and $f(\sigma_{ij}) = 0$ is replaced by an expression of the type $f(\sigma_{ij}) - g(\beta) = 0$. This results in a growth of the elasticity domain as loading increases, hence the pleasant property enjoyed by the material from an industrial viewpoint: its elastic limit can be improved with further loading. Another possibility is that the yield surface is displaced by a shift in its origin as in an expression of the type

$$f(\sigma_{ij} - a_{ij}) = 0, \tag{8.17}$$

where the tensor of component a_{ij} may be a simple expression of the actual plastic strain itself. This is called "kinematic hardening" (cf. Prager 1949—further expanded by H. Ziegler, also Mroz 1967) and $a_{ij} = Ce^p_{ij}$ is a typical *backstress* with a positive constant C that is characteristic of the considered material. Such a scheme may allow, at least qualitatively, a representation of the Bauschinger effect. Both isotropic and kinematic hardenings may be combined. Combined with loading path and pre-stress, temperature may also have a non-negligible effect on the yield surface, as clearly exhibited in the curves provided by Phillips (1968) and Phillips and Tang (1972)—and reproduced in Maugin (1992a, pp. 16–17).

8.3.3 Rate Equations

A typical rate equation for the plastic strain in perfect plasticity may be written as the normality relation

$$\dot{e}_{ij}^{p} = \dot{\lambda}\frac{\partial f}{\partial s_{ij}}, \tag{8.18}$$

where s_{ij} is the deviatoric part of the stress σ_{ij} (because of incompressibility of the plastic deformation; cf. First working hypothesis of Saint-Venant). Furthermore we have noted $\dot{\lambda}$—as if it were a time rate—the (plastic) multiplier introduced in the right-hand side of (8.18). This is to mean that the plastic process, while obviously happening in time, is not explicitly dependent on the rate at which the loading is applied so that no time scale is involved in (8.18). This equation is sometimes referred to as a model of "*time-independent plasticity*". This is an awkward coinage since a time evolution is involved in (8.18) but any parameter (e.g., an angle of torsion that grows monotonously during loading) can be used as selected time scale. Though, this essential property that is verified for sufficiently low rates of stress loading [as well established by Manjoine (1944) for mild steel] has a dramatic influence when we consider dissipation. An expression such as (8.10) will then necessarily be homogeneous of *order one* in the strain rate, while in classical irreversible processes this expression would be homogeneous of order two (like the famous Rayleigh dissipation function). When the response of the material is sensitive to the rate of loading—as happens for very malleable materials—it means that some viscosity starts to be felt in the system and we have to envisage a mixed *visco-plastic* form of deformation.

It is Melan (1938) who introduced the notion of *plastic potential* $\chi(s_{ij})$ such that (8.18) may be rewritten as

$$\dot{e}_{ij}^{p} = \frac{\partial \chi}{\partial s_{ij}}. \tag{8.19}$$

This is sometimes referred to as Mises' theory of plastic potential after a work of von Mises (1928)—cited by Freudenthal and Geiringer (1955, p. 326). We shall return to (8.19) in Sect. 8.4.

8.3.4 Prandtl-Reuss Relations

This is the small-strain case in which the elastic part satisfies standard linear elasticity and plastic deformation evolution is such that—cf. Eq. (8.18)—it is simply proportional to s_{ij}. Then we can write for the total strain rate in elasto-plasticity in shear:

$$\dot{e}_{ij} = \dot{e}^e_{ij} + \dot{e}^p_{ij} = S_{ijkl}\dot{\sigma}_{kl} + \dot{\lambda} s_{ij}, \tag{8.20}$$

where S_{ijkl} is the tensor of elastic compliances (inverse of the elasticity tensor). This was proposed by **Prandtl** (1924) for plane problems[3] and **Reuss** (1930) in the general case. Such a form where S_{ijkl} reduces to $(2\mu)^{-1}$—where μ is the shear modulus—in the case of isotropy allows relatively easily solutions of elasto-plasticity problems.

Equation (8.20) can formally be re-written as

$$\dot{\mathbf{e}} = (\mathbf{S} + \mathbf{H})\dot{\mathbf{s}}, \tag{8.21}$$

where $\mathbf{H}$ is a function of the stress. In the absence of elastic contribution, this reduces to the Lévy-Mises relations. But if *all* strains are small, one is tempted to use directly the strain, rather than the strain rate, and stress, instead of the stress rate. Thus, the equations proposed by Hencky (1924) and Nadai (1937) mimic (8.21) simply by

$$\mathbf{e} = (\mathbf{S} + 2\varphi)\mathbf{s}, \tag{8.22}$$

where φ is essentially positive during continued loading and zero during unloading. This simplified relationship was used by Nadai, and many Russian engineers—including Ilyushin (1943, 1948, 1954)—in the 1940s and 1950s. This is called the *deformation theory* of elastoplasticity (or Hencky-Nadai theory). In the 1D case this is like replacing μ by the secant modulus $\bar{\mu}$ of a nonlinear stress-strain relation. As a trivial remark, there is no need to emphasize that replacing the derivative of a function, e.g., dy/dx, by the ratio of the function itself to the argument, y/x, is strictly valid when y/x is a constant! More subtly, the replacement of (8.21) by (8.22) is theoretically possible only if the latter can be obtained from the former by a *path-independent integration*. This is the case in so-called proportional (i.e., when the stress components increase during each test proportionally to a single parameter) or "radial" loading with fixed stress axes and principal stress ratios, or for segment-wise plane yield surface, as emphasized by Freudenthal and Geiringer (1958, pp. 281, 287). Still this simplified but disputable theory, considered as an expedient, greatly facilitates the solution of boundary-value problems. This explains its acceptance and success in a time when there were no computers. It is now considered obsolete and mentioned only for the sake of curiosity.

[3]Prandtl had one of his students to work out 2D problems (H. Herbert, Thesis, Göttingen, 1909; published in the first issue of ZAMM, vol.1. p. 15, 1921, a journal founded by Prandtl and von Mises in 1921).

8.3.5 Hypo-elasticity as a Path to Elasto-Plasticity

Inspired by the work of the Polish mechanician Stanislaw. Zaremba (1903), **Clifford A. Truesdell** (1919–2000) proposed in Truesdell (1955) to postulate a relation between the rate of strain D_{ij} and the rate of stress in the general form

$$\frac{d}{dt}\sigma_{ij} = f(D_{ij}) \ or \ \frac{d}{dt}\sigma_{ij} = f(D_{ij}, \sigma_{ij}) \tag{8.23}$$

depending on whether the response is independent of or dependent on the stress field itself. This equation, for finite deformation, defines a "*hypo-elastic*" medium as a body which has no preferred state of strain or stress and thus permits the existence of an arbitrary stressed initial state.

The integration over time of (8.23) to obtain a stress-strain relation should be performed along the loading path and over deformation. But in finite strain this can produce a large variety of non-linear responses. Among these, we may have strain-hardening and strain-softening as possibilities. This hints at the possible existence of a yield limit in certain asymptotic conditions. This is the argument that led to a possible approach to elastoplasticity via hypo-elasticity as a "rational" deduction in the Truesdell-Coleman-Noll sense. There were lengthy discussions about the objective time (Jaumann co-rotational or Oldroyd convected) derivative to write a properly invariant form of (8.23), a discussion that will recur with the writing down of evolution equations in finite-strain elasto-plasticity. Hypo-elasticity, as advertised by Truesdell, was fashionable in the late 1950s and early 1960s, so much so that Eringen (1962) in one of the first books published on modern continuum mechanics devoted a large number of pages to the subject, something that disappeared altogether from the actualized book of Eringen of 1967. The same is true of Prager's (1961; Chap. 8) book that deals at length with hypoelasticity. Retrospectively, in spite of some support from hard-line Truesdellians—including in Japan with Tatsuo Tokuoka (1929–1985)—, we think that this path to plasticity was doomed. One reason for this was that the theory involved some elastic instability and does not acknowledge that yield may indeed starts with rather small strain, and thus leading to "stress-strain relations that resemble in shape though not in scale the deformation conditions of yielding" (Freudenthal and Geiringer 1958, pp. 262–263). To these much knowledgeable authors, hypo-elasticity is unrelated to real plasticity. We have to agree with them.

8.4 Mathematical Plasticity and Convexity

8.4.1 Variational Principles

Mathematics was obviously involved in some form in previously exposed developments. But we call "mathematical plasticity" those developments that started essentially with **Ernst Melan** (1890–1963) in Austria, **Rodney M. Hill** (1921–2011) in the UK, **William Prager** (1903–1980) initially in Germany and his colleagues at Brown University (USA), and also **Alexey A. Ilyushin** (1911–1995) in Moscow and **Lazar M. Kachanov** (1914–1993) in Leningrad/St Petersburg (cf. Kachanov 1960, Golovin 2008), and made extensive use of energy arguments, variational formulations (inequalities), and mathematical properties such as convexity. This format is reflected in our own textbook (Maugin 1992a).

Melan (1938) was instrumental in introducing the notion of plastic potential—cf. Equation (8.19)—and elements of shakedown theory (formation of a limit stress-strain path under cyclically imposed strain). Hill (1948, 1950) brought a fresh view by introducing seminal ideas while providing a remarkable synthetic approach for the period. The most powerful idea probably is that of *maximum plastic work* (Hill 1948) now classically referred to as the principle of *maximal dissipation* (the French call it the Hill-Mandel maximal-dissipation principle). In modern form this principle can be expressed by the following *variational inequality*:

$$(\sigma - \sigma^*) : \dot{\mathbf{e}}^p \geq 0, \quad \forall \sigma^* \in C, \tag{8.24}$$

where σ represents the stress tensor whose representative point in the appropriate space should remain in a convex set C, and $\dot{\mathbf{e}}^p$ is the rate of plastic strain. Equation (8.24) is a so-called variational inequality—since $\sigma*$ is at our disposal—what, in essence, means that plastic dissipation possibly—but not necessarily—occurs only when the stress has reached the yield surface. The important feature of Eq. (8.24) is that together with the convexity of the strain energy, it guarantees the validity of Drucker's inequality and Drucker's postulate (see below). In other words, it provides a solid foundation for the plasticity of materials that accept the normality law contained in (8.24). Such works eventually led to general studies of uniqueness and stability in nonlinear continuum mechanics. This makes Rodney Hill the father of modern plasticity theory in a thermodynamic context. This was expanded by Hill in his book of 1950, published while he was just reaching age twenty nine. This book remains *the* classic opus and indispensable reference in plasticity theory. It was written with an economy of thought and words typical of Hill. The book also generously presented the treatment of typical plasticity problems by means of the theory of slip lines, a then recent introduction for solutions of plasticity problems before the advent of computers. This applies in quasi-static loadings to plane strain problems in rigid-plastic bodies. Hill gave such a solution for the exemplary problem of a rigid punch indenting a rigid-plastic half-space. With the publication of his book at the early age of 29, Hill was recognized as a leading authority in the field.

At about the same time Prager (1949) was introducing a format of plasticity with *kinematic hardening* (plasticity surface moving with the evolving state of plastic strains), the notion of locking materials (i.e., materials exhibiting a saturation in strain; cf. Prager 1957)), and publishing deeply thought books (Prager 1955, 1961; Prager and Hodge 1954) in the field of plasticity and general continuum mechanics. Morever, after his move to Brown University, Prager became the driving force behind the expansion of a true school of elasto-plasticity, that place becoming a kind of "Mecca" for related studies; Brown indeed almost became the centre of the World for studies on elasticity and plasticity starting in the period 1940s–1960s. The following list of professors and PhD students at some time at Brown speaks for itself, looking like a real "dream team": W. Prager, D.C. Drucker, R.S. Rivlin, P.S. Symonds, E. Sternberg, J. Kestin, J.R. Rice, J.H. Weiner, L.B. Freund, R.J. Clifton, A. Needleman, B. Budiansky (Ph.D 1950), and in applied mathematics dealing with problems of continuum mechanics M.E. Gurtin (Ph.D 1961) and C. M. Dafermos. Many foreign visitors came to Brown to get acquainted with the then most recent developments in plasticity (among them, Paul Germain from France in 1952–1953, and Hans Ziegler from Switzerland).

Among successful advances made in Brown we must cite the work of **Daniel C. Drucker** (1918–2001) who established the so-called *Drucker inequality* (non negative product of stress rate and plastic strain rate; Drucker 1951), i.e.,

$$\dot{\sigma}_{ji}\dot{e}^{p}_{ij} \geq 0, \tag{8.25}$$

and "Drucker's stability postulate" in the self explanatory form (Drucker et al. 1952)

$$W = \int_{0}^{t} \left(\sigma_{ji} - \sigma^{0}_{ji}\right) \dot{e}^{p}_{ij} dt \geq 0, \tag{8.26}$$

where σ^0 is the original state of stresses and $[0, t]$ is a closed time-cycle of loading and unloading. Equation (8.25) reminds us of the Duhem inequality (8.9): it is the expression of a *local stability*, while (8.26) is more in the spirit of Hill's inequality (8.24). **H. G. Greenberg**, also at Brown, was one of those who proposed in Greenberg (1949) a variational formulation of plasticity (minimum principle exploiting a convexity argument). This was improved by **Budiansky** and Pearson (1956/57)—minimization of the total potential energy written in terms of velocities among virtual velocities compatible with kinematic conditions at the boundary. This was complemented by the Hodge and Prager (1949) principle that corresponds to a minimization of the total complementary energy among plastically and statically admissible stresses (P. G. Hodge, 1920–2014). **Symonds** (1951) introduced the ingenious notion of *plastic hinges* that allows one to treat the collapse of truss structures, and set forth the elements of shake down (or limit) analysis (cf. König 1987, Melan further up). **James R. Rice** (born 1940) produced his famous works on path-independent integrals (Rice 1968) and the thermodynamics of plasticity (Rice 1970, 1971), while **Joseph Kestin** (1993–1993) became the most acute observer and critic of the thermo-mechanics of continua (see his celebrated treatise on thermodynamics: Kestin 1966) including anelasticity described by means of internal variables of state. **J.H. Weiner** expanded the statistical theory of elasticity (see his wonderful book of 1983). **L.B. Freund** produced his theoretical works on dynamic fracture, while **R.J. Clifton** complemented theoretical approaches by performing landmark experiments in the same.

In all this was a disproportionately important contribution of Brown University to engineering and continuum mechanics while other relevant studies were conducted in other US places like Brooklyn Polytechnic, the Illinois Institute of Technology, Columbia, and Stanford University. However, some similar progress was sometimes attained in the Soviet Union in spite of the difficulties of the period. Here we must cite **Kachanov** (1942) who proposed a variational principle for elasto-plastic materials. His book (Kachanov 1956) became a classic of plasticity theory as also his successful book of 1974. In Moscow, **Ilyushin** contributed tho elasto-plasticity as soon as the mid 1930s. During World War II, he proposed (Ilyushin 1943) his celebrated method of iterative solutions. This method provided the basis of many numerical works. Somewhat in competition with Hill's book, Ilyushin published in 1948 an innovative book on plasticity with a new version as Ilyushin (1963). But in Ilyushin (1961, 1963) he proposed the so-called "Ilyushin's postulate". This belongs in the flourishing period when minimum (or maximum) principles started to play a fundamental role, directly reflecting the thermodynamic irreversibility of the evolution of the plasticity phenomenon. This postulate states that for any strain cycle $\mathbf{e}(t)$, $t \in [0, 1]$ with $\mathbf{e}(0) = \mathbf{e}(1)$ in small strains and stress σ, the strain power is positive or zero:

$$\int_0^1 \sigma_{ij}\dot{e}_{ji}(t)dt \geq 0. \tag{8.27}$$

This applies to materials with hardening. This Eq. (8.27) can be viewed as a *global stability criterion* which says that any closed response loop in a strain (abscissa)-stress (ordinate) diagram is always followed *clockwise*. It can be compared to Drucker's inequality (8.25) where $\mathbf{e}$ is reduced to the plastic strain $\mathbf{e}^p$. It is clear that Drucker's postulate (8.25), Ilysushin's postulate (8.27), the maximal dissipation rule of Hill (8.24), and Duhem's condition (8.8) belong in the same class of mathematical reasoning based on the second law of thermodynamics.

8.4.2 Application of Convex Analysis

The period of the 1950s–1970s saw the introduction of much functional analysis (energy norm, Sobolev spaces) in continuum mechanics as illustrated by the book of Duvaut and Lions (1972) among others. Furthermore, convex analysis [e.g., Rockafellar (1970), also the works of Moreau (1966, 1971) in France], the study of maximal monotonous operators (e.g., Brézis 1973) and nonlinear programming (e.g., Luenberger 1973, 1984) provided the required tools to make theoretical advances in the mathematics of somewhat critical mechanical behaviours (friction, plasticity, hysteresis), in fact the "nonsensical branches of mechanics"of Duhem (1903). This was beautifully implemented in the Thesis work of Nguyen Quoc Son (1973) where he combined these mathematical tools with the thermomechanics with *internal variables of state* [of which the anelastic strain is one example and the scalar quantities defined in Eqs. (8.15) and (8.16) above are other examples]. We exposed Nguyen Quoc Son's theory in our textbook (Maugin 1992a). We remember that the strain-rate independent plasticity can be expressed in a somewhat abstract form (A: driving force; α: variable) as

$$\dot{\alpha} = \dot{\lambda}\frac{\partial f}{\partial A}, \tag{8.28}$$

where $\dot{\lambda}$ is a so-called plastic multiplier and $f(A) = 0$ is the equation of the elasticity surface delimiting the convex domain C in A space. Here we have

$$\dot{\lambda} \geq 0 \ \ if \ \ f = 0 \ \text{ and } \ \dot{f} = 0, \tag{8.29a}$$

$$\dot{\lambda} = 0 \ \ if \ \ f < 0, \text{or } \ f = 0 \ \text{ and } \ \dot{f} < 0. \tag{8.29b}$$

This can also be expressed in the formalism of *convex analysis* as

$$\dot{\alpha} \in N_C(A) \Rightarrow \dot{\alpha} \cdot (A - A^*) \geq 0, \quad \forall A^* \in C. \tag{8.30}$$

The first of these reads: "the time rate of α belongs to the cone of outward normals to the convex set C". This is a *law of normality* which dictates the direction of the time evolution once the elasticity limit is reached. With $\alpha = \mathbf{e}^p$ and $A = \sigma$ we recognize the usual case of plasticity. This general formulation allows for the existence of angular or apex points along the surface (or limiting curve in a plane representation) of C. This is the case of Tresca's criterion of plasticity. The second of (8.30) is none other than (8.24) in elastoplasticity. If there exists a potential of dissipation Φ in the manner of Rayleigh (but expressed in terms of the force A, hence sometimes called a *pseudo-potential*), we will have the required evolution equation for α in the form [compare to Melan's form (8.19)]

$$\dot{\alpha} = \frac{\partial \Phi}{\partial A}. \tag{8.31}$$

Contrary to Rayleigh's dissipation potential, we may select the degree of homogeneity of the function Φ at our convenience in so far as the dissipation inequality is satisfied. This is what allows one to construct a good theory of plasticity (even with hardening) when the dissipation potential is homogeneous of degree one only—something which surprises to the utmost tenants of the standard theory of linear irreversible processes. This possibility allows for a strain-rate independent plasticity, a valid approximation for many materials. In fact, it is the theory of so-called *generalized standard materials* due to Halphen and Nguyen Quoc Son (1975) that is applied where the existence of a *convex* dissipation (pseudo) potential is assumed together with the *convexity* of the energy density. In the framework delineated by these authors, some of the essential above-recalled mathematical theorems of plasticity theory such as Drucker's inequality, Drucker's stability postulate, and Ilyushin's postulate follow almost automatically (cf. Maugin 1992a, Chaps. 5 and 6). The theory of fracture itself can be put in a similar general abstract framework. In contrast to the functional theory embedded in the rational thermodynamics of B. D. Coleman and W. Noll, the theory with internal variables of state as applied above yields a mathematical problem (in time) in terms of *evolution* (rate) *equations*. This is much more adapted to numerical computations as it directly fits techniques and a good mathematical frame known in other fields such as nonlinear programming. This explains the remarkable efficacy and success met by this theory within the last thirty years.

8.4.3 Uniqueness and Existence of Solutions

In 1D the uniqueness in response in stress is obvious for perfectly plastic bodies, while plastic flows indicates the non-uniqueness in the strain response. In 3D one needs to apply a standard energy argument to establish the uniqueness in stress—cf. Maugin 1992a, pp. 76–80. For a work-(positive) hardening material, uniqueness can be proved for both stress and time rates (cf. Maugin 1992a, p. 114). The matter of *existence* for perfect plastic solutions in displacement calls for a higher of difficulty. This comes from the fact that solutions often exhibit surfaces of discontinuity. Several authors have contributed to a solution of this problem, among them C. Johnson (1976), Roger Temam and Gilbert Strang (1980)—also Temam (1988) —, and above all **Pierre Suquet** (1979, 1982). The clue lies in the use of the space of functions of *bounded variation* (BD functional spaces) as this allows the representation of the observed discontinuities (slip surfaces, plastic hinges).

8.5 Physical Plasticity and Dislocations

Plasticity is one example of an exploitable continuum phenomenological theory that could be successfully expanded before the real physical mechanisms for its existence were uncovered.

After plastic deformation slip bands are usually observed at the previously polished surfaces of single crystals. It is **Geoffrey I. Taylor** (1886–1975) who in Taylor (1934) first proposed an explanation of this observed effect by the sliding mechanism by *crystal defects*, which he identified with the mathematically conceived *dislocations* of Vito Volterra (1907). This idea was practically advanced simultaneously by **Michael Polanyi** (1891–1976)—originally from Hungary, but with a chair of physical chemistry at the University of Manchester after he left

Germany in 1933—and **Egon Orowan** (1902–1989)—also originally from Hungary and with a Ph.D obtained in Berlin who moved to the UK after working for sometime in Germany and Hungary, first at the University of Birmingham and then Cambridge before finally moving to the MIT in the USA in 1950. But Taylor is a scientist of another calibre. A Cambridgian like many others, he is known above all for his outstanding contributions to dynamical meteorology and fluid mechanics, especially in turbulent flows (Rayleigh-Taylor instability and Taylor vortices). He was a man of many scientific interests so that his excursion in a field that belongs to the mechanics of materials—unknown to most fluid dynamicists—should not come as a surprise.

The insight of these three scientists is all the more remarkable that the experimental proof of the existence of dislocations as individual objects had to await the 1950s after the invention of electron microscopy. But the time evolution of dislocations indeed is the basic mechanism by which plastic flow takes place. Dislocations and disclinations are structural defects, i.e., they are the manifestation of distortions in the perfect (periodic) ordering prevailing in a crystalline structure. But this matter must be examined in both discrete-crystal and continuous pictures. The mathematical notion was introduced by **Vito Volterra** (1907)—without this naming—as manifestation of a discontinuity in elastic displacement and was illustrated by an astute thought experiment ("cut, displace, and glue back"). This creates a disturbance of the mechanical fields in the neighbourhood of this discontinuity. However, it is in the discrete approach that a more vivid picture can be obtained. For instance, for a so-called *edge* dislocation, it is the edge of the extra half-plane of atoms introduced in a regular crystal lattice which is called the *edge* dislocation. More generally, if we go along a circuit around the dislocation line of unit tangent $\boldsymbol{\tau}$, one observes a lack of closure that is a finite number of the basis vectors of the lattice. This is called the *Burgers vector* usually noted $\mathbf{b}$. This vector is parallel to $\boldsymbol{\tau}$ for a *screw* dislocation and perpendicular to $\boldsymbol{\tau}$ for an *edge* dislocation. In practice many dislocations are mixed (see, e.g., Friedel 1964) when the line direction and the Burgers vector are at any angle. As indicated above, the relationship of the notion of *dislocation* (name given by Taylor) and the phenomenon of *plasticity* was established almost simultaneously by Taylor, Polanyi and Orowan. They proposed that *shear* could be caused by the propagation of elementary linear defects they called *dislocations*.

Dislocations are observed to move under the influence of a change of the local state of stresses caused by the application of external loads. Dislocations can move by *slip* (in the plane formed by the Burgers vector and the line tangent) or by *climb* (motion outside the slip plane). Their motion propagates the plastic deformation, i.e., they ease the ductility of materials. They can interact between them, but also with the lattice and point defects. In the case of the lattice that is periodic the displacement of a dislocation needs to overcome energy barriers, a phenomenon akin to friction. The corresponding friction force is called the *Peierls-Nabarro force*. In the case of mobile point defects (foreign atoms, impurities), dislocations attract them to form so-called *Cottrell clouds*. This hinders the motion of dislocations. In turn this explains why pure metals are more ductile than alloys. When a dislocation is strongly pinned on immobile atoms, one observes a curving of the dislocation line ultimately forming a free circular dislocation and thereby a multiplication of dislocations. This is the *mechanism of Frank-Read*. The *Portevin-Lechatelier phenomenon*, observed as oscillations on a traction curve, is related to the successive pinning and unpinning of slowly moving dislocations on mobile atoms at different sites. Dislocations can also interact with precipitates. In mathematical terms dislocations can be viewed as topological defects akin to *solitons*.

Disclinations in principle are related to a discontinuity in a rotation angle instead of a displacement. They are particularly frequent in liquid crystals.

The main relationship between the description of dislocations and plastic deformation is provided by the notion of *resolved shear stress*. This is the appropriate projection of the stress onto the two characteristic directions associated with a dislocation, namely the unit normal $\mathbf{m}$ to the family of planes where slip takes place, and the slip direction of unit vector $\mathbf{s}$ within such a plane, so that $\mathbf{m}^{(k)} \cdot \mathbf{s}^{(k)} = 0$ for each slip system k in the crystal. With $\gamma^{(k)}$ the amount of slip for this system, one obtain for the plastic strain when there exists a finite number N of slip systems in the crystal, the so-called *Taylor's fundamental equation* (Taylor 1938)

$$\dot{\mathbf{e}}^p = \sum_{k=1}^{N} \dot{\gamma}^{(k)} \left(\mathbf{s}^{(k)} \otimes \mathbf{m}^{(k)} \right)_S. \tag{8.32}$$

***Schmid*'s law** (Schmid 1924; Schmid and Boas 1935) is an empirical yield criterion applied to the resolved shear stress

$$\tau_S^{(k)} := \underline{\sigma} : \left(\mathbf{s}^{(k)} \otimes \mathbf{m}^{(k)} \right)_S. \tag{8.33}$$

It reads

$$\left| \tau_S^{(k)} \right| \leq \tau_c^{(k)}, \tag{8.34}$$

where $\tau_c^{(k)}$ is a critical value of activation for the slip system k. This criterion is of the Coulomb or Tresca type (in 1D). The overall situation with crystal plasticity in the mid 1950s was examined by Seeger (1958) in his *Handbuch* article; for a more recent scientifically accurate conspectus, see Havner (1992).

Before closing this section, we note that the notion of *geometrically necessary dislocations* (for short GNDs) is of importance. Indeed, if one envisages the bending of a regular lattice, this can be realized only if dislocations are introduced to justify the angle difference between the original atomic planes. These are GNDs. This lattice curvature involves a gradient of strain, and will ultimately yield a necessary relationship between GNDs, plasticity and a strain-gradient theory. The basic relation here is that originally introduced by Nye (1953) between lattice curvature and the density tensor of dislocations. It plays a crucial role in the gradient-theory of plasticity (see below).

8.6 Finite-Strain Plasticity

Despite the writing of Eq. (8.32) in small strains, crystal plasticity is the true realm of finite-strain plasticity. This is also the case of applications and computations related to technologies that involve large deformations such as metal forming, etc. There was thus need to examine this matter more closely, so that this occupied the mind of many bright mechanicians of the continuum in the period 1970–2000. But then one has to come to grips with various matters. One is the formulation of deformation theory in such a framework. The second is the difficulty in selecting the most appropriate relevant internal variables, in particular the so-called "plastic deformation" (as emphasized by Casey and Naghdi 1992). The third is the problem of constructing invariant (i.e., objective) rate laws. Concerning the first point, in the logic of analysis, the additive decomposition of small-strain theory has to be replaced by a *multiplicative* decomposition. One such decomposition may be traced back to Bilby et al. (1955, 1957), but it was made popular later on by Lee (1969) and his co-workers. With a general deformation $\mathbf{x} = \bar{\mathbf{x}}(\mathbf{X}, t)$ characterized by the tangent map $\mathbf{F}$, a material gradient, $\mathbf{F} = \partial\bar{\mathbf{x}}/\partial\mathbf{X}|_t = \nabla_R\bar{\mathbf{x}}$, between the reference configuration K_R (of material point $\mathbf{X}$) and the current configuration K_t of placement $\mathbf{x}$, if there exist *elastic* (thermodynamically recoverable) finite deformations and *anelastic* (dissipative) deformations, they should compose the total, and only true, gradient $\mathbf{F}$ in a multiplicative manner. That is, with an obvious notation,

$$\mathbf{F} = \mathbf{F}^e\mathbf{F}^a, \tag{8.35}$$

where none of the contributors $\mathbf{F}^e$ and $\mathbf{F}^a$ of the decomposition is a true gradient. It can be said that $\mathbf{F}^a$ defines from K_R a so-called *intermediate* or *elastically released configuration* K_{relax}, as it is also obtained from K_t by applying the inverse elastic deformation $(\mathbf{F}^e)^{-1}$. Then it is clear that the decomposition (8.35) is invariant by rotations $\mathbf{Q} \in SO(3)$ of K_{relax} as we can write

$$\mathbf{F}^e\mathbf{F}^a = \hat{\mathbf{F}}^e\hat{\mathbf{F}}^a, \quad \hat{\mathbf{F}}^e = \mathbf{F}^e\mathbf{Q},\ \hat{\mathbf{F}}^a = \mathbf{Q}^T\mathbf{F}^a, \tag{8.36}$$

with

$$\mathbf{Q}\mathbf{Q}^T = \mathbf{Q}^T\mathbf{Q} = \mathbf{1}, \quad \mathbf{Q}^T = \mathbf{Q}^{-1},\ det\,\mathbf{Q} = \pm 1. \tag{8.37}$$

In practice a definite orientation can be granted to K_{relax} through a particular crystal lattice (for this, see Mandel 1971, 1982). Although the order of the decomposition (8.35) seems quite natural, it was seriously questioned by several authors (e.g., Clifton 1972; Casey and Naghdi 1980) and an additive decomposition was also proposed involving directly finite symmetric elastic and anelastic strain tensors (Green and Naghdi 1965) or an additive decomposition of the gradient $\mathbf{H} = \mathbf{F} - \mathbf{1}$ (Nemat-Nasser 1979). We retain the more commonly accepted decomposition (8.35). For the constitutive behaviour, we note that only $\mathbf{F}^e$ is observable by elastic unloading, we may consider an energy per unit volume of the reference configuration in the form

$$W = \bar{W}(\mathbf{F}^e, \alpha, \theta), \tag{8.38}$$

and define entropy and associated internal forces by

$$S = -\frac{\partial \bar{W}}{\partial \theta}, \quad \mathbf{T}^e := \frac{\partial \bar{W}}{\partial \mathbf{F}^e}, \quad A = -\frac{\partial \bar{W}}{\partial \alpha}. \tag{8.39}$$

Then we look for a convenient and suggestive expression for the residual dissipation inequality. To reach such an expression first note that

$$\dot{\mathbf{F}}^e = \dot{\mathbf{F}} \cdot (\mathbf{F}^a)^{-1} - \mathbf{F}^e \cdot \dot{\mathbf{F}}^a \cdot (\mathbf{F}^a)^{-1}, \tag{8.40}$$

obtained by computing $\dot{\mathbf{F}}$ on the basis of (8.35) and applying $(\mathbf{F}^a)^{-1}$ to the right to the result. Furthermore, setting

$$\bar{\mathbf{T}}^v := \mathbf{F}^a \cdot \mathbf{T} - \mathbf{T}^e, \bar{\mathbf{T}} := \mathbf{T} \cdot \mathbf{F}, \tag{8.41}$$

it is readily checked from the second law of thermodynamics that we obtain the following residual dissipation inequality

$$\Phi = \Phi_{intr} + \Phi_{th} \geq 0, \tag*{(842)}$$

with

$$\Phi_{intr} = tr\left\{\bar{\mathbf{T}}^v \cdot \dot{\mathbf{F}}^e + \bar{\mathbf{T}} \cdot \dot{\mathbf{F}}^a\right\} + A\,\dot{\alpha}, \Phi_{th} = -(\mathbf{Q}/\theta).\nabla_R \theta. \tag{8.43}$$

Then we note that the original first Piola-Kirchhoff stress is recovered by computing

$$\mathbf{T} = (\mathbf{F}^a)^{-1} \cdot \left(\bar{\mathbf{T}}^e + \bar{\mathbf{T}}^v\right) = \mathbf{T}^e + \mathbf{T}^a, \tag{8.44}$$

wherein

$$\mathbf{T}^e = (\mathbf{F}^a)^{-1} \cdot (\partial \bar{W}/\partial \mathbf{F}^e), \mathbf{T}^v = (\mathbf{F}^a)^{-1} \cdot \bar{\mathbf{T}}^v. \tag{8.45}$$

The first of (8.43) is of interest because it clearly separates the dissipative effects related to $\dot{\mathbf{F}}^e$ and $\dot{\mathbf{F}}^a$ (cf. Maugin 1994). Accordingly, the first is often related to pure viscous processes, hence the notation for $\mathbf{T}^v$ and the second to pure plasticity effects, so that the superscript a will now be replaced by p. To proceed further we discard the "viscous" type of dissipation so that the first of (8.44) yields

$$\mathbf{T} = -(\mathbf{F}^p)^{-1} \cdot \bar{\mathbf{T}}^e = (\mathbf{F}^p)^{-1} \cdot (\partial \bar{W} / \partial \mathbf{F}^e). \tag{8.46}$$

It is salient to introduce strain measures in the configuration K_{relax} by

$$\mathbf{C}^e := (\mathbf{F}^e)^T \mathbf{F}^e, \quad \mathbf{E}^e = \frac{1}{2}(\mathbf{C}^e - 1_{relax}), \tag{8.47}$$

that are true covariant symmetric tensors in the configuration K_{relax} equipped with unit $\mathbf{1}_{relax}$.

We introduce the Jacobian determinants

$$J^e = det\, \mathbf{F}^e, \quad J^p = det\, \mathbf{F}^p, \quad J_F = J^e J^p > 0. \tag{8.48}$$

The following *plastic strain rates* are also useful:

$$\mathbf{D}^p = \left\{ \mathbf{F}^e \cdot \mathbf{L}^p \cdot (\mathbf{F}^e)^{-T} \right\}_S = (\mathbf{D}^p)^T \text{ in } K_t, \mathbf{L}^p = \dot{\mathbf{F}}^p (\mathbf{F}^p)^{-1}, \tag{8.49}$$

and

$$\mathbf{D}^p_{relax} = \{ \mathbf{C}^e \cdot \mathbf{L}^p \}_S = \left(\mathbf{D}^p_{relax} \right)^T \text{ in } K_{relax}, \tag{8.50}$$

while the second Piola-Kirchhoff stress relative to K_{relax} reads as

$$\mathbf{S}_{relax} = J^e (\mathbf{F}^e)^{-1} \cdot \sigma \cdot (\mathbf{F}^e)^{-T}. \tag{8.51}$$

Then the *intrinsic* dissipation per unit volume reads

$$\Phi_{intr} = tr(\sigma \cdot \mathbf{D}^p) + J_F^{-1} A\, \dot{\alpha} \text{ in } K_t, \tag{8.52}$$

and

$$(\Phi_{intr})_{relax} = tr\left(\mathbf{S}_{relax} . \mathbf{D}^p_{relax} \right) + \tilde{A}\, \dot{\alpha} \text{ in } K_{relax}, \tag{8.53}$$

together with the energy (per unit volume of K_{relax})

$$W_{relax} = \tilde{W}(\mathbf{E}^e, \alpha, \theta), \tag{8.54}$$

and the laws of state

$$\mathbf{S}_{relax} = \frac{\partial \tilde{W}}{\partial \mathbf{E}^e}, \quad \tilde{A} = -\frac{\partial \tilde{W}}{\partial \alpha}, \quad S_{relax} = -\frac{\partial \tilde{W}}{\partial \theta}. \tag{8.55}$$

Here $\mathbf{E}^e$ is the *observable* mechanical variable of state.

A *finite-strain rate-independent theory of elastoplasticity* is now completed by assuming the existence of a *yield* (hyper-) *surface* $f\left(\mathbf{S}_{relax}, \tilde{A}\right) = 0$ bounding a convex set C in $\left(\mathbf{S}_{relax},\tilde{A}\right)$ space, and considering the following plastic evolution equations

$$\mathbf{D}^{p}_{relax}=\dot{\lambda}\frac{\partial f}{\partial \mathbf{S}_{relax}},\quad \dot{\alpha}=\dot{\lambda}\frac{\partial f}{\partial \tilde{A}}, \tag{8.56}$$

where $\dot{\lambda}\geq 0$, λ being the plastic multiplier. This is the theory exposed in Maugin (1994, 2011, Chap. 2). First we note that the intrinsic dissipation due to stresses in the elastically released configuration is given by

$$tr\left(\mathbf{S}_{relax}\cdot\mathbf{D}^{p}_{relax}\right)=tr\left(\mathbf{M}_{relax}\cdot\mathbf{L}^{p}\right),\quad \mathbf{M}_{relax}:=\mathbf{S}_{relax}\cdot\mathbf{C}^{e}, \tag{8.57}$$

where $\mathbf{M}_{relax}$ is the Mandel stress relative to the configuration K_{relax}. This makes some say that the *Mandel stress is the driving force of plasticity* (cf. Mandel 1971). Second, the Taylor Eq. (8.32) of crystal plasticity is now written as (cf. Bertram 2005, p. 300)

$$\mathbf{L}^{p}=\dot{\mathbf{F}}^{p}\left(\mathbf{F}^{p}\right)^{-1}=-\sum_{k=1}^{N}\dot{\gamma}^{(k)}\operatorname{sgn}\left(\tau_{s}^{(k)}\right)\left(\mathbf{s}_{0}^{(k)}\otimes\mathbf{m}_{0}^{(k)}\right), \tag{8.58}$$

where $\mathbf{s}$ and $\mathbf{m}$, with zero subscript, are now expressed in the reference frame. The corresponding plastic dissipation power (8.57) is then estimated as

$$\Phi_{relax}=\mathbf{M}_{relax}\cdot\mathbf{L}^{p}=\sum_{k}\left|\tau_{s}^{(k)}\right|\dot{\gamma}^{(k)} \tag{8.59}$$

that is positive in states of yielding and zero in elastic states.

The exposition just given eschews the problem of objectivity of the relevant strain rate in the rate equation. But there was a much heated discussion about this matter and the related matter of the so-called *plastic spin* in the 1980s (cf. in particular, Dafalias 1983, 1984, 1985), from which it followed that the co-rotational or Jaumann derivative may be the preferred objective time derivative to avoid spurious oscillations in dynamic simulations. Some years before, Mandel (1971) had pondered the case of a yield surface $f(\mathbf{S}_{relax},\alpha,\theta)=0$ where α stands for the hardening variables. He suggested considering an evolution equation in the form

$$\left[(D_{\omega}\mathbf{F}^{p})\left(\mathbf{F}^{p}\right)^{-1}\right]_{S}=\dot{\lambda}\mathbf{B}(\mathbf{S}_{relax},\theta,\mathbf{A}\cdot\beta), \tag{8.60}$$

where

$$D_{\omega}\mathbf{F}^{p}:=\dot{\mathbf{F}}^{p}-\underline{\omega_{D}}\cdot\mathbf{F}^{p},\quad \underline{\omega_{D}}=\underline{\dot{\beta}\beta^{T}},\quad \mathbf{B}=\frac{\partial f}{\partial \mathbf{S}_{relax}}. \tag{8.61}$$

Here $\underline{\omega}_D$ is the angular velocity of a director frame. The latter is attached to the crystal structure in its released configuration. It provides a privileged frame. The elastic deformation then is none other than the lattice deformation. In a polycristal one would take as director frame anyone of the frames attached to the constituent crystals. The model is completed by the following normality law for the internal variables

$$\dot{\alpha} = \dot{\lambda}\frac{\partial f}{\partial \mathbf{A}}. \tag{8.62}$$

More is to be found on various formulations of finite-strain plasticity in the review of Cleja-Ţigoiu and Soós (1990), and the book of Bertram (2005). Of course, such a theory must be considered in the case of propagation of strong discontinuities such as shock waves. Piau (1975) was a pioneer along this line. Note in conclusion that, along a different line, Naghdi (1990) leans in favour of a yield function in strain space **E**.

8.7 Varia

8.7.1 Anisotropy

Yield criteria for *anisotropic* materials were proposed quite early by von Mises (1928) and Hill. Crystal plasticity obviously is a direct domain of application of this notion (cf. Havner 1992). But of greater interest here is the theory of artificially conceived materials such as fibre-reinforced composite materials approximated by homogeneous continua endowed with privileged directions. While the elasticity aspects were discussed in detail by Anthony J.M. Spencer (1972) in Nottingham, the case of plasticity was dully studied by his disciple and co-worker, Tryfan C. Rogers (1988, 1990) with appropriate generalizations of Mises' and Tresca's criteria (see also Maugin 1992a, pp. 304–309)—see Boehler (1988) for the state of research as on 1987. Anisotropic plasticity has been particularly well cultivated in Germany (cf. works by H. Lippmann, J. Betten, O.T. Bruhns, etc.).

8.7.2 Numerical Plasticity

Started modestly with the slip-line solution for plane plasticity (e.g., in Hill 1950; Freudenthal and Geiringer 1958, p. 353 on) and other graphical methods, and Ilyushin's iteration method, numerical plasticity has evolved to reach a very sophisticated technical level with the implementation of the finite-element method (practically created in the mid 1960s) in space—exemplified in the book of Owen (1980)—together with various algorithms for time integration for the evolution problem. This progress benefited from mathematical advances in convex analysis and nonlinear programming. This was particularly worked out by scientists such as J.-J. Moreau, Nguyen Quoc Son, and O. Desbordes in France in the 1970s and 1980s, and by Taylor, Simo and Hughes (e.g., Simo and Hughes 1998) with their so-called "radial return algorithm" (exploitation of constant tangent operators in rate-independent elastoplasticity) in the USA. A more recent book on the subject is by Han and Reddy (2013). Some previous books came from Japan where numerical computations were very active (cf. Washizu 1982; Miyoshi 1985).

8.7.3 Homogenization in Elastoplasticity

In some circumstances there is need for a procedure of *homogenization* (finding an equivalent—in some mathematical sense—homogeneous material to a highly heterogeneous material). This is needed in polycrystals and in elasto-plastic composites. Advances along this line were first achieved by Hill (1965)—and also Mandel (1971). These authors exploited the so-called *Hill-Mandel principle of macro-homogeneity* which established an energy equivalence between the average (over a representative volume element) of the energy of *statically* admissible (micro) stress field in a kinematically admissible (micro) stress field, and the energy of the obtained macro stress and (macro) strain that result from volume homogenization. The case of elastoplastic constituents (without hardening) was expertedly established by Suquet (1985). As to flowing polycrystals, homogenization was constructed by means of a generalization of E. Kröner's self-consistent scheme by Berveiller and Zaoui (1978).

8.7.4 Viscoplasticity

Viscoplasticty is an admixture of viscous and plastic behaviours. While the first involves a characteristic time through viscosity, the second involves the notion of threshold (yield). In vivid terms, we can describe viscoplasticity as the possibility to take an excursion outside the elasticity domain C but then there is a kind of recall to it with the involved characteristic time. A first model of this behaviour was introduced by **Eugene C. Bingham** (1922) when the elastic part is totally absent and replaced by a rigid behaviour. Mathematically, we can write the corresponding evolution equation for (small) plastic strain the following form (this in fact is the Duvaut-Lions formulation, 1972)

$$\dot{\mathbf{e}}^p = \frac{1}{2\eta}(\underline{\sigma} - \Pi_C \underline{\sigma}), \tag{8.63}$$

where $\eta > 0$ is a viscosity coefficient and the symbol Π_C indicates the projection on C. Obviously, when $\underline{\sigma}$ is in C, the projection $\Pi_C\underline{\sigma}$ is $\underline{\sigma}$ itself so that the last factor in (8.63) then vanishes and there remains only the elastic strain (or a rigid behaviour). Perzyna (1966) deduced (8.63) from a dissipation potential in the manner of Eq. (8.31) by writing (on adopting the Duvaut-Lions formulation)

$$\dot{\mathbf{e}}^p = \frac{\partial \Omega}{\partial \underline{\sigma}}; \quad \Omega = \frac{1}{4\eta}\|\underline{\sigma} - \Pi_C\underline{\sigma}\|^2. \tag{8.64}$$

Bingham gave its original model for simple shear only. This we often illustrate with the emerging flow of tooth paste from a squeezed tube. It is Hohenemser and Prager (1932) who formulated the general case—also Ilyushin in 1940 and Oldroy in 1947 as cited by Freundenthal and Geiringer (1958, p. 255). In more recent times, many numerous empirical and semi-empirical flow stress models have been used in computational viscoplasticity. One of the first books fully devoted to viscoplasticity was by Critescu and Suliciu (1982) with a special interest in mathematical properties. As to Micunovic (2009), he gives first hand both theoretical and experimental aspects in a modern formalism. An original viewpoint on viscoplasticity was exposed by Valanis (1971).

8.7.5 *Coupling with Other Properties (Porosity, Damage, Magnetism)*

The influence of pressure was discarded in above-given examples of criteria. Nonetheless, pressure may have a non-negligible effect, so that the mean stress, defined by

$$\sigma_m = \frac{1}{3}(\sigma_1 + \sigma_{\dot{e}} + \sigma_3) = \frac{1}{3}I_1(\underline{\sigma}), \tag{8.65}$$

may intervene in the yield criterion. For instance, generalizing the Huber-Mises and Tresca criteria, we may write

$$I_2(\mathbf{s}) = K(\sigma_m), \tag{8.66}$$

or

$$Sup_{\alpha,\beta}|\sigma_\alpha - \sigma_\beta| = 2K(\sigma_m), \tag{8.67}$$

with

$$K(\sigma_m) = A|\sigma_m|^\gamma, \tag{8.68}$$

and exponent γ slightly smaller than 1 for rocks. Criterion (8.66) was proposed by von Schleicher in 1926 (cf. Nadai 1950). Criteria (8.66) and (8.67) show that the mean stress practically acts like a hardening parameter for the Huber-Mises and Tresca criteria.

Moreover, the first of Saint-Venant's working hypotheses (isochoric plastic deformations) has to be released when the material presents some *porosity* as in the presence of many small cavities. The corresponding generalization of the Huber-Mises yield for isotropic bodies was established in a celebrated paper by Gurson (1977). It reads

$$f\left(\sigma_{ij}, p\right) = \frac{\sigma_{eq}^2}{\sigma_Y^2} - \left(1 + p^2\right) + 2p\,q\,\cosh\left(\frac{3}{2}\frac{\sigma_m}{\sigma_Y}\right) = 0, \tag{8.69}$$

where σ_Y is a material characteristic constant (yield stress in the Huber-Mises case), σ_m is the already defined mean stress, p is the porosity (volume fraction of cavities), and q is a numerical factor to be determined. J.B. Leblond in Paris and his co-workers (e.g., Cologanu et al. 1997) have improved on Gurson's result. A recent development concerns so-called *microporomechanics* (cf. Dormieux et al. 2006) and the related yield criteria.

Another mechanical phenomenon that can couple with plasticity is *damage*. The latter is commonly described as the loss of elasticity due to the growth of micro-cavities and microcracks present in the material. This was expanded by J. Lemaitre (1985) in Paris-Cachan—see also Maugin (1992a, Chap. 10), Voyiadjis and Kattan (2005, Chap. 7). An interesting point is that damage can introduce some anisotropy.

In ferromagnetic bodies domain walls that separate magnetic domains are pinned on structural defects. They move by successive pining and unpinning. This induces couplings between the irreversible process of magnetization and viscoplasticity of the material. This was examined in finite strains by Maugin and Fomèthe (1982)—see also Micunovic (2009, Chap. 6). In small strains Maugin et al. (1987) have proposed a theory of coupled irreversible magneto-mechanical processes including the influence of plasticity on the formation of magnetic hysteresis curves.[4]

8.7.6 Gradient Plasticity

The most recent avatar of elasto-plasticity is called *gradient plasticity*. The interest for such a theory stems from the following facts. In some situations of deformation like in torsion, a marked spatially non-uniform strain is exhibited. Furthermore, Nye (1953) has shown that macroscopic distortion of the non-uniform shear type is related to the presence of *geometrically necessary dislocations* (GNDs)—also Ashby (1970). From these remarks there follows the conception of plasticity flow rules that involve not only the stress (like in classical plasticity) but also the

[4]A strong analogy between dissipation in magnetism and solid mechanics and between irreversible magnetization and viscoplasticity of crystals is given in Tables 9.1 and 9.2 in Maugin (1999).

gradient of the stress with a characteristic length (Aifantis 1984, 1987)—of which the thermodynamic formulation can be disputed; cf. Maugin (1990)—or a real expansion of a phenomenological theory of *strain-gradient plasticity* (Fleck and Hutchinson 1993, 1997, 2001) comforted by experiments on the torsion of very thin copper wires (cf. Fleck et al., 1994), and made more "rational" and mathematical by Fleck and Willis (2009). Another much relevant work is Gurtin and Anand (2009).

From Nye's theory and using Kröner's synthetic notation, a tensor of dislocation density, usually noted α_{ij} in Cartesian components is introduced, with the symbolic relationship $\underline{\alpha} \approx \nabla \times \underline{\gamma}$ with the distortion $\underline{\gamma}$. Fleck and Hutchinson (1993, also 1997). They go one step further by incorporating their approach in a Mindlin type of gradient elasticity-plasticity by introducing the second gradient of the displacement $\eta_{ijk} = u_{k,ij} = \eta_{jik}$ and the energy in terms of the usual strain and of this gradient by the energy variation

$$\delta W = \sigma_{ij}\delta e_{ij} + \tau_{ijk}\delta\eta_{ijk}, \tag{8.70}$$

from which there follow the constitutive equations

$$\sigma_{ij} = \frac{\partial W}{\partial e_{ij}}, \quad \tau_{ijk} = \frac{\partial W}{\partial \eta_{ijk}}, \tag{8.71}$$

where σ_{ij} is the second-order symmetric stress and τ_{ijk} is a third-order stress (or hyperstress). Note that the notion of applied couples and double forces will necessary appear at boundaries in addition to usual traction forces. For an *incompressible* displacement field (a current hypothesis in elasto-plasticity), only the "deviatoric" parts σ'_{ij} and τ'_{ijk} are retained. The associated space of relevant generalized stresses of vector $\Sigma = (\underline{\sigma}', \underline{\tau}')$ is therefore a five +eighteen = 23-dimensional space, so is the space of dual stresses $E = \left(\mathbf{e}', \underline{\eta}'\right)$. This allows for the formulation of a generalized flow rule in the Σ-space and a variational inequality formulation in the Drucker-Hill form (cf. Fleck and Hutchinson 1997), a normality law, as

$$\left(\sigma_{ij} - \sigma^*_{ij}\right)\dot{e}^p_{ij} + \left(\tau_{ijk} - \tau^*_{ijk}\right)\dot{\eta}^p_{ijk} \geq 0, \tag{8.72}$$

where the stress state $(\underline{\sigma}, \underline{\tau})$ is associated with the plastic strain rate $\left(\dot{\mathbf{e}}, \dot{\underline{\eta}}\right)$ and $(\underline{\sigma}^*, \underline{\tau}^*)$ is any other stress state on or within the yield surface. Equation (8.72) is a direct generalization of the Hill- Mandel maximal-dissipation principle. Because of the presence of the third-order stress in the yield criterion, a length scale is involved and is characteristic of the strain-gradient theory.

A finite-strain theory has been expanded with a generalization of Nye's relation in the self-explaining form $\underline{\alpha} \approx \nabla_R \times \mathbf{F}^p$, where ∇_R denotes a referential gradient. The corresponding theory rapidly acquires a high degree of complexity with which we shall not deal.

8.8 Conclusion

From the above conspectus of plasticity we can distinguish three periods or essential steps. Considering that really serious things started with the work of Tresca (hence the chosen date of 1864 as the beginning of the perused period), we first observe a period of initiation with the works of a few remarkable individuals, namely, Tresca, Saint-Venant, Lévy, Huber, von Mises, Prandtl, Reuss, and Bingham. Whatever will be studied in the future will base on their work with unavoidable reference to their names, proposed criteria of yield, basic equations and singular behaviours for the sake of comparison, generalization or simple argued explanation. That is why all mechanical engineers and mechanician scientists know or should know their name and the related relatively simple equations (criteria, evolution).

With the second period started just before the Second World War and extending until circa 1970, we have a period of consolidation with the promotion of extensions and formulation of principles intimately connected with irreversible thermodynamics. The names of Nadai, Hencky, Prager, Ilyushin, Kachanov, Hill, Drucker, Westergaard, Freudenthal, Koiter, Odqvist, Budiansky, and Naghdi are principally associated with these developments that started to exploit serious arguments of thermodynamics brought in by Bridgman, Eckart, Kestin and others (e.g., Rice, Mandel, Ziegler). Practically in parallel we witness a conjunction between mechanical engineering and some aspects of materials science illustrated by the initial work of Taylor on dislocations and plasticity, and the original work of Frenkel and Kontorova (1938). This really kindled a serious study of crystal plasticity with works by, e.g., Schmid, Nye, Bilby, Kröner, Mandel, Havner, etc. This recently expanded along some unexpected lines such as gradient plasticity.

The third period, roughly started in 1970, uses more abstract mathematical tools —e.g., convex analysis—and more inclusive geometrical concepts while simultaneously allowing for a true 3D computational plasticity. On the one hand, this allowed the proof of powerful theorems of uniqueness and existence, and the proposal of efficient schemes of computation, including in large strains. On the other hand, a clarification of the theory of finite deformations with adapted geometrical notions also contributed to this immense progress in these computations required by industrial applications. We have gone a long way since the pioneers of the 1960s such as Lippmann and Mahrenholtz (1967).

Born with rather crude experiments missing highly precise measuring devices but also with the remarkable insight of Saint-Venant, Huber and Mises, plasticity evolved to reach an insuperable degree of sophistication in both physical and mathematical aspects, and a highly efficient predictive power in computation. It succeeded to resolve the initial conflict between science and engineering, and between microscopic and macroscopic levels of understanding. This is now materialized in a corresponding gathering of a multitude of dislocation motions and continuum theory in a unified multi-scale approach. Most of the basic aspects of this evolution have been evoked perhaps in an unsophisticated, too much simplified, manner, or not cultivated to the degree expected by specialists. Thus the emphasis was placed on rate-independent plasticity, while rate effects are also of great interest in some materials. Only some of viscoplasticity was mentioned along this line. Also a rather strict framework of convexity and normality laws—so-called *associated plasticity*—was privileged although *non-associated flow rules*—for which yield function and plastic potential do not coincide and normality is no longer the rule—are feasible in some cases such as in Zirconium or rock-like materials (cf. Mroz 1963; Maier and Hueckels 1979—see Figs. 4 and 5 by these authors; Ziegler and Wehrli 1987, pp. 216–218; Houlsby and Puzin 2002, 2006). This shortcoming is only due to our blinkered vision and our lack of deeper knowledge in that direction, with an inevitable bias related to the period of our formative years (1960s–1970s) and professionally active career (1970s–2000).

8.9 Note on the Bibliography

The "classic" books on plasticity are definitely those of the period 1930–1960, by Nadai (1930, 1950), Sokolowsky (1946), Hill (1950), Ilyushin (1948), Zener (1948), Westergaard (1952), Prager (1955, 1961), Prager and Hodge (1954), and Kachanov (1956, 1960, 1974). They may seem outdated to most readers in the 2010s, but their convincing power and ready style are unforgettable. This is also the case of the synthetic encyclopaedia article of Freudenthal and Geiringer (1958) that really gave a complete overview of the field as in the mid 1950s, and of Bell (1973) for an incredible collection of experimental records since the beginning of continuum mechanics. Koiter (1960) gives the viewpoint of a true mechanician while Bridgman (1952) presents interesting experimental observations by a grand physicist. Thomas (1961) is much preoccupied by the propagation of discontinuity waves in the tradition of Hadamard. Much more recently, among the many books published for pedagogical purposes we single out the opus of Ziegler (1977) and the brilliant exposition by Lubliner (1990). As a textbook oriented to students with an inclination towards mathematics, Maugin (1992a) may be of interest. We do not list books with an engineering bias with the exception of the efficacious books of Mendelson (1968) and Chakrabarty (1987). Past reviews of some interest are those of Osakada (2008) and Chaboche (2008). The *International Journal of Plasticity* founded in 1985 by Akhtar Khan (Baltimore) and the associated series of international conferences organized by the same scientist provide a privileged forum for research papers on plasticity in the last thirty years.

References

Aifantis EC (1984) On the microstructural origin of certain inelastic models. Trans ASME J Eng Mater Technol 106:326–330
Aifantis EC (1987) The physics of plastic deformation. Int J Plasticity 3:211–247
Appell P (1921) Traité de mécanique rationnelle, Third edition. Gauthier-Villars, Paris [facsimilé reprint by Gabay, Paris, 1991]
Ashby MF (1970) The deformation of plastically non-homogeneous crystals. Phil Mag 21:399–424
Barré de Saint-Venant A (1871a) Rapport sur un mémoire de M. Maurice Lévy relatif aux équations générales des mouvements intérieurs des corps solides ductiles, au delà des limites où l'élasticité pourrait les ramener à leur premier état. C R Acad Sci Paris 73:86–91
Barré de Sain-Venant A (1871b) Sur la mécanique des corps ductiles. C R Acad Sci Paris 73:1181–1184
Barré de Saint-Venant A (1871c) Mémoire sur l'établissement des équations différentielles des mouvements intérieurs opérés dans les corps solides ductiles au delà des limites où l'élasticité pourrait les ramener à leur premier état. J Math Pures et Appl 16:308–316
Bauschinger J (1886) Mitt. Mech. Lab Munich (Yearly report) (cited by Timoshenko, 1953, p. 280)
Bell JF (1973) The experimental foundations of solid mechanics. In: Handbuch der Physik, Vol. VIa/1, Eds Flügge S, Truesdell CA. Springer, Berlin (also, 1984), pp. 1–813
Bertram A (2005) Elasticity and Plasticity of large deformations: an introduction. Springer, Berlin
Berveiller M, Zaoui A (1978) Extension of the self-consistent scheme to plastically flowing polycrystals. J Mech Phys Solids 26:325–344
Bilby BA, Bullough R, Smith E (1955) Continuous distributions of dislocations: new application of the methods of non-Riemannian geometry. Proc Roy Soc Lond A231:263–273
Bilby BA, Gardner LRT, Stroh AN (1957) Continuous distributions of dislocations and the theory of plasticity. Actes du Xième Congrès International de Mécanique Appliquée (Bruxelles, 1956), vol VIII. Université de Bruxelles, Belgique, pp 35–44
Bingham EC (1922) Fluidity and plasticity. McGraw-Hill, New York
Boehler A (ed) (1988) Plastic behaviour of anisotropic solids. Editions du CNRS, Paris
Bouc R (1971) Modèle mathématique d'hystérésis. Acustica Acta 24:16–25
Brézis M (1973) Opérateurs maximaux monotones et semi-groupes non linéaires dans les espaces de Hilbert. North-Holland, Amsterdam
Bridgman PW (1952) Studies in large plastic flow and fracture. Harvard University Press, Cambridge, Mass
Budiansky B, Pearson CE (1956/57) The variational principle and Galerkin's procedure for nonlinear elasticity. Quart Appl Math 14:328–331
Casey J, Naghdi PM (1980) A remark on the use of the decomposition F = Fe Fp in Plasticity. J Appl Mech 47:672–675
Casey J, Naghdi PM (1992) A prescription for the identification of finite plastic strain. Int J Engng Sci 10:1257–1278
Cauchy AL (1828) Sur les équations qui expriment l'équilibre ou les lois du mouvement intérieur d'un corps solide élastique ou non élastique. In:Exercices de mathématiques, vol. 3, pp. 160–187, Sept. 1828 [This is the fons et origo of modern continuum mechanics, presenting in print the ideas originally submitted to the Paris Academy of Sciences on September 30, 1822]
Chaboche JL (2008) A review of some plasticity and viscoplasticity constitutive theories. Int J Plasticity 24:1642–1693
Chakrabarty J (1987) Theory of plasticity. McGraw-Hill, New York
Cleja-Ţigoiu S, Soós E (1990) Elastoplastic models with relaxed configurations and internal state variables. Appl Mech Rev (ASME) 43:131–151
Clifton RJ (1972) On the equivalence of $\mathbf{F}^e\mathbf{F}^p$ and $\mathbf{F}^p\mathbf{F}^e$. Trans ASME J Appl Mech 39:287–289

Cologanu M, Leblond J-B, Perrin G, Devaux J (1997) Recent extensions of Gurson's model for porous ductile metals. In: Suquet P (ed) Continuum micromechanics. Springer, New, York, pp 61–130
Coulomb CA (1784) Recherches théoriques et expérimentales sur la force de torsion et sur l'élasticité des fils de métal. Hist Acad Roy Sci Paris, pp. 229–269
Critescu N, Suliciu I (1982) Viscoplasticity. Martinus Nijhoff Publ, The Hague
Dafalias YF (1983) Corotational rates for kinematic hardening at large plastic Deformations. J Appl Mech 50:561–565
Dafalias YF (1984) The plastic spin concept and a simple illustration of its role in finite plastic transformation. Mech Materials 3:223–233
Dafalias YF (1985) The plastic spin. J Appl Mech 52:865–871
Dormieux L, Kondo J, Ulm F-J (2006) Microporomechanics. John Wiley & Sons, UK
Drucker DC (1951) A more fundamental approach to plastic stress-strain relations. First US national congress of applied mechanics. ASME, New York, pp 487–491
Drucker DC, Prager W, Greenberg HJ (1952) Extended limit design theorems for continuous media. Quart J Appl Math 9:301–309
Duhem P (1903) L'évolution de la mécanique (published in seven parts in: Revue générale des sciences, Paris; as a book: A. Joanin, Paris) [English translation: The evolution of mechanics, Sijthoof and Noordhoff, 1980]
Duvaut G, Lions JL (1972) Les inéquations en mécanique et en physique. Dunod, Paris [English translation: Inequations in mechanics and physics, Springer, Berlin, 1979]
Eckart C (1948) The thermodynamics of irreversible processes IV, The theory of elasticity and anelasticity. Phys Rev 73:373–382
Eringen AC (1962) Nonlinear theory of continuous media. McGraw Hill, New York
Fleck NA, Hutchison JW (1993) A phenomenological theory for strain gradient effects in plasticity. J Mech Phys Solids 41(12):1825–1857
Fleck NA, Hutchinson JW (1997) Strain gradient plasticity. Advances in applied mechanics, vol 33. Academic Press, New York, pp 295–361
Fleck NA, Hutchinson JW (2001) A reformulation of strain gradient plasticity. J Mech Phys Solids 49:2245–2271
Fleck NA, Muller GM, Ashby MF, Hutchinson JW (1994) Strain gradient plasticity: theory and experiment. Acta Metall Mater 42(2):475
Fleck NA, Willis JR (2009) A mathematical basis for strain-gradient plasticity theory- Part I. J Mech Phys Solids 57:161–177; Part II, ibid, 57:1045-1057
Frenkel YI, Kontorova TA (1938) On the theory of plastic deformation and twinning. Physik Sowjet Union 13:1–15
Freudenthal A, Geiringer H (1958) The mathmatical theories of inelastic continuum. In: Handbuch der ¨Physik, Ed. S. Flügger, Bd. VI. Springer, Berlin, p. 229–433
Friedel J (1964) Dislocations. Pergamon Press, Oxford, UK [French original, Gauthier- Villars, Paris, 1956]
Gerstner FJ (1831) Handbuch der Mechanik, vol 1. Herbig, Leipzig
Golovin SA (2008) On the history of the Russian school of anelasticity in solids. Solid State Phenomena (English translation) 137:231–236
Green AE, Naghdi PM (1965) A general theory of an elastic-plastic continuum. Arch Rat Mech Anal 18:251–281
Greenberg HJ (1949) On the variational principles of plasticity. Grad. Div. Appl. Math., Brown Univ. Report, A-11-54, March 1949
Groot SR de, Mazur P (1962) Non-equilibrium thermodynamics, North-Holland, Amsterdam [Dover reprint, New York, 1986]
Guest J (1900) On the strength of ductile materials under combined stress. Phil Mag Fifth series 50:369–413
Gurson AL (1977) Continuum theory of ductile rupture by void nucleation and growth, Part I: yield criteria and flow rules for porous ductile materials. J Eng Mater Technol 99:2–15

Gurtin ME, Anand L (2009) Thermodynamics applied to gradient theories involving the accumulated plastic strain: the theories of Aifantis and Fleck and Hutchinson and their generalization. J Mech Phys Solids 57:405–421
Halphen H, Son Nguyen Quoc (1975) Sur les matériaux standards généralisés. J Mécanique (Paris) 14:39–63
Han W, Reddy BD (2013) Plasticity: mathematical theory and numerical analysis, 2nd edn. Springer, New York
Havner KS (1992) Finite plastic deformation of crystalline solids. Cambridge University Press, Cambridge, UK
Hencky H (1923) Über die einige statisch bestimmete Fälle des Gleichgewichts in plastischen Körpen. Zeit Angew Math Mech 3:241–251
Hencky H (1924) Zur Theorie plastischer Deformationen und die hierdurch Material herfogerufenen Nachspannungen, Zeit. Angew; Math. Mech. 4:323–334
Hill R (1948) A variational principle of maximum plastic work in classical plasticity. Quart J Mech Appl Math 1:18–28
Hill R (1950) Mathematical theory of plasticity. Oxford University Press, UK
Hill R (1965) Micro-mechanics of elastoplastic materials. J Mech Phys Solids 13:89–101
Hodge P G, Prager W (1949) A variational principle for plastic materials with strain Hardening. J Math and Phys. 27, 1–15, [This journal was published by MIT]
Hohenemser K, Prager W (1932) Über die Ansätze der Mechanik isotroper Kontinua. Zeit Angew Math Mech 12:216–226
Houlsby GT, Puzin AM (2002) Rate-dependent plasticity models derived from potential functions. J of Rheology 46(1):113–126
Houlsby GT, Puzin AM (2006) Principles of hyperplasticity: an approach to plasticity theory based on thermodynamic principles. Springer-Verlag, London
Huber MT (1904) Le travail de distorsion spécifique comme mesure du taux de sollicitation d'un matériau. Czasopismo Techn. Lvov, 3ème série, no.22:80–81
Huber MT (2004) Arch. Mech.(PL), 56/3:173–190 (English translation of Huber, 1904)
Ilyushin AA (1943) Some problems in the theory of plastic deformation (in Russian). Prikl Mat Mekh 7:245–272
Ilyushin AA (1948) Plasticity. Moscow (French Translation, Eyrolles, Paris, 1956)
Ilyushin AA (1954) On relations between stresses and small strains in continuum Mechanics. Prikl Mat. Mekh. (English Translation) 18:641–666
Ilyushin AA (1961) On the postulate of stability. J Appl Math Mech (PMM.) 25:746–752
Ilyushin AA (1963) Plastichnost'. Osnovy obshchey matematicheskoy teorii (Plasticity, Foundations of the General Mathematical Theory), Moscow, Acad. Sci. USSR
Johnson C (1976) Existence theorems for plasticity problems. J Math Pures Appl 55:431–444
Kachanov LM (1942) Variational principles for an elastic-plastic medium. Prikl Mat Mekh 6(2): 3–12
Kachanov LM (1956) Theory of plasticity. Moscow
Kachanov LM (1960) Theoria polzuchesti (Creep theory). Fizmatgiz, Moscow
Kachanov LM (1974) Fundamentals of the theory of plasticity, MIR Publishers, Moscow [in French: Eléments de la théorie de la plasticité, Editions de Moscou, 1975]
Kestin J (1966) A course in thermodynamics, Blaisdell, Waltham. Mass, USA
Koiter WT (1960) General theorems for elastic-plastic solids. In: Progress in solid mechanics. In: Sneddon IN, Hill RM (eds), North-Holland, Amsterdam, pp. 167–221
König JA (1987) Shakedown of elastic-plastic structures, (translation form the Polish). Elsevier, Amsterdam
Lee EH (1969) Elastic-plastic deformation at finite strain. ASME Trans J Appl Mech 36:1–6
Lemaitre J (1985) Coupled elastoplasticity and damage constitutive equations. Comp Math Appl Mech Engng 51:31–49 (Proc. Fenomech 84, Stuttgart, 1984)
Lenoble M (1900) Contribution à l'étude des déformations permanentes des fils métalliques; allongement par traction. Doctoral Thesis, Bordeaux

Lévy M (1871) Extrait du mémoire sur les équations générales des mouvements intérieurs des corps solides ductiles au delà des limites où l'élasticité pourrait les ramener à leur premier état. J Math Pures Appl 16:369–372

Lippmann H, Mahrenholtz O (1967) Plastomechanik der Umformung metallischer Werkstoffe. Springer-Verlag, Berlin

Lubliner J (1990) Plasticity theory. Mcmillan, New York

Luenberger DG (1973) An approach to nonlinear programming. J Optimization, Theory and Applications 11(3):219–227

Luenberger DG (1984) Linear and nonlinear programming. Addison-Wesley, Mass

Maier G, Hueckel T (1979) Nonassociated and coupled flow rules of elastoplasticity for rock-like materials. Int J Rock Mech Sci & Geomech Abstr 16:77–92

Mandel J (1971) Plasticité classique et viscoplasticité (CISM Lecture Notes, Udine, Italy). Springer-Verlag, Vienna

Mandel J (1982) Définition d'un repère privilégié pour l'étude des transformations anélastiques du polycristal. J Méca Théor Appl. 1(1):7–23

Manjoine MJ (1944) Influence of rate of strain and temperature on yield stresses of mild Steel. Trans ASME J Appl Mech 11:1–11

Manville O (1927) La physique de Pierre Duhem. Mém Soc Sci Phys Nat Bordeaux, 7ème Série, 171–636 [The most comprehensive and unsurpassed analysis of Duhem's scientific works by one of his direct disciples]

Maugin GA (1990) Internal variables and Dissipative Structures. J Non-Equilibrium Thermodynamics 15:173–192

Maugin GA (1992a) The thermomechanics of plasticity and fracture. Cambridge University Press, U.K.

Maugin GA (1992b) Thermodynamics of hysteresis, in: Advances in Thermodynamics, vol. 7, "Extended thermodynamic systems", Eds. S. Sieniutycz and P. Salamon. Taylor and Francis, New York p. 25–52

Maugin GA (1994) Eshelby stress in elastoplasticity and ductile fracture. Int J Plasticity 10:393–408

Maugin GA (1999) The thermomechanics of nonlinear irreversible behaviors: An Introduction. World Scientific, Singapore

Maugin GA (2011) Configurational forces. CRC Press, Chapman & Hall, Boca Raton, Fl

Maugin GA (2013) Continuum mechanics through the twentieth century: a concise historical perspective. Springer, Dordrecht

Maugin GA (2014) Continuum mechanics through the eighteenth and nineteenth centuries. Springer, Dordrecht

Maugin GA, Fomethe A (1982) On the viscoplasticity of ferromagnetic crystals. Int J Engng Sci 20:885–908

Maugin GA, Sabir M, Chambon M (1987) Coupled Magneto-mechanical Hysteresis in Ferromagnets: Application to Nondestructive Testing. In: Yamamoto Y, Miya K (eds) Electromagneto-mechanical Interactions in deformable solids and structures (IUTAM, Tokyo, 1986). North-Holland, Amsterdam, pp 255–264

Melan E (1938) Zur Plastizität des raümlichen Kontinuums. Ing Arch 9:116–125

Mendelson A (1968) Plasticity theory and applications. Macmillan, London

Mićunović MV (2009) Thermomechanics of viscoplasticity. Springer, New York

Mises R von (1913) Mechanik der festen Körper im plastisch deformalen Zustand. Gött Nach Math.-Phys Kl 582–592

Miyoshi T (1985) Formulation of the numerical analysis of plasticity. North- Holland, Amsterdam

Mohr O (1900) Z. Ver. Deut. Ing., p. 1524 (cited by Timoshenko, 1953, p. 268)

Moreau J-J (1966) Fonctionnelles convexes. Multigraphed, Collège de France, Paris

Moreau JJ (1971) On unilateral constraints, friction and plasticity (Lecture notes, C.I.M.E; Bressanone, Italy). Edizioni Cremonese, Rome

Mroz Z (1963) Non-associated flow laws in plasticity. J Mécanique 2:21–42

Mroz Z (1967) On the description of anisotropic work hardening. J Mech Phys Solids 15:163–175

Nadai A (1930) Plasticity. McGraw-Hill, New York (Translation from the German)
Nadai A (1937) Plastic behavior of metals in the strain-hardening range, Parts I. J Appl Phys 8:205–213
Nadai A (1950) Theory of flow and fracture of solids, 2nd edn. McGraw-Hill, New York
Naghdi PM (1990) A critical review of the state of finite plasticity. Zeit Angew Math Phys 41:315–394
Nemat-Nasser S (1979) Decomposition of strain measures and their rates in finite deformation elastoplasticity. Int J Solids Structures 15:155–156
Nguyen Quoc Son (1973) Doctoral Thesis in Mathematics, Université Pierre et Marie Curie (then called Paris 6). France, Paris
Nye JM (1953) Some geometrical relations in dislocated crystals. Acta metal 1(2):153–162
Odqvist FKG (1963) Die Verfestigung von flusseisenahnlichen Körern. Zeit Angew Math Mech 13:360–363
Odqvist FKG (1966) Mathematical theory of creep and creep rupture. Oxford University Press, UK
Osakada K (2008) History of plasticity and metal forming analysis. In: Proceedings of the 9th International Conference on Technology of Plasticity, pp. 22–43
Owen DJ (1980) Finite elements in plasticity: theory and practice. Pineridge Press Ltd, Swansea, UK
Perzyna P (1966) Fundamental problems in viscoplasticity. In: Ed. Yih CS (ed) Advances in applied mechanics, vol. 9. Academic Press, New York, pp. 243–377
Phillips A (1968) Yield surfaces of pure aluminium at elevated temperatures. In: Boley BA (ed) Thermoinelasticity, IUTAM Symposium, East Kilbride, 1968). Springer, Vienna, pp. 241–258
Phillips A, Tang JL (1972) The effect of loading path on the yield surface at elevated Temperatures. Int J Solids Structures 8:563–474
Piau M (1975) Acceleration waves in viscoplastic elastoplastic media. J Mécanique 14:1–38
Prager W (1949) Recent developments in the mathematical theory of plasticity. J Appl Phys 20:235–241
Prager W (1955) Probleme der Plastizitätstheorie. Birkhäuser, Basel
Prager W (1957) On ideal locking materials. Trans Soc Rheology 1:169–175
Prager W (1961) Einführung in die Kontinuumsmechanik. Birkhäuser, Basel [English translation: Introduction to mechanics of continua (1961), Ginn and Co., Boston]
Prager W, Hodge PG (1954) Theory of perfectly plastic solids. Wiley, New York
Prandtl LT (1924) Spannungsverteilung in plastischen Körpern. In: Proceedings of the 1st International Congress on Mechanics, Delft, p. 43–54
Reuss A (1930) Berücksichtigung der elastischen Formanderung indr Plastizitätstheorie. Zeit Angew Math Mech 10:266–274
Rice JM (1968) A path-independent integral and the approximate analysis of strain concentration by notches and cracks. Trans ASME J Appl Mech 33:379–385
Rice JM (1970) On the structure of stress-strain relations for time-dependent plastic deformations in metals. Trans ASME J Applied Mech 37:728–737
Rice JM (1971) Inelastic constitutive equations for solids: An internal variable theory and its application to metal plasticity. J Mech Phys Solids 19:433–455
Rockafellar RT (1970) Convex analysis. Princeton University Press, Princeton, NJ
Rogers TC (1988) Yield criteria, flow rules and hardening in anisotropic plasticity. In: Proceeding of the IUTAM-ICM symposium in yielding, damage and failure of anisotropic solids, Boehler JP (ed), Mech. Eng. Publ. London (1989)
Rogers TC (1990) Plasticity of fibre-reinforced materials. In: Maugin GA (ed) Continuum models and discrete systems, vol 1. Longman, London, pp 87–102
Schmid E (1924) "Streckgrenze" von Kristallen Schubspannungsgesetz (Yield point of crystals, Critical shear stress law). In: Proceedings of the first international congress of applied mechanics Delft, 1924
Schmid E, Boas W (1935) Kristallplastizität. Julius Springer, Berlin

Seeger A (1958) Kristallplastizität. In: Handbuch der Physik, Ed. S. Flügge, Bd. VII/2, Springer, Berlin
Simo JC, Hughes TJR (1998) Computational inelasticity. Springer, New York
Spencer AJM (1972) Deformation of fibre-reinforced materials. Clarendon Press, Oxford, UK
Sokolowsky WW (1946) Theory of plasticity (in Russian). Moscow
Suquet P (1979) Un espace fonctionnel pour les équations de la plasticité. Ann Fac Sci Toulouse 1:77–87
Suquet P (1982) Plasticité et homogénisation. Université Pierre et Marie Curie, Paris, Thèse de Doctorat ès Sciences Mathématiques
Suquet P (1985) Homogénisation de matériaux élastoplastiques. Doctoral Thesis in Mathematics, Université Pierre et Marie Curie, Paris
Symonds PS (1951) Shakedown in continuous media. Trans ASME J Appl Mech 18:18–35
Taylor GI (1934) The mechanism of plastic deformation of crystals. Part I-Theoretical. Proc Roy Soc Lond A145:362–404
Taylor GI (1938) Plastic strain in metals. J Inst Metals 68:307–324
Taylor GI (1947) A connexion between the criterion of yield and the strain ratio relationships in plastic solids. Proc Roy Soc Lond A191:441–446
Taylor GI, Quinney H (1931) The plastic distorsion of metals. Phil Trans Roy Soc Lond 230: 323–362
Temam R (1988) Mathematical problems in plasticity (translation from the French). Gauthier-Villars, Paris
Teman R, Strang G (1980) Functions of bounded variations. Arch Rat Mech Anal 75:7–21
Thomas TY (1961) Plastic flow and fracture of solids. Academic Press, New York
Timoshenko SP (1953) History of the Strength of Materials. McGraw Hill, New York [Dover reprint, New York, 1983] [Review of this book by Truesdell reprinted in Truesdell, 1984, p. 251–253.]
Tresca HE (1872) Mémoire sur l'écoulement des corps solides à de fortes pressions. Mémoires présentés par divers savants Acad Sci Paris 20:75–135
Truesdell CA (1955) Hypo-elasticity. J Rat Mech Anal 4:83–133
Truesdell CA (1984) An idiot's fugiiive essays on science. Springer, New York
Valanis KC (1971) A theory of viscoplasticity without a yield surface. Arch Mech 23:517–533
Volterra V (1907) Sur l'équilibre des corps élastiques multiplement connexes. Ann Sci de l'Ecole Norm Sup 24:401–517
von Mises R (1928) Zeit Angew Math Mech 8:161–185
Voyiadjis GZ, Kattan PI (2005) Damage mechanics. Taylor and Francis, Boca Raton. Fl, USA
Washizu K (1982) Variational methods in elasticity and plasticity. Pergamon, Oxford
Westergaard HM (1952) Theory of elasticity and plasticity. Wiley, New York
Zaremba S(1903) Bull Int Acad Sci Cracovie (Krakow) 85:380, 594, 614 [Also: Mémorial des Sciences Mathématiques, No. 82, Gauthier-Villars, Paris, 1937]
Zener C (1948) Elasticity and anelasticity of metals. Chicago University Press, Chicago
Ziegler H (1977) An introduction to thermomehnaics. North-Holland, Amsterdam (Second edition, 1986)
Ziegler H, Wehrli C (1987) The derivation of constitutive equations from the free energy and the dissipation functions. In: Yu TY, Hutchinson JW (eds) Advances in applied mechanics, vol. 25. Academic Press, New York, p. 183–238

Chapter 9
Fracture: To Crack or Not to Crack. That Is the Question

If only I could break a leg, what a lot of scientific work I could do Attributed to Thomas H. Huxley (1825–1895)

Abstract The expansion of the theory of fracture is one of the most exciting and thought provoking intellectual construct in continuum mechanics. In a period of almost a century, it required the combination of ingredients of differing nature: acute experimental observations, clever reasoning both in elasticity and thermodynamics, introduction of plasticity theory and of new notions related to the material toughness, application of powerful theorems (e.g., Noether's), development of the theory of configurational forces, and expansion of appropriate numerical strategies to deal with field singularities. In these adventures the most acknowledged heroes certainly are Inglis, Griffith, Westergaard, Snedddon, Irwin, Dugdale, Barenblatt, Wells, Eshelby, Cherepanov, Rice, and more recently, Freund, Nishioka, Atluri, Bazant, and many less known characters. The story which became "history" here is told with an emphasis on the most subtly inventive and landmark contributions.

9.1 Introduction

Nowadays the term fracture is most often heard in medical media (e.g., the femoral neck fracture among elderly patients, the hip fracture) or in politics (e.g., the social fracture between different classes of a population, etc.). But here we indeed understand this term as it was introduced in the strength (or resistance) of materials,

---

Caution. This is not a course on fracture (for which there exists a multitude of good books with varied amount of engineering, physical or mathematical emphasis); neither does it intend to peruse a complete history of the field. Only the most striking advances are reported in the period covering the twentieth century. These may be considered as critical landmarks.

G.A. Maugin, *Continuum Mechanics through the Ages - From the Renaissance to the Twentieth Century*, Solid Mechanics and Its Applications 223,
DOI 10.1007/978-3-319-26593-3_9

Fig. 9.1 Charles E. Inglis (1875–1952)

signifying the ruin by separation into parts of an initially unique body—or one that looked like this at the engineering scale of observation -, so that no more strength can be transmitted between the parts. It looks quite natural now to view this phenomenon as due to the irreversible propagation of defects such as cracks. But this view is relatively new on the time scale of the history of mechanics even though Leonardo da Vinci may have had some qualitative idea about the matter. In truth, the first precise connection between crack extension and fracture seems to be mathematically expressed in an original work by Inglis[1] (1913)—cf. Fig. 9.1. With this exemplary start we can say that fracture mechanics really exclusively belongs in the twentieth-century development of continuum mechanics.

[1]Charles Edward Inglis (1875–1952) was one of the most famous British civil engineers of the early twentieth century. He noticeably expanded civil engineering at Cambridge where he was professor and head of the Engineering Department. He was considered the greatest teacher of engineering of his time (according to the obituary of Fellows of the Royal Society published in 1953). His main works deal with the construction and vibrations of bridges, although here we emphasize his breakthrough contribution to the study of the effects of defects on the strength of structural elements.

9.2 The Birth of Fracture Theory: Inglis, Griffith

Inglis' contribution to the Spring Meeting of the Fifty-fourth Session of the Institution of Naval Architects in March 1913 was immediately acknowledged as a much awaited breakthrough in the field as witnessed in the vivid discussion (in pp. 231–241 of the Proceedings of the Meeting), that followed his presentation. Indeed we have here one of the comparatively few attempts made to adapt the mathematical theory of elasticity to the practical problems met in Naval Architecture—as emphasized by one contributor to the discussion, L. Woollard (p. 240). In his report, Inglis first recalls that "the destructive influence of a crack is a matter of common knowledge, and is particularly pronounced in the case of brittle non-ductile materials. This influence is turned to useful account in the process of glass cutting". He mentions ductile materials and the possible effect of plastic yielding but his analysis is confined to elastic materials along the standards set forth by A.E.H. Love in his celebrated treatise on elasticity. The main object of his paper is the determination of the stresses around a hole in a plate, the hole being elliptic in form; by making one axis of the ellipse very small the stresses due to the existence of a fine crack can then be investigated. The best résumé of the scientific results of Inglis' investigation is given by P. Hopkinson (himself a famous mechanical engineer) during the discussion: "Mr Inglis has shown us exactly how the stress at the end of the crack varies with its curvature and size; he has shown that it is proportional to the square root of the length of the crack and inversely proportional to its radius of curvature" (within the limits of elasticity of the material). It was a fortunate happening that a paper by E.G. Coker on the photo-elastic experiments on the stress field at a crack tip was presented at the same session and favourably concurred with Inglis' theoretical considerations. Inglis' analysis particularly applied to the stress concentration produced by holes in decks of ships (reinforcement had to be implemented) as noticed by Timoshenko (1953, p. 435), himself a contributor to this problem in the 1920s.

The next step on the path to a true "theory" of fracture was taken by Alan A. Griffith[2]—cf. Fig. 9.2—who introduced the notion of *energy* in the possible formulation of a criterion for fracture. This justifies that he be considered the "father of the modern theory of fracture". In his epoch-making paper published as (Griffith 1921) Griffith singlehandedly opened the way for a true mechanical and thermodynamic theory of fracture in the elastic regime (so-called brittle fracture). Of course he knew of Inglis' work and exploited his solution, but he realized that "the weakening of a material by a crack could be treated as an equilibrium problem in

[2]Alan A Griffith (1893–1963) was an English mechanical engineer. Educated (B. Eng., M. Eng., D. Eng.) at the school of Mechanical engineering of the University of Liverpool, he had a fruitful career in aeronautics, culminating in the aerodynamic theory of turbine design that was to lead to the successful technical development of the jet engine. He had some interaction with (Sir) Geoffrey I. Taylor (membrane-soap film—analogy as a method of solving complex—e.g., torsion–elasticity problem) before producing his breakthrough contribution to fracture in 1920.

Fig. 9.2 Alan A. Griffith (1893–1963)

which the reduction of strain elastic energy, when the crack extends, could be equated to the increase in surface energy due to the increase in surface area at the crack" (words of G.I. Taylor reported by Gilman 1998). Indeed, the beauty of Griffith's theory stems from the fact that "it uses the elegance of a thermodynamic argument to deal with the singularity that appears at the tip of a crack in linear elasticity" (Taylor's words). Technically, the contents of Griffith's paper can be summarized thus (cf. Timoshenko 1953, pp. 358–360; and Roylance 2001). The value of the critical crack length can be found by setting the derivative of the total involved energy to zero. The usual strain energy per unit volume—in fact unit thickness of the specimen—(in a linear elastic material and one-dimensional setting) is given by $U = E\varepsilon^2/2 = \sigma^2/2E$ where σ is the stress and E is Young's modulus. But during extension of the crack of length a, it is observed that a region of material adjacent to the free surfaces of the crack is unloaded and its strain is released. This region is essentially made of two triangular regions near the crack flanks of width a and height βa. Parameter β is chosen so as to agree with Inglis' solution (in plane stress loading this gives $\beta = \pi$). Accordingly, the total strain energy released is given by the strain energy per unit volume times the volume in both triangular regions, that is (in our notation, not Griffith's):

$$U = -\left(\frac{\sigma^2}{2E}\right) \times \left(2 \times \frac{\pi a \cdot a}{2}\right) = -\frac{\pi a^2 \sigma^2}{2E}. \tag{9.1}$$

In forming the crack, bonds must be broken and the corresponding energy will be absorbed by the material. With γ the surface energy (in Joule/meter^2) and accounting for the two faces of the crack, this provides (still per unit depth in the specimen) the involved surface energy as $S = 2\gamma a$. The vanishing of the derivative of $U + S$ thus yields the *critical stress* as

$$\sigma_f = (2E\gamma/\pi a)^{1/2} \tag{9.2}$$

Experiments show that the product of the square root of the flaw length a and the stress at fracture σ_f was nearly constant, i.e.,

$$\sigma_f \sqrt{a} \approx C. \tag{9.3}$$

Comparison of (9.2) and (9.3) leads to the remarkable result of Griffith:

$$C = (2E\gamma/\pi)^{1/2}. \tag{9.4}$$

With $E = 62\,\mathrm{GPa}$ and $\gamma = 1\,\mathrm{J/m^2}$ this gives excellent agreement of Griffith's predicted fracture stress with experimental results for *glass*, a notably brittle material. Griffith's theory was largely unrecognized before the Second World War, as it could not deal with the most interesting industrial case of ductile materials (e.g., metals)—in both ship building and aeronautic industries—where plasticity must play a decisive role. Nonetheless, both Inglis' and Griffith's works decidedly exhibit the importance of a detailed study of cracks in elasticity or more complex mechanical behaviour.

9.3 The Analysis of the Stress Field at Cracks

As a somewhat distraction from our main argument we must also report on essential developments in the mathematical analysis of the stress field in the neighbourhood of a crack tip. Two works are of utmost importance in this context, both relying on the mathematical exploitation of complex variables. One is the formidable work of Westergaard[3] (1939). In this masterpiece this author exploits the complex-variable representation for the mechanical fields of strains and stresses in plane problems of the elasticity of isotropic bodies introduced by G.V. Kolossov in 1909 and

[3]Harald M. Westergaard (1888–1950) was a Danish structural engineer who obtained his PhD (1916) at the University of Illinois in Urbana after some training in Germany, then taught at this university and moved to Harvard as a professor of civil engineering in 1936. His main scientific interests were in the resistance of ductile materials to combined stresses, stability of elastic structures, strength of concrete structures and highways, and bridges during earthquakes, etc. In 1939 he published his celebrated paper on cracks.

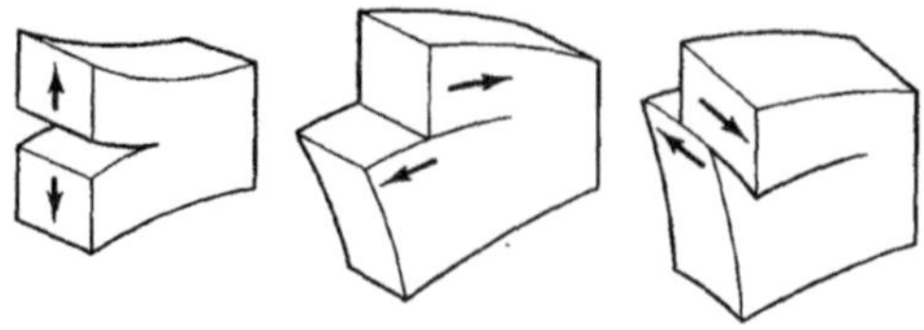

Fig. 9.3 From *left* to *right* Mode I, II, III (From Maugin 1992, Fig. 7.4, p. 138)

developed to a large extent by N.I. Muskhelisvili (as illustrated in detail by this author in Muskhelishvili 1953). The problem solved is the evaluation of the behaviour of mechanical fields in the vicinity of the tip of a straight crack in an infinite plane subjected to the action of a normal symmetric load. This is achieved for the three standard modes of cracking (so-called *mode I*: opening mode, *mode II*: shear plane mode (sliding), and *mode III*: shear antiplane mode; cf. Fig. 9.3) and provides for each mode a typical behaviour in the characteristic form

$$\sigma \approx \frac{K}{\sqrt{2\pi r}} f(\theta), \tag{9.5}$$

where r is the radial distance from the crack tip, $f(\theta)$ is a nondimensional function of the angle made by this radius vector with the direction of extension of the crack, and K is a numerical factor commonly called the *stress intensity factor* that differs from one mode of cracking to another[4] (see Fig. 9.4). Factor $(2\pi r)^{-1/2}$ in expression (9.5) shows the theoretical singular nature of the stress distribution as we approach the crack tip. In practice one as to evaluate K for each case of crack, structure and load geometry. Quantity K may be used in design and analysis by arguing that the material can withstand crack tip stresses up to a critical value of K, noted K_C, beyond which the crack propagates rapidly without control. This provides a measure of the *material toughness*. In turn, this allows one to relate the fracture stress σ_f and K by the simple formula

$$\sigma_f = \frac{K_C}{\alpha\sqrt{\pi a}}, \tag{9.6}$$

for a crack of length a with a factor α equal to one for an edge crack and generally of the order of unity in other situations. This viewpoint was to be related to the energy approach by Irwin (Sect. 9.4 below).

[4]We have given the detail of this analysis in Maugin (1992, Appendix 4).

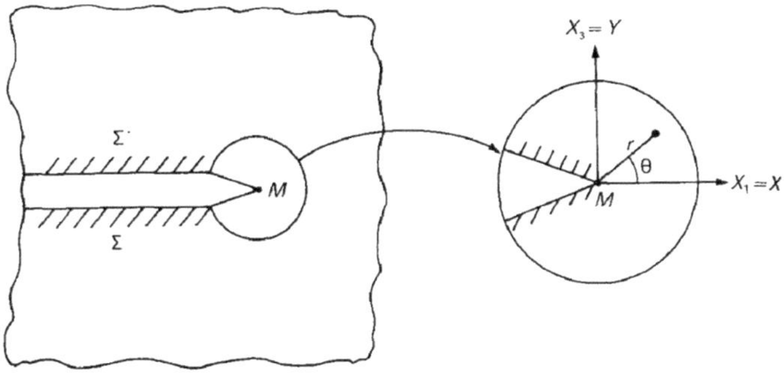

Fig. 9.4 Crack tip (r, θ) parametrization (From Maugin 1993, Fig. 7.1, p. 148)

The second work of importance in crack studies was produced by Sneddon (1946)[5] and showed the influence of war related research on the expansion of some theoretical studies. As a matter of fact, while being a Junior Scientific Officer at the British Ministry of Supply in the period 1942–1945 during which he spent some time at the Cavendish Laboratory, he was asked by (Sir) Nevill F. Mott (1905–1996; Nobel Prize in Physics in 1977), to examine the determination of the stress field in a infinite body containing a disc-shaped crack. In Mott's mind this was related to the mechanical resistance of steel tank armour plates where bubbles of gas flattened in the roll process transformed into disc-shaped cracks. This may be viewed as an expansion of Larmor's (1892) ideas. Sneddon solved the full three-dimensional problem in elasticity involving a crack by introducing a system of dual integral equations, a mathematical technique in which he became a supreme master. From here on the role of crack-tip solution and plasticity will take over the field of fracture in metals.

9.4 Irwin and Energy-Release Rate

Since Griffith's solution applied only to brittle materials while metals were taking much importance in various constructions (ship-building, aeronautics), it became necessary to involve plasticity and dissipation of some kind ahead of the crack tip in the examination of the fracture phenomenon. This forward step was taken in a

[5]Ian N. Sneddon (1919–2000) was a Scottish applied mathematician, educated in Glasgow and Cambridge. He was a professor in Keele and then Glasgow (1956–1984). He is famous as an insuperable master of integral transforms in the solution of mixed boundary-value problems in potential theory. Among others, he mentored Anthony M. Spencer and Peter Chadwick. A charming personality, he had broad scientific interests, even producing a pioneering book on wave mechanics in 1948 with N.F. Mott.

group dealing with ductile materials and working under Irwin[6] at the US Naval Research Laboratory (NRL) during World War II. While the phenomenon of extension of a plastic zone was a recognized one for a long time, Irving's group realized that if elastic strain energy is released as a crack grows, dissipated energy includes both surface energy (as already accounted for by Griffith) and plastic dissipation. While elastic energy provides the *driving force* for fracture, dissipated energy provides the *thermodynamic resistance* to fracture. That is, an amount G_p of plastic dissipation (and perhaps dissipation from other sources) should be added to the surface energy 2γ. Plastic dissipation will largely dominates surface energy (perhaps a few hundreds times more)—cf. Irwin (1948, 1957), Irwin and Kies (1952). With this consideration, Griffith's criterion (9.4) is replaced by the new criterion

$$C := \sigma_f \sqrt{a} = \left(\frac{E(2\gamma + G_p)}{\pi} \right)^{1/2}, \tag{9.7}$$

which becomes significant for ductile materials with G_p of the order of 1000 J/m^2.

Another fruitful notion introduced by Irwin's group (Irwin and Kies 1952, 1954) is that of *energy release rate* (ERR). This is not a rate per unit time, but the rate per unit lengthening of a crack. This can be evaluated as the change of total elastic energy of the system per unit area of crack growth. That is,

$$G := \left[\frac{\partial U}{\partial a} \right]_P = - \left[\frac{\partial U}{\partial a} \right]_u, \tag{9.8}$$

where subscripts P and u mean with load P or displacement u kept fixed in the evaluation of the derivatives. For instance, for *mode I* of cracking (opening mode) Irwin found that (this is Irwin famous relation)

$$G = G_I = K_I^2/E \quad \text{or} \quad \left(1 - \nu^2\right) K_I^2/E, \tag{9.9}$$

whether the crack problem is treated in plane stress or in plane strain. Here ν is Poisson's ratio and K_I is the stress intensity factor in mode I. Equations (9.9) show that Irwin's criterion and Griffith's one are essentially identical. Irwin also found the general expression for a planar crack in linear elasticity for the most general loading conditions. Then the energy-release rate is expressed in terms of the three stress intensity factors. On assuming that the size and shape of the energy-dissipation zone remains approximately constant during brittle fracture, he suggested that the energy needed to create a unit fracture surface is a constant that depends only on the

[6]George R. Irwin (1907–1998) was an American mechanical engineer initially educated in humanities and physics who had a long career at the US Naval Research Laboratory (NRL), from 1937 to 1967, and then taught at Lehigh University and next at the University of Maryland. His innovative work on fracture was kindled by his interest in penetration problems in armoured plates and damage to aircraft structures.

material, and is thus a material property that may be called the *fracture toughness*. Rather than this critical G_{Ic}, one nowadays uses the notion of critical stress intensity factor, K_{Ic} already mentioned in Eq. (9.6). This is accepted as a defining property within the frame of the linear elastic theory of fracture. With this notion we can formulate the following criterion:

(i) If $K < K_C$, there is no propagation, and we can symbolically write: $\dot{a} = 0$;
(ii) If $K = K_C$, propagation is possible $(\dot{a} \geq 0)$ but we do not know whether it will actually happen or not (so that $\dot{a} = 0$ is not excluded).

Note that K_C is difficult to determine experimentally in modes II and III, while it is easily determined in mode I (as done in Irwin's group).

The state of the art in fracture studies as of the late 1940s was reported by Orowan (1948). As a complement, it is worth mentioning a fracture criterion proposed by Leguillon (2002). This fracture *onset* criterion uses two parameters. Indeed, it seems that both energy (based on G) and strength (based on σ) criteria are necessary conditions for fracture, but neither one nor the other are sufficient. But the consistency between the two conditions provides a general form of criterion for crack nucleation involving a length, the energy criterion providing a lower bound of admissible length crack while the stress criterion provides an upper bound. The analytical justification of this criterion exploits the knowledge of the singular stress field around the notch tip—with an open angle between zero (case of a crack) to π (case of a straight edge)—obtained by asymptotic reasoning. The relevance and possible criticism of this mixed criterion are commented by Gross and Seelig (2001, but p. 126 on in the English translation of 2011).

One of the most recent advances dealing essentially with Griffith's approach is represented by an original mathematically founded variational approach to fracture as developed at length by Bourdin et al. (2008). This comprehensive paper is extremely mathematical (e.g., exploiting the full formalism of functional spaces, convex analysis, and the notion of Hausdorff measure of a set). It finds its root in the pioneering work of Francfort and Marigo (1998) but it matter-of-factly incorporates eighty years of mathematical experience since Griffith's work. Local or even global minimality is the mathematical argument that replaces the standard stationarity. The approach is first applied to "Griffith-like" energies, but it is also shown how to account for Barenblatt's idea of a cohesive zone (see next Section for this notion). The pure Griffith case is clear-cut form the point of view of irreversibility. The case including a "cohesive energy is more challenging because there is no obvious threshold for irreversibility" (the authors' words, 2008, p. 9). Another interesting result is the possible prevision of the path of the crack but without uniqueness result. A few numerical examples with multi-cracking and the appearance of multiple cracks illuminate this original powerful approach.

Globally, therefore, Irwin's group improved on both linear elastic fracture and the plastic analysis. Further progress would be achieved by paying more attention to the plasticized region about the crack tip in extension. For this we have to turn to works by Wells, Dugdale and Barenblatt.

9.5 Accounting for the Plastic Zone

It is Dugdale[7] (1960) who stimulated the development of so-called *cohesion models of cracking* by his introduction of a narrow yielding zone on the continuation of a through crack in a metal sheet. This matter is discussed in many books on fracture (e.g., Anderson 1995). The plastic zone introduced at a crack tip may have a size that is of the order of the crack size, and both size and shape of this zone can evolve with the increase of the applied load and the lengthening of the crack. This implies that a rather general theory of crack growth should account for local conditions for the initial crack growth, including nucleation, growth, coalescence of voids or decohesion at a crack tip, and a global energy balance criterion should be formulated for further crack growth and instable fracture.

Another well known contributor to this problem was Barenblatt[8] (1962; cf. Fig. 9.5) who accepts that ahead of the crack front—expanding in mode I—there exits a zone in which the "atoms" can be pushed aside at a variable distance δ and that this separation leads to cohesive stresses which are opposed to a clear separation. These cohesive forces vary from zero—where $\delta = 0$-as a function of δ, $\sigma(\delta)$, according to a law that is characteristic of the material (cf. Fig. 9.6). It is assumed that the zone of loss of cohesion is independent of the change and remains equal to itself, during the crack's extension. Willis (1967) and Rice (1968) have shown the equivalence of the Dugdale-Barenblatt approach with Griffith's results; in particular, one can write (compare Eq. (9.8))

$$G = -\frac{\partial U}{\partial a} = \int_0^{\delta_t} \sigma(\delta) d\delta. \tag{9.10}$$

This section would not be complete without mention of an interesting crack parameter known as the *Crack Tip Opening Displacement* (CTOD). This notion was introduced by Wells (1961, 1963) in his studies of structural steels and can be described thus. High toughness of steel sheets cannot be characterized by the standard linear theory of fracture mechanics. Wells observed in his experiments that

[7]Donald Stephen Dugdale is a British mechanical engineer (PhD Bristol 1952 for his studies in the field of plasticity and indentation hardness of metals under the supervision of J. Morrison and Rodney Hill). He set up a research school at the University College of Swansea, South Wales (UK). He then had a career in various institutions, including UK Government laboratories, the Illinois Institute of Technology in Chicago, the University of Sheffield and at Trinidad at the University of the West Indies and at the Sultanate of Oman. He is an acknowledged specialist of fatigue, plasticity and cracks.

[8]Grigori I. Barenblatt (born 1927) is a Russian mechanician-applied mathematician who is most famous for his works in the theory of fluid and gas flows in porous media, the mechanics of non-classical deformable solids, turbulence, self-similarity and so-called intermediate asymptotics. He obtained his doctoral degree in Moscow under the guidance of A.N. Kolmogorov. His long fruitful career was spent in Moscow, Cambridge and Berkeley. He mentored, among others, Genady P. Cherepanov, who became a master in crack studies.

Fig. 9.5 Grigori I. Barenblatt (born 1927)

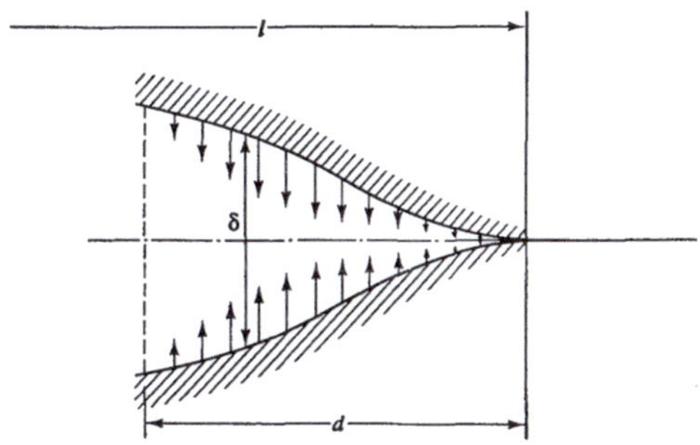

Fig. 9.6 Barenblatt's cohesive forces and crack-face displacement at the crack tip (From Maugin 1992, Fig. 7.10, p. 155)

before fracture the walls of a crack get separated and the crack tip evolves from a sharp shape to a rounded off form due to plastic deformation. This rounding is all the more pronounced that the material is tougher. The CTOD is usually defined as the displacement of the two parallel crack faces at the original crack tip with a typical semicircle blunt of the crack at the originally sharp tip. Interest in this crack parameter principally stems from its relatively easy measurement.

9.6 Invariant Integrals as Measures of Toughness

Although J. D Eshelby had paved the way in the exploitation of so-called *integral invariants* by applying E. Noether's invariance theorem in elasticity, it is in the mid 1960s with Cherepanov[9] (1967; cf. Fig. 9.7) in Russia and

---

[9]Genady P. Cherepanov (born 1937) was probably the most brilliant student of G.I. Barenblatt. He was the youngest ever DSc in Mechanics (1964) in the former Soviet Union. He is internationally known for his seminal work in the theory of deformation and fracture of materials and structures He can be considered one of the founders of *configurational mechanics* with the introduction of invariant and path-independent integrals in fracture science. He immigrated to the USA in 1990. He published a marvellous book (Cherepanov 1979).

Fig. 9.7 Genady P. Cherepanov (born 1937)

Rice[10] (1968; cf. Fig. 9.8) in the USA that a new measure of toughness was introduced in the form of a *path integral* around a crack tip. This was achieved in the framework of a nonlinear elastic theory—precisely a monotonic deformation theory of plasticity (no unloading!) under the now accepted name of the "*J*-integral" with critical value at fracture denoted J_{Ic}. This new approach reduces to Griffith's theory in the pure linear elastic case, and it is readily shown that J_{Ic} can be converted to the critical K_{Ic} introduced above in Sect. 9.4. The *J*-integral is mathematically defined by the expression

$$J = \int_{\Gamma} \left(W dy - T_i \frac{\partial u_i}{\partial x} ds \right), \quad W = \int_0^{\varepsilon_{ij}} \sigma_{ij} d\varepsilon_{ij}, \tag{9.11}$$

where W is the volume density of strain energy, T_i are the components of the traction vector, at Γ, ds is an element of length along the path Γ, which path is arbitrary

[10]James R. Rice (born 1940) is one of the most brilliant engineering scientists of the second half of the twentieth century. Not only did he contribute the celebrated *J*-integral but he also achieved seminal works in other branches of theoretical mechanics, civil-environmental engineering, materials physics, and geophysics. In particular we note his breakthrough works that dealt with the structure of inelastic constitutive equations, microscopic mechanisms of cleavage and ductile or creep rupture, localization into shear zones, with geophysical applications to landslides, fault systems and earthquakes. Having obtained his PhD at Lehigh in 1964, he became a professor first at Brown University and then at Harvard (since 1981).

Fig. 9.8 James R. Rice (born 1940)

clockwise around the tip of the crack of which the two faces are free of load ($x = x_1$ and $y = x_2$ are Cartesian coordinates in the plane, x being along the crack length). The scalar defined in (9.11) can be seen as the component along axis x, denoted with index *1*, of a vectorial quantity

$$J_i = \int_{\Gamma} \left(W n_i - \mathbf{n}.\sigma.\frac{\partial \mathbf{u}}{\partial x_i} \right) d\Gamma, \tag{9.12}$$

where $\mathbf{n}$ is the unit normal pointing to the outside of the domain encircled by Γ in the counter clockwise direction (see Fig. 9.9).

In a thermodynamic approach J is the conjugate of the rate of increase of the crack length so that the irreversible progress of the crack is characterized by the dissipation (cf. Maugin 1992, pp. 144–147)

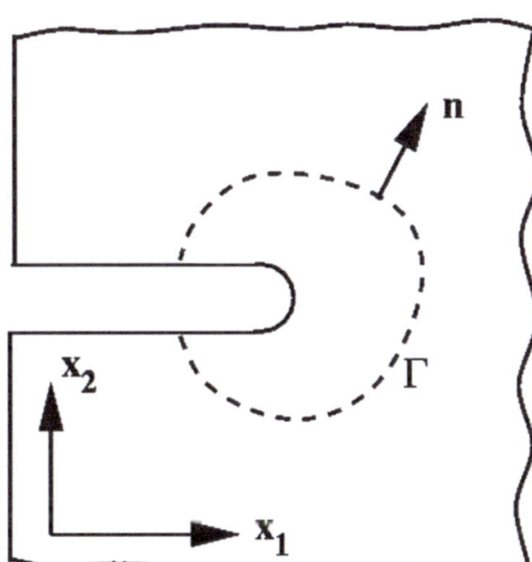

Fig. 9.9 Computation of the *J*-integral

Fig. 9.10 John D. Eshelby (1916–1981)

$$\Phi = J\dot{a} \geq 0. \tag{9.13}$$

It can also be noted that (9.12) is none other than the integral

$$J_i = \int_\Gamma n_j b_{ji} d\Gamma, \tag{9.14}$$

where tensor of Cartesian components b_{ji} is defined by

$$b_{ji} := W\delta_{ji} - \sigma_{jk} u_{k,i}. \tag{9.15}$$

This is called the (quasi-static) *Eshelby stress tensor* after Eshelby (1951). This tensor—which is not symmetric—is involved in all considerations of material inhomogeneity (cf. Maugin 1993, 2011) as initially introduced by Eshelby[11] (1951; cf. Fig. 9.10), a field singularity being a manifestation of some inhomogeneity.

[11]John (≪Jock≫) D. Eshelby (1916–1981) was a British physicist best known for his original work on dislocation motion, the driving force on a material inhomogeneity and on a field singularity, the continuum theory of lattice defects, and the celebrated "Eshelby" inclusion problem. He is considered the father of, and the main contributor to, the theory of configurational forces. Educated in Bristol (PhD 1949), he worked in Cambridge and taught in Sheffield as a reader and then as a professor (1971–1981). Among his direct co-workers we note B.A. Bilby, C. Atkinson and A.N. Stroh.

For completeness we note that the dual I-integral of the linear theory of fracture was introduced by Bui (1973) who based his reasoning on the complementary elastic energy. Other generalizations concern the fracture mechanics of materials described by means of a microstructure such as in Cosserat or polar media (cf. Atkinson and Leppington 1974; Maugin 1998b) or materials with gradients of strains (cf. Maugin and Trimarco 1992).

9.7 The Realm of Configurational-Material Forces

Expressions (9.11) through (9.15) belong in what is now called the theory of *configurational* or *material* forces. Originally (with Eshelby's works of the 1950s) this developed from an application, to non-dissipative systems, of the most famous theorem of mathematical physics of the twentieth century, the invariance theorem proved by the German mathematician Noether (1918). Without entering details (cf. Maugin 1993, 1999, 2011), this theorem essentially says that with a Lagrangian-Hamiltonian principle that delivers field equations of a physical theory there is associated a system of conservation laws, each of the latter being generated by a element of the invariance group of the theory. The energy conservation is such a law that is generated by invariance under time translation in classical mechanics. In the case of small-strain inhomogeneous elasticity, we have the following scheme at all regular material points with equations in Cartesian tensor notation:

$$\frac{\partial}{\partial t}(\rho_0 \dot{u}_i) - \frac{\partial}{\partial x_j}\sigma_{ji} = 0, \quad \sigma_{ji} = \frac{\partial}{\partial e_{ij}} W\left(e_{pq}; x_k\right) = \sigma_{ij}, \tag{9.16}$$

$$\frac{\partial}{\partial t}(K + W) - \frac{\partial}{\partial x_j}\left(\sigma_{ji}\dot{u}_i\right) = 0, \tag{9.17}$$

$$\frac{\partial}{\partial t}P_i - \frac{\partial}{\partial x_j}b_{ji} = f_i^{inh}, \tag{9.18}$$

$$K = \frac{1}{2}\rho_0 \dot{u}_i \dot{u}_i, \quad P_i = -\rho_0 \dot{u}_j u_{j,i}, \tag{9.19}$$

$$b_{ji} = -\left(L\delta_{ji} + \sigma_{jk} u_{k,i}\right), \quad L = K - W, \quad f_i^{inh} = \left.\frac{\partial L}{\partial x_i}\right|_{\dot{\mathbf{u}},\mathbf{e}\,fixed}, \tag{9.20}$$

$$\left(\frac{\partial}{\partial t}(\rho_0 \dot{u}_i) - \frac{\partial}{\partial x_j}\sigma_{ji}\right)u_{i,k} + \left(\frac{\partial}{\partial t}P_k - \frac{\partial}{\partial x_j}b_{jk} - f_k^{inh}\right) = 0. \tag{9.21}$$

Here Eqs. (9.16) are the field equations with strain energy W and kinetic energy K, per unit volume—with L the corresponding Lagrangian density -, Eqs. (9.17) and (9.18) are the associated conservation laws of energy and material momentum (respectively, invariance under time translation and spatial translation in material space), vector of components P_i is called the *material* or *field* momentum (in contrast with $p_i := \rho_0 \dot{u}_i$, the classical physical momentum), b_{ji} is the *dynamical* Eshelby stress, and f_i^{inh} is the "material force of inhomogeneity" computed by taking the explicit derivative of L with respect to space. Equation (9.21) is a so-called *Noether identity* which shows how (9.17) can be deduced from the first of (9.16) for non-vanishing displacement gradient. Integration of (9.18) in the plane over the volume encircled by the Γ path of Sect. 9.6 will yield the expression of the J-integral (9.14) for quasi-statics and in absence of distributed material inhomogeneities. The interest of the identity (9.21) is that its shows by mimicking it how to deal with more general cases that are not a priori deductible from a variational formulation (e.g., thermo-elasticity of conductors, visco-elasticity, plasticity, damage, etc.).

In this general framework, we note the elegant exact "analytical theory of brittle fracture" formulated by Dascalu and Maugin (1993) which, in full dynamics and finite deformations, yields compatible formulas for the driving force on the crack and the associated energy-release rate in terms of the Lagrangian L and Hamiltonian H densities—one being the Legendre transform of the other -, of the material momentum and of the velocity of progress of the crack tip in the material.

9.8 Dynamic Fracture

Dynamic fracture is well documented in Freund (1990) who has been one of the most active contributors to this field; see also the review of Nishioka (1998). By dynamic fracture it must be understood the inclusion of inertial terms. This important domain of fracture mechanics relates to *fast* fracture processes and *impact* fracture problems where effects of the crack velocities play a significant role. Episodes of growth initiation, acceleration, deceleration, arrest, etc., are all concerned with it. Of necessity the question of generalizing the J-integral approach to such a case is posed. Nishioka and Atluri[12] (1983) have pioneered in this direction of obtaining a path-independent "dynamic" J-integral. In the present framework, this is arrived at by exploiting the dynamic equation of conservation of material momentum (9.19). By integration of this Eq. (9.18) in the plane over the volume $V_\Gamma - V_\varepsilon$ encircled by the $\Gamma + \Gamma_\varepsilon$ path of Sect. 9.6—where Γ is sufficiently far from

[12]Starting in the early 1980s Satya N. Atluri—originally from India, now at Irvine, California—and Toshihisa Nishioka (Kobe, Japan) and their co-workers have been among the most productive contributors to dynamic fracture and its computational aspects (cf. Nishihoka's contribution in Cherepanov 1998) and Atluri's (1986) book. In the USA one must also emphasize the generous contributions of Lambert B. Freund (Brown University) and Ares J. Rosakis (CALTECH)—in particular, Freund's landmark book of 1990. In the former Soviet Union we note the works of V.Z. Parton and L.I. Slepyan, while in France H.D. Bui was the leading figure in the field.

Fig. 9.11 Huy Duong Bui (1937–2013)

the crack tip and Γ_ε is very close to the crack tip (see Fig. 9.11), this yields instead of (9.14):

$$J_i = \lim_{\varepsilon \to 0} \left\{ \int_{\Gamma + \Gamma_\varepsilon} n_j b_{ji} d\Gamma - \int_{V_\Gamma - V_\varepsilon} \dot{P}_i dV \right\}. \tag{9.22}$$

with b_{ji} defined in (9.19). Equivalent forms involve the Hamiltonian H rather than L, such as (cf. Nishioka 1998; pp. 589–590)

$$J_i = \lim_{\varepsilon \to 0} \left\{ \int_{\Gamma + \Gamma_\varepsilon} \left(H n_i - n_k \sigma_{kp} u_{p,i} \right) + \int_{V_\Gamma - V\varepsilon} \left(\rho \ddot{u}_k u_{k,i} - \rho \dot{u}_k \dot{u}_{k,i} \right) dV \right\}. \tag{9.23}$$

Both forms (9.21) and (9.22) are obtained for fixed *far-field* paths.

Bui[13] (1977b)—see Fig. 9.11—had given before a path-independent that moves with the crack tip, i.e., what we would get by integrating (9.18) over Γ encircling

[13]Huy Duong Bui (1937–2013) was a French engineer of Vietnamese origin. Having graduaded from the prestigious Ecole Polytechique and the School of Mines in Paris and mentored by Jean Mandel, he rapidly became a highly regarded specialist of fracture mechanics, with a special taste for invariance properties. He worked at the French Electricity National Company (EdF) and the Laboratory of Solid Mechanics (LMS) of the Ecole Polytechnique developing many aspects of fracture theory in relation to metal structures and the construction of atomic power plants. His book of 1977 was a landmark in the field.

the volume V_Γ about the crack tip and moving with it, that is (cf. Maugin 1993, pp. 167–169):

$$J_1 = \int\limits_{\Gamma(moving)} \left(-Ln_1 - n_k\sigma_{kp}u_{p,1} + v_C P_1 n_1\right)d\Gamma - \frac{d}{dt}\int\limits_{V_\Gamma(moving)} P_1 dV, \quad (9.24)$$

where v_C is the magnitude of the crack-tip velocity. Clearly this expression (9.24) is the most natural and direct dynamic generalization of the *1*-component of (9.14), but the computation to evaluate an integral over a path that moves with the crack tip may be difficult unless one employs a moving-mesh procedure. Much more on dynamic fracture mechanics can be found in Freund (1990) and the many works devoted to this specialized field in the last twenty years and usually reported at sessions of the periodic International Congress of Fracture, a very successful series of symposia founded in 1965 by the Japanese scientist Takeo Yokobori (born 1917) —cf. the general view of this author on the physical mechanisms involved in the mechanics of fracture in Cherepanov (1998, pp. 698–708). Anyway, various criteria of fracture have been proposed for elastodynamic fracture in which toughness may be a function of the time evolutions $(\dot{a}, \ddot{a}, ..)$ of the crack length. In particular, the dynamic *J*-integral may be used as a possible critical parameter (cf. Nishioka 1998, p. 588). Another prominent contributor to the dynamics of fracture was the Swedish scientist Bertram Broberg (cf. Broberg 1999). In Russia, Parton (cf. Parton and Boriskovsky 2002) published an interesting book on the subject.

9.9 Extensions and More Recent Developments

9.9.1 Electro-Magneto-Elastic Generalizations

It was not long before some of the above results were generalized to the case of *piezo-electricity*, yielding, for instance, the following "piezo-electric" *J*-integral in quasi-electro-elastostatics (Suo et al. 1992; but this was not the first obtaining of this form)

$$J = \int\limits_{\Gamma} \left(Wn_1 - \mathbf{n}\cdot\sigma\cdot\mathbf{u}_{,1} - \mathbf{n}\cdot\mathbf{D}\varphi_{,1}\right)d\Gamma, \quad (9.25)$$

where $\mathbf{D}$ is the electric displacement vector and φ is the electrostatic potential. Here W is jointly quadratic in the deformation and the electric field is such that $\mathbf{E} = -\nabla\varphi$. However, more exact results in the nonlinear linear framework had been given before by Pak and Herrmann (1986a, b), with an expression in the reference configuration such as

$$J(\Gamma) = \int_{\Gamma} \left\{ \tilde{W} N_1 - \mathbf{N} \cdot \left(\mathbf{T}^E + \mathbf{T}^F \right) \cdot \frac{\partial \mathbf{u}}{\partial X_1} - \left(\hat{\mathbf{D}} \cdot \mathbf{N} \right) \frac{\partial \hat{\varphi}}{\partial X_1} \right\} dS, \tag{9.26}$$

where $\mathbf{T}^E$ and $\mathbf{T}^F$ are "elastic" and "electric-field "contributions to the first Piola-Kirchhoff stress of finite-strain continuum mechanics and $\hat{\mathbf{D}}$ is the material electric displacement. Another approach was proposed by Maugin and Epstein (1991) on the basis of their theory of material forces. This electro-elastic case applies particularly to brittle electro-ceramics. A pioneering work on cracks in piezoelectricity was published by Parton (1976). The case of soft magnetic and paramagnetic bodies was similarly formulated by Sabir and Maugin (1996) and the much more complicated case of elastic ferromagnets (exhibiting a magnetic-spin internal degree of freedom) was settled by Fomethe and Maugin (1998).

9.9.2 The Consideration of Generalized Functions

The natural mathematical language for correctly dealing with singular fields in a continuum is the theory of *generalized functions* or *distributions* (in the sense of the French mathematician L. Schwartz) such as Dirac's delta function. However, while the theory is more than seventy years old, very few applications have been given in continuum mechanics. First applications to the theory of running cracks were in fact given by H.D. Bui and his direct co-workers (e.g., Bui et al. 1987). But a true exposition with applications to dislocations, cracks, and other defects was belatedly expanded only by Dascalu and Maugin (1994) and Maugin (1998a). In this last work the original field equations (e.g., balance of linear momentum and energy) are expressed in terms of distributions so that they automatically include terms related to singular points and surfaces. In this type of approach the energy is the sum of the potential energy of the (regular) defective body and of the energy associated with the extension domain of the defects—see also Maugin (2011), Chap. 8, pp. 220–226 and 234–238. As an example of such "distributional" conservation law we can cite the following conservation law of energy for a steady progressing straight crack:

$$D_t(W + F_{cr} H(l - x_1) \cdot \delta(x_2)) = Div(\mathbf{T} \cdot \mathbf{v}) + \overline{\mathbf{T}}^d \cdot \mathbf{v} \delta(\partial B), \tag{9.27}$$

where operators D_t and Div are to be understood in the sense of distribution theory. $H(l - x_1)$ is the Heaviside function of the interval [0, l], $\delta(x_2)$ is a Dirac function, and the dot in Eqs. (9.27) denotes the tensor product of distributions. F_{cr} is the critical value of the driving force (J-integral in the Griffith's vision where $F = F_{cr}$ at fracture) corresponding to a dissipation $\Phi = G^* = cF$ where c is the steady evolution velocity of the crack tip along axis x_1. The quantity

$$W_D := F_{cr}H(l - x_1) \cdot \delta(x_2) \tag{9.28}$$

may be called the ≪energy of the crack≫. With F having reached the fixed value F_{cr} and evaluating the time derivative of the Heaviside function, the left-hand side of (9.27) is none other than

$$D_t W + G^* \delta(x_1 = l, x_2 = 0), \tag{9.29}$$

where G^* is the localized energy-release rate. Equation (9.27) shows, as announced, that the *material-plus-defect* system is a *conservative* one. Moreover, it automatically accounts for the boundary data in stress at the boundary ∂B. This whole interpretation may have been too mathematical for most engineering scientists interested in fracture.

9.9.3 Computational Mechanics of Material Forces

Section 9.7 already emphasized the close relationship between fracture mechanics and the theory of material forces when dealing with path-independent integrals. But there is more to this. In a short paper of far-insight Braun (1997) pointed out a fact of deep computational interest that can be simply illustrated thus. For instance, the theory of material forces leads to the simultaneous existence of two forms of the equation of linear momentum or equilibrium, e.g., Eqs. (9.16) and (9.18). That is, in quasi-statics and for materially homogenous elastic bodies,

$$\frac{\partial}{\partial x_j}\sigma_{ji} = 0 \quad \text{and} \quad \frac{\partial}{\partial x_j}b_{ji} = 0. \tag{9.30}$$

But fields σ_{ji} and b_{ji} are not of the same degree in the elastic displacement gradient (cf. the definition of b_{ji}), the field b_{ji} presenting potentially the same singularity as the energy. Imagine that the first of (9.30) is solved quite exactly by using a sufficiently well designed discretization grid in the FEM (finite-element method). On evaluating from this solution the quantity in the left hand side of the second of (9.30)—which should vanish in the absence of material inhomogeneities—we might well obtain a non vanishing value, i.e., symbolically

$$\left[\frac{\partial}{\partial x_j}b_{ji}\right]_{computed} = f_i^{comp} \neq 0. \tag{9.31}$$

The minimization of the field of spurious material forces $\mathbf{f}^{comp}$ will cause a coordinated displacement of the initial computational points. This can be achieved in several iterative steps. This constitutes the essence of the creative remark made by Braun (1997) in his insightful paper. This was followed by an implementation in the evolution of cracks in inhomogeneous materials by Müller and Maugin (2002) and many other papers by Dietmar Gross and his co-workers in Darmstadt and Paul Steinmann and his co-workers in Kaiserslautern (e.g., Steinmann et al. 2001) and other people in different countries. Per force, this has become a very technical matter. Müller and Maugin (2002) exhibit the change of direction of the progressing crack in a materially inhomogeneous elastic bodies (with localized soft and hard inclusions ahead of the crack), showing how the crack can be attracted or repelled by these inclusions depending on the relative elasticity of the inclusions compared to that of the ambient material.

Here we must mention some of the numerous works devoted to the study of the *branching* of cracks, obviously a problem of high importance but unfortunately accompanied by a high technicality in analysis. Such works are exemplified by Amestoy et al. (1979) and Amestoy and Leblond (1992) [see also Leblond 2003, Sects. 4.2 and 6.3; Leblond and Frelat 2000]. Along the same analytical difficulties, we note the treatment of the stability and perturbation in the crack front by Leblond and his co-workers in the years 2000–2015.

Simultaneously, with the deduction of the conservation (or rather, non-conservation) of material momentum in the presence of dissipative effects such as heat conduction and plasticity, the world of material forces—exposed by Kienzler and Herrmann (2000) with applications to the strength of materials—expanded considerably and it became possible to envisage numerical schemes capable of treating these more complex cases, with application to the propagation of cracks and the development of irreversible fracture. The notion of material forces due to heat conduction was introduced by Epstein and Maugin (1995)—also Bui (1977a, b) in small strains—while that of material force due to the presence of an internal variable of state in the free energy density was introduced by the writer in the early 2000s and formalized, e.g., in Maugin (2011, Chaps. 5 and 6). Typically, in quasi-statics, Eq. (9.18) is replaced by

$$\frac{\partial}{\partial x_j} b_{ji} + f_i^{inh} + f_i^{th} + f_i^{intr} = 0, \tag{9.32}$$

wherein

$$b_{ji} = W\delta_{ji} - \sigma_{jk} u_{k,i}, \quad W = W\left(e_{ji}, \theta, \alpha; \mathbf{x}\right), \tag{9.33}$$

$$\sigma_{ji} = \frac{\partial W}{\partial e_{ji}}, \; f_i^{inh} = -\left.\frac{\partial W}{\partial x_i}\right|_{expl}, \; f_i^{th} = S\frac{\partial \theta}{\partial x_i}, \; f_i^{intr} = A\frac{\partial \alpha}{\partial x_i}, \tag{9.34}$$

and

$$S = -\frac{\partial W}{\partial \theta}, \quad A = -\frac{\partial W}{\partial \alpha}. \tag{9.35}$$

The last two quantities are the entropy density and the thermodynamic force associated with the internal variable α. In elasto-plasticity α may be the plastic strain e_{ij}^{p} and a hardening parameter κ. Thus (9.32) yields the local equilibrium equation at regular points in material space as

$$\frac{\partial}{\partial x_j} b_{ji} = \left.\frac{\partial W}{\partial x_i}\right|_{expl} - S\frac{\partial \theta}{\partial x_i} + \sigma_{pq}\frac{\partial e_{pq}^{p}}{\partial x_i} + q\frac{\partial \kappa}{\partial x_i}, \quad q = -\frac{\partial W}{\partial \kappa}. \tag{9.36}$$

Here $W = W\left(e_{ij}^{e} = e_{ij} - e_{ij}^{p}, \theta, e_{ij}^{p}, \kappa; x_i\right)$ where the strain tensor has an additive decomposition in elastic and plastic parts. This is easily generalized to the case of large strains (cf. Näser et al. 2007) where the total deformation gradient admits a multiplicative decomposition. The numerical exploitation of Eq. (9.36) for cracks extending with a plastic process zone ahead is shown by Näser et al. (2007), and case studies of crack propagation—including the evolution in the path of cracking—are given in the ANSYS (2015) software, thus demonstrating the applicative power of the method of configurational/material forces, a most interesting method in fracture mechanical investigations.

9.9.4 Peridynamics

The word "peridynamics" is a neologism created from the Greek word "peri" meaning "the surrounding" and "dynamics" that needs no explanation. It in fact is a special version of the *nonlocal theory of continua*. As such, it will replace the usual partial differential equations by integro-differential equations. In truth, it essentially provides a bridge between standard local continuum models and nonlocal atomistic models. It is an extension of classical continuum mechanics suitable for modelling discontinuous phenomena such as discontinuous displacement. Fracture indeed is the most frequent field of application of this new technique which attempts at modelling continuous media, discrete particles, and defects within the same set of equations (cf. Ha and Boberu 2011). The main idea is to avoid the difficulty presented by the existence of field singularities such as happens in the study of fracture for which the partial differential equations of classical mechanics are not well equipped. It directly provides an efficient numerical method to deal with these cases. The corresponding material model includes an explicit length scale. The early developments of this approach are principally due to a small group of authors around Silling (2000)—for greater detail see the review by Silling and Lehoucq (2010).

Fig. 9.12 Zdenek P. Bazant (born 1937)

9.9.5 *Size Effects*

Size effects are of fundamental importance in fracture. This has been a recurring subject of study in works by Zdenek P. Bazant[14] (e.g., Bazant 2002; Bazant and Planas 1998; see Fig. 9.12) at Northwestern University, Alberto Carpinteri in Torino, and others, both in mechanical/civil engineering and applied physics. Size effects may cause the nominal strength to depend on structure size, e.g., through boundary layer effects, diffusion phenomena, statistical size effects, fracture-mechanics size effects and fractal nature of crack surfaces. Size effects can be defined by comparison of geometrically similar structures of different sizes (Bazant). Linear elastic fracture mechanics exhibits a rather strong size effect. Another size effect is due to ductility (cf. Bazant and Planas 1998). A rigorous approach to this matter should involve an asymptotic analysis of some kind with a typical scaling (cf. Bazant 1997, 2002). Carpinteri's numerous works (e.g., Carpinteri 1997) are characterized by a strong interest in the relationship between crack propagation and catastrophe theory, applications of fractional calculus, stability of cracks and size effects. Of direct relation to this is the question of scaling

---

[14]Zdenek P. Bazant (born 1937) is an American engineer of Czech origin—with a long career at Northwestern University, Evanston—who has been one of the most productive and prolific authors in various fields of mechanical/civil engineering: creep of concrete, the stability of structures, scale effects in mechanics, and a nonlocal theory of damage.

properties of cracks [cf. the review of Bouchaud (1997) from the viewpoint of a physicist interested in the morphology of fracture surfaces].

9.10 Summary and Conclusion

In the second half of the twentieth century fracture mechanics has expanded to become one of the most exciting research domains in continuum mechanics. The essential reason for this remarkable development is that fracture mechanics intimately relates to a basic query of dramatic content: will a catastrophic issue such as ruin happen to man-made or natural (in geophysics) structures in certain physical situations? Fracture mechanics is a *failure theory*. In contrast with a deterioration happening simultaneously through an entire zone or surface (like in distributed damage or chemical decohesion—like zinc pest—in some alloys) the failure by fracture is one that takes place by propagation throughout a structure, hence the emphasis placed on the possible extension of cracks, the main object of interest in the theory. Along this line of thought one naturally distinguishes between fracture by sharp cracks that characterizes what happens in brittle or quasi-brittle elastic bodies and the fracture in ductile elastic-plastic bodies. This is at first the privileged application domain of so-called *linear elastic fracture mechanics* as exemplified by the early works of Inglis and Griffith, and the introduction of the notion of *energy-release rate* (Irwin) and of the J-integral by Cherepanov and Rice. This is applicable insofar as the material is elastic except perhaps in a vanishingly small region at the crack tip under the assumption of small scale yielding, with a stable or unstable crack growth. The innovative works of Dugdale and Barenblatt were useful in bringing in the picture the notion of cohesive zone. But *elastic-plastic fracture mechanics* is the realistic theory that applies in the conditions of ductile fracture that is accompanied by formation of a relatively large plastic zone at the crack tip and fracture happens through stable crack growth. Crack tips are no longer sharp, but are naturally rounded, and the notion of crack-tip-opening-displacement (Wells) acquires its whole practical importance. In all we have emphasized the role played by the rich notion of configurational or material forces, first illustrated by the celebrated J-integral.

Dynamical fracture became important with the necessary consideration of *fast* fracture processes and *impact* fracture problems (cf. contributions of Freund, Nishioka, Atluri, Broberg, Bui, Morozov, Parton). Most recent developments included extension to the fracture of materials with coupled electro-magneto-mechanical behaviours (in particular, piezo-electricity) and materials admitting a microstructure. These generalizations were facilitated by the inclusive view brought in by material forces and the canonical background that this notion naturally offers with an automatically compatible formulation in time (energy considerations) and material space (configurational forces). The latter also benefited to efficient numerical computation methods while generalized functions present the best theoretical mathematical background for such approaches. Scale effects are obviously of prevailing importance in all problems involving cracks (cf. Bazant), while new numerical schemes such as meshless ones (cf. Belytschko and Chen 2013; Rao and Rahman 2000; Chen et al. 2006) and peridynamics are well adapted to fracture problems, indeed accounting for these scale effects and the occurrence of field singularities.

Not being a specialist of numerics, I probably did not pay sufficient due attention to numerical procedures useful in fracture studies. In addition to the already mentioned recently developed techniques involving configurational forces, meshless methods, and peridynamics, we should note the strong-discontinuity method and Extended Finite Element Method (ex-FEM—cf. Belytschko and co-workers), and the more classical FEM and Boundary Element Method (BEM) as particularly illustrated in older books by Parton and Morozov (1978) and Owen (1983) and Aliabadi and Rooke (1991), respectively, without mentioning what is handily available in commercial software packs.

In conclusion, we note that the technical literature on fracture, both theoretical and experimental, is overwhelming and this extremely high production rightly reflects the tremendous activity in the field. The book by Anderson (1995) has become an inevitable reference in English for engineers while that of Gross and Seelig (2001) prevails on the German market with an excellent updated English translation (2011). Other books of interest are those of Atluri (1986), Broberg (1999), Freund (1990), Bui (1977a, b), Lawn (1993) and Leblond (2003), which put accent on different aspects of the phenomenon. Innumerable technical papers are nowadays published in specialized journals (such as the *International Journal of Fracture* or *Engineering Fracture Mechanics*) but they also abound in general journals of mechanical sciences and numerical computations. A generous—but already obsolete—bibliography is given in Maugin (2011, Chap. 8).

References

Aliabadi MH, Rooke DP (1991) Numerical fracture mechanics. Springer, Dordrecht

Amestoy M, Leblond JB (1992) Crack paths in plane situations-II-detailed form of the expansion of the stress intensity factors. Int J Solids Struct 29:465–501

Amestoy M, Bui HD, Van Dang K (1979) Déviation infinitésimale d'une fissure dans une direction arbitraire. C R Acad Sci Paris IIb-289:99–102

Anderson TL (1995) Fracture mechanics: fundamentals and applications. CRC Press, Boca Raton

ANSYS (2015) Material forces: a novel approach to fracture mechanics (white paper; available on the web)

Atkinson C, Leppington FG (1974) Some calculations of the energy-release rate G for cracks in micropolar and couple-stress elastic media. Int J Fract 10:599–602

Atluri SN (1986) Computational methods in the mechanics of fracture. North-Holland, Amsterdam

Barenblatt GI (1962) The mathematical theory of equilibrium cracks in brittle fracture. In: Sih GS (ed) Advances in applied mechanics, vol 7. Academic Press, New York, pp 55–129

Bazant ZP (1997) Scaling of quasi-brittle fracture: asymptotic analysis. Int J Fracture 83(1):19–40

Bazant ZP (2002) Scaling of structural strength. Hermes-Penton, London [Second edition (2005), Elsevier, London]

Bazant ZP, Planas J (1998) Fractural size effects in concrete and other quasi-brittle materials. CRC, Boca Raton

Belytschko T, Chen JS (2013) Meshless and particle methods. Wiley-Blackwell, UK
Bouchaud E (1997) Scaling properties of cracks. J Phys Condens Matter 9:4319–4344
Bourdin B, Francfort GA, Marigo JJ (2008) The variational approach to fracture. J Elasticity 91:3–148 [also as a separate book, Springer. New York, 2008]
Braun M (1997) Configurational forces induced by finite-element discretization. Proc Estonian Acad Sci Phys Math 46:24–31
Broberg B (1999) Cracks and fracture. Academic Press, New York
Bui HD (1973) Duality entre intégrales indépendantes du contour dans la théorie des corps fissurés. C R Acad Sci Paris 276A:1425–1428
Bui HD (1977a) Mécanique de la rupture fragile. Masson, Paris
Bui HD (1977b). Stress and crack displacement intensity factors in elastodynamics. In: Proceeding of the Internatuional congress fracture 4 (Waterloo, Canada), vol 3, pp 91–95
Bui HD, Nguyen QS, Ehrlacher A (1987) Thermomechanical coupling in fracture mechanics. In: Bui HD, Nguyen QS (eds) Thermomechnaical couplings in solids. North-Holland, Amsterdam, pp 327–341
Carpinteri A (1997) Structural mechanics: a unified approach. Chapman & Hall, London
Chen Y, Lee JD, Eskandarian A (2006) Meshless methods in solid mechanics. Springer, New York
Cherepanov GP (1967) Crack propagation in continuous media. Prikl Mat Mekh (PMM) 31 (3):476–488 (in Russian) [reproduced in English translation in Cherepanov, 1998, pp. 41–53]
Cherepanov GP (1979) Mechanics of brittle fracture (translation form the Russian). McGraw-Hill, NewYork
Cherepanov GP (ed) (1998) Fracture: a topical encyclopedia of current knowledge. Krieger Publ. Co., Malabar
Dascalu C, Maugin GA (1993) Forces matérielles et taux de restitution de l'énergie dans les corps élastiques homogènes avec défauts. C R Acad Sci Paris II-317:1135–1140
Dascalu C, Maugin GA (1994) The energy of elastic defects: a distributional approach. Proc Roy Soc London A445:23–37
Dugdale DS (1960) Yielding of steel sheets containing slits. J Mech Phys Solids 8(2):100–104
Epstein M, Maugin GA (1995) Thermoelastic material forces: definition and geometric aspects. C R Acad Sci Paris II-320:63–68
Eshelby JD (1951) The force on an elastic singularity. Phil Trans Roy Soc Lond A244:87–112 [reproduced in Cherepanov, 1998, pp. 452–475]
Fomèthe A, Maugin GA (1998) On the crack mechanics of hard ferromagnets. Int J Non-linear Mech 33:85–95
Francfort GA, Marigo JJ (1998) Revisiting brittle fracture as an energy-minimization problem. J Mech Phys Solids 46(8):1319–1342
Freund LB (1990) Dynamic fracture mechanics. Cambridge University Press, UK
Gilman JJ (1998) Alan Arnold Griffith, An appreciation. In: Cherepanov GP (ed) Fracture–a topical encyclopedia of current knowledge. Krieger, Malabar, pp xix–xxii
Griffith AA (1921) The phenomena of rupture and flow in solids. Phil Trans Roy Soc Lond A221:163–198 [reproduced in Cherepanov, 1998, pp. 2–29]
Gross D, Seelig TH (2001) Bruchmechanik, 3rd edn. Springer, Berlin [Updated English translation, Springer, Berlin, 2011]
Ha YD, Boberu F (2011) Characteristics of dynamic brittle fracture with peridynamics. Engn Fract Mech 78(5):1156–1168
Inglis CE (1913) Stresses in a plate due to the presence of cracks and sharp corners. Trans Inst Naval Archit 55:219–230 (discussion of paper in pp 231–241)
Irwin GR (1948) Fracture dynamics. In: Fracturing of metals, American Society for Metals, Cleveland, Ohio
Irwin GR (1957) Analysis of stresses and strains near the end of a crack traversing a plate. J Appl Mech 24:361–364
Irwin GR, Kies JA (1952) Fracturing and fracture dynamics. Weld J 31:95–100 (Research Suppl)

Irwin GR, Kies JA (1954) Critical energy rate analysis of fracture strength of large welded structures. Weld J 33:193–198 (Research Suppl)
Kienzler R, Herrmann G (2000) Mechanics in material space. Springer, Heidelberg
Larmor J (1892) The influence of flaws and air-cavities on the strength of materials. Phil Mag 33:70
Lawn BR (1993) Fracture of brittle solids. Cambridge Solid State Science Series, 2nd edn. CUP, Cambridge
Leblond JB (2003). Mécanique de la rupture fragile et ductile. Hermès, Paris (197 pages)
Leblond JB, Frelat J (2000) Crack kinking from an initially closed crack. Int J Solids Struct 37:1595–1614
Leguillon D (2002) Strength or toughness? A criterion for crack onset at a notch. Eur J Mech A/Solids 21:61–72
Maugin GA (1992) The thermomechanics of plasticity and fracture. Cambridge University Press, UK
Maugin GA (1993) Material inhomogeneities in elasticity. Chapman & Hall, London
Maugin GA (1998a) Variations on a theme of A.A. Griffith. In: Cherepanov GP (ed) Fracture: a topical encyclopedia of current knowledge. Krieger Publ. Co., Malabar, pp 494–509
Maugin GA (1998b) On the structure of the theory of polar elasticity. Phil Trans Roy Soc Lond A356:1367–1395
Maugin GA (1999) The thermomechanics of nonlinear irreversible behaviors: an Introduction. World Scientific, Singapore and New Jersey [See, in particular, Chapter 8]
Maugin GA (2011) Configurational forces. CRC/Chapman & Hall/Taylor & Francis, Boca Raton
Maugin GA, Trimarco C (1992) Pseudo-momentum and material forces in nonlinear elasticity: variational formulations and application to brittle fracture. Acta Mech 94:1–28
Müller R, Maugin GA (2002) On material forces and finite element discretization. Comp Mech 29:52–60
Muskhelishvili NI (1953) Some basic problem in the mathematical theory of elasticity. Noordhoff, Groningen
Näser B, Kaliske M, Müller R (2007) Material forces for inelastic models at large strains: application to fracture mechanics. Comp Mech 40:1005–1013
Nishioka T (1998) On the dynamic J-integral in dynamic fracture mechanics. In: Cherepanov GP (ed) Fracture: a topical encyclopedia of current knowledge. Krieger Publ. Co., Malabar, pp 575–617
Nishioka T, Atluri SN (1983) Path-independent integrals, energy-release rates, and general solutions of near-tip fields in mixed-mode dynamic fracture mechanics. Eng Fract Mech 18:1–22
Noether E (1918) Invariante Variationsproblem. Klg-Ges Wiss Nach Göttingen. Math Phys Kl.2:235–257
Orowan E (1948) Fracture and strength of solids. Rep Prog Phys 12:185–232
Owen DRJ (1983) Engineering fracture mechanics: numerical methods and applications. Pineridge Press Ltd, Swansea
Pak YE, Herrmann G (1986a) Conservation laws and the material momentum tensor for the elastic dielectric. Int J Engng Sci 24:1365–1374
Pak YE, Herrmann G (1986b) Crack extension force in elastic dielectrics. Int J Engng Sci 24:1375–1388
Parton VZ (1976) Fracture mechanics of piezoelecric materials. Acta Astronaut 3:671–683
Parton VZ, Boriskovsky VG (2002) Dynamic fracture mechanics, Two volumes, 2nd edn. CRC Press, Boca Raton (translation for the Russian)
Parton VZ, Morozov EM (1978) Elastic-plastic fracture mechanics. MIR Publ, Moscow (in English)
Rao BN, Rahman S (2000) An efficient meshless method for fracture analysis of cracks. Comput Mech 26:398–408
Rice JR (1968) A path independent integral and the approximate analysis of strain concentration by notches ad cracks. J Appl Mech 35:379–386

Roylance D (2001) Introduction to fracture mechanics, Lecture notes, MIT, Department of materials science and engineering, June 14, 2001 (available on the www)
Sabir M, Maugin GA (1996) On the fracture of paramagnets and soft ferromagnets. Int J Non-linear Mech 31:425–440
Silling SA (2000) Reformulation of elasticity theory for discontinuities and long-range Forces. J Mech Phys Solids 48:175–209 [the first paper in this field]
Silling SA, Lehoucq RB (2010) Peridynamic theory of solid mechanics. Advances in applied mechanics, vol 44. Academic Press, New York, pp 73–168
Sneddon IN (1946). The distribution of stresses in the neighbourhood of a crack in an elastic solid. In: Proceedings of the Royal Society of London A: Mathematical, Physical and Engineering Sciences, pp 187, 229
Steinmann P, Ackerman D, Barth HJ (2001) Application of material forces to hyperelastostatic fracture mechanics, Part 2: computational setting. Int J Solids Structures 38:5509–5526
Suo Z, Kuo CM, Barnett DM, Willis JR (1992) Fracture mechanics for piezo- electric ceramics. J Mech Phys Solids 40:739–765
Timoshenko SP (1953) History of the strength of materials. McGraw-Hill, New York [Dover reprint, 1983, New York]
Wells AA (1961) Unstable crack propagation in metals: cleavage and fast fracture. In: Proceedings of the crack propagation Symposium, College of Aeronautics and the Royal Aeronautical Society, Cranfield, vol I, pp 210–230
Wells AA (1963) Application of fracture mechanics at and beyond general yielding. Brit Weld J 10(11):563–570 [introduction of the COD]
Westergaard HM (1939) Bearing pressure and cracks. Trans ASME J Appl Mech 6A:49–53
Willis JR (1967) A comparison of the fracture criteria of Griffith and Barenblatt. J Mech Phys Solids 15(3):151–162

Chapter 10
Geometry and Continuum Mechanics: An Essay

Abstract Geometry, analysis, and numerics all apply to a good modelling of continua. Mathematics is the natural language of physics (Galileo Galilei), geometry is the natural language of continuum mechanics. This essay emphasizes the more or less elementary notions of differential geometry that are hidden in the bases of classical continuum mechanics (Killing's theorem, covariance, Riemannian curvature). Then it examines the intervening of more modern and sophisticated notions that have been introduced for pedagogical purpose in harmony with present day mathematics and others of which the need appeared in the twentieth century development of this science: connections, torsion, Cartan's forms and spaces. The influence of Einstein's theory of gravitation on this increased geometrization and the role played by the formulation of a geometric theory of evolving structural rearrangements and defects such as dislocations and material inhomogeneities is of prime importance. The main actors in this historical perspective appear to be Pfaff, Lie, Riemann, Killing, Cartan, Kondo, Kröner, Bilby, and Noll.

10.1 Introduction

According to the mid nineteenth-century clichés listed in his Dictionary of "idées reçues", Gustave Flaubert recalls that "mechanics is the lowest part of mathematics". Jean Dieudonné (one of the creators of the Bourbaki group in twentieth-century mathematics), in a contribution in the Bulletin of the French Mathematical Society, places "applied mathematics at the lowest level in the classification of mathematics" (not a compliment in his snob's mouth)—something which must have pleased Jacques-Louis Lions! With his usual hubris Dieudonné claims that "(pure) mathematics is the honour of the human species" (this was the title of his

---

Dedicated to Marcelo Epstein on the occasion of his 70th anniversary (December 03, 2014; lecture delivered at the university of Calgary, Canada, at the Society of Natural Philosophy Meeting, August 2015).

G.A. Maugin, *Continuum Mechanics through the Ages - From the Renaissance to the Twentieth Century*, Solid Mechanics and Its Applications 223,
DOI 10.1007/978-3-319-26593-3_10

“modest” autobiography). Then it seems salient to ask the following two questions: (1) what about analysis and continuum mechanics? and (2) what about geometry and continuum mechanics? It would be difficult to put continuum mechanics at the bottom of analysis since both fertilized each other during the eighteenth and nineteenth centuries and their parallel progress often is the result of research by the same scientists, of necessity. This is markedly illustrated by the case of Joseph Louis Lagrange and Augustin Louis Cauchy. Furthermore, the very concept of “continuum” partakes of the same vision (and philology) as “continuity”, a basic notion of analysis if we avoid pathological cases (i.e., singularities). For geometry and continuum mechanics the matter must be pondered. This is the object of the present essay in a somewhat historical perspective where the most creative characters appear to be J.F. Pfaff, G.B. Riemann, W Killing, E. Cartan, K. Kondo, E. Kröner, and W. Noll.

10.2 A Fundamental Theorem by Killing

The answer to the posed question (relationship between geometry and continuum mechanics?) depends on the required degree of sophistication. With the early developments of continuum mechanics in the eighteenth century (cf. Maugin 2013), it is clear that R^3 and E^3 are completely identified as only three-dimensional Euclidean space is considered for physical space, the arena of physical events, and only Cartesian systems of coordinates are envisaged. But it was soon realized that the practical solution of specific problems in two or three dimensions requires the introduction of more adapted systems of coordinates, especially in problems of elasticity. The “engineer-scientist” Gabriel Lamé in France and Italian analysts (e.g., Enrico Betti and Eugenio Beltrami) were among those who elaborated such adapted systems of coordinates (e.g., curvilinear coordinates for spherical shells; cf. Lamé 1859), and wrote equations of elasticity in them. Underlying this useful advance is the idea of some embryonic concept of “invariance”: the fact that fundamentally physically identical problems can be solved in formally different frameworks, with the choice of the most efficient system for a solution that takes account of the shape of objects, the problem symmetries, and the type of boundary conditions. This is not far from our modern notion of covariance, and the introduction of tensor analysis. In fact, these works certainly kindled the development of tensor analysis by the Italian mathematicians Gregorio Ricci (1853–1925) and Tullio Levi-Civita (1873–1941) who introduced the spot-on expression of “absolute differential calculus” (Ricci and Levi-Civita 1900). As a matter of fact, Ricci had both Betti and Beltrami as professors. But we cannot ignore the strong influence exerted by Georg F. Bernhard Riemann (1826–1866) and Elwin B. Christoffel (1829–1900), where the former is responsible for a good introduction of curvature in R^n (after the innovative works of Gauss on surfaces) and the latter for that of covariant derivative (Christoffel 1869; also Ricci-Curbastro 1887). This matter of “covariance” plays a critical role in the pedagogy of continuum mechanics: should we teach the basic equations of continuum mechanics in covariant form without mention of a specific system of coordinates (or even the nature of the considered space, not to speak of an arbitrary number of dimensions)? We do not dwell with this here.[1]

At a less aerial view it is of interest to consider a basic question of continuum mechanics, what will help us to connect the present discursive approach with *group theory*. The inevitable scientist here is seldom cited by mechanicians: it is Wilhelm Killing (1847–1923), a German mathematician who had a rather quiet life where religion and mathematics intermixed harmoniously.[2] His main mathematical contributions are to the theories of Lie algebras, Lie groups and non-Euclidean geometry. He in fact invented Lie algebras independently of Sophus Lie (1842–1899). It is in these studies (see, e.g., Killing 1888) that he was led to the introduction of what are now called *Killing vector fields*. Such fields on a Riemannian manifold preserve the metric: we can say that they are infinitesimal generators of *isometries*. In modern mathematical terms, a vector field X is a Killing field if the Lie derivative with respect to X of the metric g vanishes. This is written as

$$L_X g = 0. \tag{10.1}$$

The connection with standard continuum mechanics is now clear, especially in elasticity. In this context one needs to compare a generally deformed state of a continuum with a standard provided by a rigid body. This will provide a relationship between continuum mechanics and Euler's rigid-body mechanics. If $\mathbf{v}$ denotes the usual velocity field of Euclidean components v_i, $i = 1, 2, 3$ in E^3, then the differential condition

$$\frac{\partial v_i}{\partial x_j} + \frac{\partial v_j}{\partial x_i} = 0, \tag{10.2}$$

is the local condition for the continuum motion to be that of a rigid-body, as the integral of (10.2) is none than the velocity field of a rigid-body motion written as

[1]However, our own experience is that teaching in a simple, perhaps naïve, format using a direct intrinsic notation or a Cartesian tensor notation with indices of one kind only is most often sufficient and rewarding in an introductory course to engineering students. It is not so bad for students in mathematics either! For advanced courses, one must appeal to the precise language of mathematics as a vehicle for formulating concepts and carrying out logical arguments. Modern differential geometry then is the required tool. However, one then faces the insuperable problem to ask from students a simultaneous excellent knowledge of modern geometry, continuum thermodynamics, and, often, materials science and solid-state physics. Only a few versatile students will possess this curriculum and a few researchers will dare to acquire this knowledge just by themselves.

[2]Wilhelm Killing wrote a dissertation under the supervision of Karl Weierstrass and Ernst Kummer at Berlin in 1872. He was for some time a professor and administrator at a Seminary College, but finally became a professor of mathematics at the University of Münster (his Alma Mater) while he and his wife entered the Third Order (i.e., as lay members) of Franciscans.

$$v_i(\mathbf{x}, t) = \Omega_{ij}(t)x_j + V_i(t), \tag{10.3}$$

with spatially uniform skew symmetric Ω_{ij} and spatially uniform translational velocity V_i, i.e.

$$\Omega_{ij} = -\Omega_{ji},\ \Omega_{ij,k} = 0,\ V_{i,k} = 0. \tag{10.4}$$

Equation (10.3) is the velocity field associated with a motion

$$\mathbf{x} = \mathbf{Q}(t) \cdot \mathbf{x}^* + \mathbf{d}, \tag{10.5}$$

such that (T = transposed)

$$\underline{\Omega} = \dot{\mathbf{Q}}\mathbf{Q}^T = -\underline{\Omega}^T, \quad \mathbf{Q}^{-1} = \mathbf{Q}^T, \quad \det \mathbf{Q} = +1, \tag{10.6}$$

where $\mathbf{Q}$ is a spatially uniform orthogonal matrix and $\mathbf{d}$ is a spatially uniform vector.[3] That is, the motion is composed of a spatially uniform (but possibly time-dependent) rotation and a spatially uniform translation. The contact with the general Killing formula (10.1) is as follows. A true deformation of a deformable continuum is described by a nonvanishing symmetric strain tensor that is none other than a metric on the appropriate material manifold. The invariance in the distance between two infinitesimally close material points (i.e., no deformation) is an isometry with the condition (10.2), i.e.,

$$L_{\mathbf{v}}\mathbf{C} = \mathbf{D} = \mathbf{0}, \tag{10.7}$$

where $\mathbf{C}$ is a strain tensor and $\mathbf{D}$, the symmetric part of the velocity gradient, is the rate-of-strain tensor. Together with the notion of squared infinitesimal distance (the famous ds^2) and that of strain (metric), Eqs. (10.2) are in fact the most basic relationships of continuum mechanics with geometry. In other words, *Killing's theorem is the basis of elasticity*. There is no need to emphasize the obvious group structure of orthogonal transformations in E^3, i.e., $SO(3)$.

Three remarks are in order.

First of all we can now say that the importance of group theory for continuum mechanics was in the air in the late 1890s as proved by the first incursion in this domain in the groundbreaking book of Eugène and François (Cosserat and Cosserat 1909). In this original opus, perhaps inspired by works of Sophus Lie, the Cosserats applied to a variational formulation an invariance which they call *Euclidean action*. This is the first manifestation of an application of the invariance under rigid-body displacements (10.5) written in infinitesimal form, and putting on equal footing rotations and translations, and thus emphasizing the group form as well as the

[3]The various invariances considered in continuum mechanics that exploit orthogonal transformations are reviewed in Maugin (2014).

justified consideration of a general deformation involving both the usual degree of translation (the usual displacement) and an internal rotation (the future "Cosserat continuum"). This paved the way for the construction of generalized continuum mechanics. This move and its underlying notion of group of transformations were applauded by Elie Cartan, a specialist of Lie groups.

The second remark pertains to this famous geometer, Elie Cartan (1869–1951). It is said that Cartan's (1894) dissertation was essentially a rigorous rewriting of Killing's paper. But Cartan, with a long active scientific life, was to go much farther than Killing and Lie (see below).

Finally, because of personal interest in the matter, we cannot avoid mentioning the problem of the definition of rigid-body motions in relativity. In the period 1908–1911, one had to envisage a mechanical behaviour more complex than pure fluidity, for instance, elasticity, in the relativistic framework. As already pointed out, elasticity is meaningful only if defined with respect to a standard (a comparison medium) that is normally rigid-body behaviour. But here there is a caveat. The inherent limitation to physical speeds imposed by relativity makes it inadmissible to have a physical object of arbitrary large dimension in rigid-body rotation. Still this was the subject of vivid discussions in the early 1910s. It was proposed by several authors, in particular Born (1911) and Herglotz (1911), that the condition of rigid-body motion be defined *locally* in Minkowski's four dimensional space-time by $(\alpha, \beta = 1, 2, 3, 4;\ x^4 timelike)$

$$\partial_\alpha u_\beta + \partial_\beta u_\alpha = 0, \tag{10.8}$$

in a local spatial section of space-time. This is Killing's theorem written in covariant formalism, but applied to the essentially spatial part of the four-dimensional Minkowskian metric. More precisely (compare to (10.1)):

$$L_u \underline{P} = 0, \quad P_{\alpha\beta} := g_{\alpha\beta} + c^{-2} u_\alpha u_\beta. \tag{10.9}$$

Here $\boldsymbol{P}$ is the so-called spatial projector, $\boldsymbol{g}$ is the space-time metric—of signature $(+, +, +, -)$-, and $\boldsymbol{u}$ is the fourth-velocity. The Born-Herglotz condition (10.8) is a direct space-time generalization of the classical condition (10.2).

At this point, before proceeding to further complications, we must also mention the *Navier-Saint-Venant conditions of compatibility* in small-strain elasticity. The question is the following one: can we deduce in a unique manner the displacement vector (three components) from the knowledge of a strain tensor (six components)? There must exist a set of equations that express this condition; these are known as the Navier-Saint Venant necessary and sufficient conditions usually written in Cartesian tensor components as

$$\varepsilon_{ikr}\ \varepsilon_{jls}\ e_{ij,kl} = 0, \tag{10.10}$$

where e_{ij} denotes the strain and ε_{ikr} stands for the alternation symbol. Another intrinsic and suggestive writing of (10.10) is

$$\nabla \times (\nabla \times \mathbf{e}) = \mathbf{0}; \tag{10.11}$$

This short hand notation is written in analogy with the condition $\nabla \times \mathbf{A} = \mathbf{0}$ for the existence of a potential φ for the vector field **A** so that $\mathbf{A} = \nabla \varphi$. The writing (10.11) is probably due to Kröner (1955)—see more herein after. Equation (10.10) is deduced through some manipulations in textbooks on elasticity. But the geometrical meaning of this equation is much deeper: it means that the material manifold (the set of material points constituting the elastic body) is *flat* and remains flat in the course of deformations that do not allow for the appearance of defects. Indeed, the usual material manifold is usually considered to be a flat Euclidean space with vanishing curvature (in the sense of Riemann). The Riemann curvature tensor involves space derivatives of the Christoffel symbols—themselves defined from derivatives of the metric-, hence second-order space derivatives of the metric. In three dimensions this curvature tensor reduces to a second-order tensor called the Einstein tensor, whose six independent Cartesian components are none other than the left-hand side of Eq. (10.10) when **e** is viewed as the metric of interest (cf. Maugin 1993, pp. 54–58). This is probably as far as an engineering student is exposed to a relationship between geometry and continuum mechanics in his curriculum.

10.3 The Role of Elie Cartan

Elie Cartan is one of the most famous geometers of the first half of the twentieth century (see Chern and Chevalley 1952).[4] A continuator of Johann Friedrich Pfaff (1765–1825) and of Sophus Lie (1842–1899; see Cartan 1948), he has left a definite print in the theory of Lie groups and algebras, associative algebras, non-Riemannian geometry, Finsler spaces, the theory of forms, integral invariants, etc. But more remarkably, he introduced and developed new fields that were to find later on splendid applications in theoretical and mathematical physics. Here we must first mention the theory of *spinors* (synthesis in Cartan 1938) introduced about 1910, hence much before the discovery of spin for elementary particles in quantum physics (the discovery of the spin of the electron dates from 1926). We recall that a spinor is a complex vector that allows an expression of space rotations by a

[4]Elie Cartan was a never tired scientist. Between 1893 and 1947 he published no less than 187 works, including many long memoirs, pamphlets and full-length books. He usually was a single author with only a few exceptional papers co-authored by J.A. Schouten. The best thorough description of his works is given by himself in the Notice he wrote for his election to the Paris Academy of Sciences in 1931 (see Cartan 1974). The main heroes of Cartan are Jacobi, Lie, Riemann, Killing, Poincaré, Levi-Civita, Fubini, Einstein, and H. Weyl. Some of his direct disciples were Charles Ehresmann (1905–1979) and Georges de Rham (1903–1990). Of course one cannot neglect the influence of Hermann Grassmann's (1809–1877) "theory of extensions"—exterior or Grassmann algebra—on Cartan's inception of the calculus of differential forms.

two-dimensional representation. The other new advance made by Cartan is his innovative work in non-Riemannian spaces, including those now named *Riemann-Cartan spaces*, was spaces with both curvature *and* torsion (Cartan 1925, 1935).

Cartan knew well his mechanics as he had followed lectures by Gabriel Koenigs and Paul Appell. He replaced Paul Painlevé on his chair of rational mechanics, also teaching both analytic and celestial mechanics. But his most appropriate and well deserved affiliation was the chair of Higher Geometry that he obtained in 1924. He formed and influenced many French mathematicians in the inter-war period 1920–1940 (cf. Gispert and Leloup 2009). In particular, together with Jacques Hadamard he was instrumental in the creation of the Bourbaki group, of which his own son, Henri, together with Dieudonné, was a most active member.

Apart from a re-evaluation of Killing's works on Lie groups in 1894, Cartan published (Cartan 1899) an extensive work in which, pursuing the works of Pfaff,[5] Jacobi and Lie, he pioneered a modern theory of *differential forms* and introduced the notion of exterior derivative. This provided an approach to multivariable calculus that is independent of coordinates. This found applications in an astute condensed rewriting of the equations of electromagnetism and also continuum mechanics, and of course in an aesthetically wonderful rewriting of the generalized Stokes' theorem. This is well illustrated in the book of Harley Flanders (1925–2013; book in 1963) and that of Flanders' former student, Theodore Frankel (1997). This last book is often presented as the most successful book on the applications of modern differential geometry to physics.[6] Cartan was very much interested in Einstein's theory of gravitation where he almost immediately contributed to a refinement in the mathematical tools of that theory (Cartan 1922a, b). He was very much enthusiastic about possible physical applications of his "Einstein-Cartan" spaces. He went all the way to declare that "if physics can be geometrized at all, then all physical laws must be expressible in terms of partial differential equations governing the torsion of the relevant space" (Cartan 1931; Translation from the French by the author). Cartan certainly had the idea that implementing the Cosserats' notion of generalized continuum would require using some of the geometrical notions he had so forcefully introduced. But he could not guess that his "connection" was to become an ideal mathematical tool in the theory of dislocations, physical objects of which the existence was proved only in the 1930s. As to generalized theories of gravitation with spin, applications of Einstein-Cartan spaces and the formulation of general relativity exploiting exterior derivatives, they will have to await the 1970s.

---

[5]Remember that Pfaff's problem deals with the conditions under which a differential expression in R^n is an exact differential. This plays a basic role in Caratheodory's axiomatic approach to thermodynamics (cf. Maugin 2014, Chap. 9).

[6]See, in particular, pp. 118–122 on Maxwell's equations and pp. 617–629 on continuum mechanics in Frankel (1997). As witnessed by the writer in the mid 1960s, Paul Germain, in Paris, used to formulate the balance laws of continuum mechanics in the language of exterior derivatives. The author particularly appreciated the book of Flanders (1963). Cartan's son, Henri, even wrote a textbook on differential forms (Cartan 2006).

10.4 The Influence of the Theory of General Relativity

As a preliminary we note that contrary to the initial gravitation theory of Newton that dealt with mass points, Einstein's theory of gravitation is from the start a *continuum* theory, since the source of geometry is to be found in energy, momentum and *stresses*. It is so much the case that it will prove difficult to deal with mass points (i.e. singularities in their own right) in this framework. As to the Killing conditions (10.2) it takes the required form using covariant differentiation.

The theory of gravitation devised by Eisntein (1914, 1916), also known as the theory of *general relativity*, was rapidly accepted by his fellow mathematicians and physicists David Hilbert (1862–1943), Gustav Mie (1869–1957) and Herrmann Weyl (1865–1957). The first two of these scientists had been in competition with Einstein, while Weyl tried to extend Einstein's theory so as to incorporate electromagnetism in its geometrical framework. Astronomer (Sir) Arthur Eddington (1882–1944) and physicist Wolfgang Pauli (1900–1958) easily grasped the Riemannian tensor formalism exploited by Einstein (with the help of his friend, the mathematician Marcel Grossmann (1878–1936)). They published comprehensive books on relativity as early as the beginning 1920s (Eddington 1923; Pauli 1921). Then general relativity became a subject of vivid interest in cosmology and the beginnings of astrophysics while Einstein engaged himself in a rather vain quest for a generalized theory that should ultimately include quantum effects. Elie Cartan was more inclined to a generalization by considering a non-Riemannian space-time. He was on the right track with the notions of "Cartan's connection" and torsion (Cartan 1922a, b, 1923).

Any way, the tensor analysis and Riemannian spaces exposed in—at the time—a *reasonable* formalism (i.e., with explicit contra-variant and co-variant indices) do not seem to have been obstacles to an apprehending by many people. Many of these readers met first with these notions by reading the nicely written and readable small book by Einstein (1956). This was the case of many curious engineers and amateur physicists.[7] This must have been the case of Kazuo Kondo (1911–2001), a Japanese aeronautical engineer. Although occupied with problems of the mechanics of structural members, he seems to have been the first to propose an interpretation of the theory of structural defects (e.g. plasticity and dislocations) in geometrical terms borrowed from Einstein's theory of gravitation (Kondo 1952). It is with this revolutionary work that the sophistication of Riemannian geometry entered the field of classical continuum mechanics. Kondo was to expand further aspects of his

[7]This was the case of the author who learned his basic tensor analysis in Einstein (1956) and later on in the treatise of Landau and Lifshitz on theoretical physics (Volume 2 on the theory of fields). This is of some importance because no amateur could have learned this from more recent treatises using abstract differential geometry without first having been exposed to hard mathematics. Of course, the teaching of André Lichnerowicz in Paris and the small (in format) book (Lichnerowicz 1946) he had published on the subject were also essential for French students such as the author. Walter Noll followed Lichnerowicz's lectures when he stayed in Paris in 1949–1950. Before Lichnerowicz's book, there were the very little read treatise by Thiry (1925) and the original book by Brillouin (1938); but these two books did not reach mechanical engineers.

geometrical approach in a series of publications known as the *R.A.A.G. Memoirs* (Kondo 1955). The problem with these publications is that their author introduced too much unnecessary philosophy and too many neologisms (probably approximately translated—or even perhaps not translatable—from the Japanese). A somewhat parallel advance was made by Ekkehart Kröner (1919–2000) in Germany and a group around Bruce A. Bilby in the UK. For this see Sect. 10.5.

The influence of general relativity in the present context is also to be appreciated in the light of the remarkable book of Marsden and Hughes (1983). This book set new standards in the mathematical formulation of classical continuum mechanics by using systematically differential geometry in its new modern outlook. Of course this reflected the initial formation of Jerold Marsden (1942–2010) at Princeton (PhD 1968) in the environment of John Archibald Wheeler (1911–2008) who used some of this formulation in his course (as reasonably illustrated in the monumental book of Misner et al. 1973).[8] Among other creative works in particle physics, quantum mechanics, astrophysics and cosmology, Wheeler was a partisan of *geometrodynamics*, a program of physical and ontological reduction of every physical phenomenon, such as gravitation and electromagnetism, to the geometrical properties of a curved space-time (among his many PhD students, we note R.P. Feynman).

10.5 The Influence of the Theory of Dislocations

Here one must emphasize the role played by Kröner (1958), Bilby et al. (1955, 1957), and Walter Noll (1967). Kröner first recognized that the Navier-Saint Venant condition (10.10) or (10.11) is no longer satisfied when the elastic displacement is no more continuous as a result of the presence of dislocations. There are so many of these that a density of dislocations has to be introduced, while Johannes M. Burgers (1895–1981) in the Netherlands had introduced the notion of "Burgers" vector (measure of the discontinuity in displacement) to characterize a single dislocation in Burgers (1939). As (10.10) has now acquired a non-vanishing right-hand side the material manifold is now equipped with a non-zero curvature. The material manifold has thus become Riemannian and, accordingly, is no longer Euclidean. Kröner (1955) astutely related this misfit with the original Navier-Saint Venant compatibility condition by introducing the notion of elastic incompatibility. Equation (10.2) is replaced by a balance between the Einstein tensor and Kröner's incompatibility tensor that can be expressed in terms of the gradient of a dislocation density tensor whose integral over an elementary surface equals the Burgers vector. The dislocation density tensor satisfies a conservation law (see Maugin 1993, pp. 57–58) which can be interpreted as Bianchi's identity in differential geometry. This tensor can be identified with the linearized version of the Cartan torsion tensor—more on this in classical elasticity and its generalized-continuum versions in Lazar and Maugin (2007).

[8]As directly witnessed by the author as a graduate student at Princeton.

Kröner was aware of the possible relationship between Cosserat continua and Cartan's geometry.[9] His student, Friedrich Hehl established this relation in his Habilitation diploma obtained in Clausthal in 1969. Hehl was to become one of the best specialists in the Einstein-Cartan theory of gravitation with spin (cf. Hehl et al. 1976). We examined Kröner's scientific legacy in Maugin (2003).

About the same time as Kröner, Bilby and his co-workers in Sheffield engaged in an ambitious programme of which the purpose was to develop a geometric theory of the continuous distribution of dislocations. They emphasized that (Bilby et al. 1955) "the geometry of the continuously dislocated crystal is most conveniently analyzed by treating the manifold of lattice points in the final state as a non-Riemannian one with a single asymmetric connexion. The coefficients of that connexion are then expressed in terms of the generating deformations that relate the dislocated crystal to the reference lattice". It is at this point that the local density of dislocations can be identified with the torsion tensor associated with the connection. The latter possesses the property of so-called "distant (or absolute) parallelism" in the sense of Elie Cartan (1931). Bilby et al. developed their theory in a series of six long papers between 1955 and 1966. The result is aesthetically rewarding but still now of limited use.

Both Noll (1967) and Wang (1967) published mathematical contributions to the subject matter invoking a general approach to the theory of uniformity and homogeneity of simple bodies. These contributions, although often cited, were very little exploited as being probably too abstract and not legible by engineers or materials scientists. They indeed involved a Cartan connection. These two papers were much admired by Marcelo Epstein and the present author. It was due to these two authors (Epstein and Maugin 1990) to show that the Cartan connection based on the anelastic part $\mathbf{F}^a$—as a matter of fact, its inverse $\mathbf{K}$—of the multiplicative decomposition of the deformation gradient was intimately related with the *material Eshelby stress* (sometimes called Eshelby energy-momentum stress) of J.D. Eshelby's theory of material inhomogeneties, and the associated fully material balance of material (or configurational) forces. This is dealt with at length in a book (Maugin 2011).

10.6 The Theory of Local Structural Rearrangements

Following along the line of the preceding section, we simply note the following expressions:

$$\mathbf{F} = \mathbf{F}^e\mathbf{F}^a, \quad \mathbf{K} = (\mathbf{F}^a)^{-1}, \quad \Gamma = \left(\nabla_R \mathbf{K}^{-1}\right) \cdot \mathbf{K}, \tag{10.12}$$

[9]Kröner associated the memories of both the Cosserats and Cartan in the dedication of the IUTAM Symposium he organized in the Black Forest in 1967 (Kröner Editor 1968).

where $\mathbf{F}$ is the total deformation gradient between the reference configuration and the actual one, $\mathbf{F}^e$ is the elastic part and $\mathbf{F}^a$ the anelastic part - both non-integrable separately in a displacement[10] -, of $\mathbf{F}$, and Γ is a Cartan connection on the material manifold; notice that this is not defined from the metric.

This opened the way for a synthesis where all configurational forces are seen as contributing jointly to a balance of *material* momentum. This we called the *theory of local structural rearrangements*, since it includes the local—in general thermodynamically irreversible—transformations of the material manifold due to effects of macro- and micro-structural defects, foreign material inclusions (inhomogeneities), temperature conduction, phase transformations, and material growth. An especially interesting case is the last one mentioned: growth.

This must concern open thermodynamic systems as the import of nutrients is necessary. Therefore, the complexity of the geometrical description already reached for closed systems is normally increased by the coupling with the diffusion of mass and the transport of species. It was shown by Epstein and Maugin (2000) that a first-order gradient theory based on Eq. (6.1) was not sufficiently well equipped to deal with this case. In particular, thermodynamic consistency with mass diffusion required considering a second-gradient theory. This was achieved in a work started in Ciarletta and Maugin (2011). The reached geometric complexity is such that first-order and second-order connections must be introduced together with material Eshelby stress tensors of first and second orders. This complexity allows one to examine the problem posed by the existence of singular lines that may appear during the process of growth of soft biological tissues.

10.7 Modern Differential Geometry and Its Use in Continuum Mechanics

Within about sixty years the relationship between differential geometry and continuum mechanics has continuously strengthened reaching an incredible level of sophistication and complexity with the necessary consideration of a non-Riemannian structure for the material manifold. This we owe essentially to the combined efforts of Kondo, Kröner, Bilby, Noll, Wang and, more recently, Epstein and co-workers. Rather recent works using generalized Riemannian spaces or non-Riemannian spaces in dislocated elastic bodies are by Rakotomanana (2003), Zubov (1997) and Yavari and Goriely (2012). A short retrospective is appropriate at the end of this contribution.

No doubt that geometry is the basis on which the kinematics and deformation theory of continuum mechanics rely. We agree with Epstein (2010) that the language of continuum mechanics is geometry. Until recently only geometry as made analytical by René Descartes and considering the three-dimensional Euclidean

[10] Accordingly they are related to Pfaffian differentials.

space as the normal arena of continuum mechanics was acknowledged as the standard background. Differential geometry as formulated by nineteenth century mathematicians (above all Gauss and Riemann) was the tool that introduced the notions of metric and eventually curvature.[11] Two facts have complicated the picture. One was the influence of the consideration of non-Euclidean spaces in gravitational theory following Einstein and others. The second, in fact related to the first, was the recognition that taking account of the presence of many structural defects requires abandoning the peculiarity of the Euclidean nature of material space in favour of more general concepts introduced by modern differential geometers such as Elie Cartan: non-Riemannian spaces and the allied incursion of group theoretical concepts, following pioneering works by Pfaff, Lie and Killing in the nineteenth century.

The author remembers that in the 1950s–1970s the three main books that offered an in-depth presentation of all necessary tools to foster progress in differential geometry applied to physics were celebrated books by Jan A. Schouten (1883–1971)—Book of 1954—and Luther P. Eisenhart (1876–1965)—books of 1966 (in fact 1924) and 1927. Both these geometers were certainly stimulated by Einstein's theory of general relativity. They were true contributors to advances in tensor calculus for which they are considered "masters of tensors". These books are still cited as the recommended ones in the field by Ericksen (1960) and Eringen (1971) in their summaries of tensor analysis for mechanicians. The title of Schouten's book smells good of the initial development of the field by Ricci. The long period of scientific activity by these two mathematicians explains the durable influence that they had. They provided a continuous thread between the origins of tensor calculus and the 1960s. But their books, although extremely rich and worth a perusal in depth, will certainly look much antiquated to all students of the early twenty-first century. In the mean time, since the 1960s the Bourbaki tradition in general mathematics and Elie Cartan and his disciples have left a definite print that transpires in all modern expositions. Each teacher has thought good to publish his own book on the matter, so that we have a large choice at our disposal. One nice example is provided by Chapters III through V in the book of Choquet-Bruhat et al. (1977) aimed at mathematical physicists. For memory we recall the book of Marsden and Hughes (1983) and that of Frankel (1997). All these authors had contributed in one way or another to general-relativistic studies or quantum-field theory. Finally, the most recent addition in this line at the time of writing seems to be the book by our almost twin-brother[12] Marcelo Epstein (2014)[13] to whom we are happy to dedicate the present essay on the occasion of his seventieth anniversary.

---

[11]And therefore, by negation, a good definition of flatness; think of the Navier-Saint Venant equations of compatibility.

[12]M. Epstein was born on the third of December 1944, while the author was born on the second of December of the same year, however in quite distant places in the world.

[13]Even Epstein tried his hand at some relativistic formulation of anelastic continua (see Epstein et al. 2006).

Gallery of Portraits

Johann F. Pfaff (1765–1825)

Sophus Lie (1842–1899)

G. Bernhard Riemann (1826–1866)

Elwin B. Christoffel (1829–1900)

Wilhelm Killing (1847–1925)

Elie Cartan (1869–1951)

Kazuo Kondo (1911–2011)

Ekkehart Kröner (1919–2000)

References

Bilby BA, Bullough R, Smith E (1955) Continuous distributions of dislocations: new application of the methods of non-Riemannian geometry. Proc Roy Soc Lond A231:263–273

Bilby BA, Gardner LRT, Stroh AN (1957) Continuous distributions of dislocations and the theory of plasticity. Actes du Xième Congrès International de Mécanique Appliquée (Bruxelles, 1956), vol VIII. Université de Bruxelles, Belgique, pp 35–44

Born M (1911) Elastizitätstheorie und Relativitätstheorie. Phys Zeit 12:569–575

Brillouin L (1938) Les tenseurs en mécanique et en élasticité (Vol. I of a course on mathematical physics) Masson, Paris [Reprint by Dover, New York, 1946; English translation, Academic Press, New York, 1963]

Burgers JM (1939) Some considerations of the field of stress connected with dislocations in a regular crystal lattice. Proc Kon Nederl Kad Wissensch 42(293–325):378–399

Cartan E (1894) Sur la structure des groupes de transformations finis et continus. Thesis Univ. Paris, Editions Nony (Second edition, Vuibert, Paris, 1933)

Cartan E (1899) Sur certaines expressions différentielles et le problème de Pfaff. Ann Scient Ecole Normale Supérieure 239–332

Cartan E (1922a) Sur les équations de la gravitation d'Einstein. J Math Pures Appl 39:141–203

Cartan E (1922b) Sur une généralisation de la notion de courbure de Riemann et les espaces à torsion. C R Acad Sci Paris 174:593–596

Cartan E (1923) Sur les variétés à connexion affine et la théorie de la relativité généralisée. J Math Pures Appl 40:325–412 [also Ann Ecole Normale 42:17–88]

Cartan E (1925) La géométrie des espaces de Riemann. Hermann, Paris

Cartan E (1931) Le parallélisme absolu et la théorie unitaire du champ. Revue de Métaphysique et de Morale, pp 13–28

Cartan E (1935) La théorie des groupes continus et des espaces généralisés. Hermann, Paris

Cartan E (1938) La théorie des spineurs. Hermann, Paris

Cartan E (1948) Un centenaire: Sophus Lie. In: Les grands courants de la pensée mathématique, Ed. F. Le Lionnais. Hermann, Paris (New Edition, 1998)

Cartan E (1974) Notice sur les travaux scientifiques. Gauthier-Villars, Paris [with a complete list of publications and his "Notice" written in 1931, in pp 15–112, and complemented in footnotes for the works after this date]

Cartan H (2006) Differential forms. Dover, New York (First French edition, Hermann, Paris, p 1967

Chern SS, Chevalley C (1952) Obituary: Elie Cartan and his mathematical work. Bull Am Math Soc 58:217–250

Choquet-Bruat Y, DeWitt-Morette C (1977) Analysis, manifolds and physics. Elsevier, Amsterdam

Christoffel EB (1869) Ueber die transformation der homogener differentialaustdrücke zweiten grades. Journal für d reine und angewandte Mathematik 70:46–70

Ciarletta P, Maugin GA (2011) Elements of a finite-strain-gradient thermomechanical theory for material growth and remodelling. Int J Non-Linear Mech 46:1341–1346

Cosserat E, Cosserat F (1909) Théorie des corps déformables. Hermann, Paris, 226 Pages (Reprint by Editions Gabay, Paris, 2008; Reprint by Hermann Archives, Paris, 2009) [English translation: N68–15456: Clearinghouse Federal Scientific and Technical Information, Springfield, Virginia NASA, TT F-11561 (February 1968); another translation by D. Delphenich, 2007]

Eddington A (1923) The mathematical theory of relativity, 1st edn. Cambridge University Press, UK

Einstein A (1914) Die formale Grundlage der allgemein Relativitätstheorie. Sitz der Preuss Akad Wiss Berlin, 1030–1085

Einstein A (1916) Die Grundlage der allgemeine Relativitätstheorie. Ann der Phys 49:769–822 [English translation in: Lorentz HA, Einstein A, Minkowski H, Weyl H (1923). The principle

of relativity: a collection of original memoirs. Methuen, London; reprinted by Dover, New York, 1952]
Einstein A (1956) The meaning of relativity. Princeton University Press, Princeton, NJ [First edition, P.U.P., 1923; Text of lectures delivered at Princeton University, May 1921; 118 pages]
Eisenhart LP (1927) Non-Riemannian geometry. American Mathematical Society, New York
Eisenhart LP (1966) Riemannian geometry. Princeton University Press, NJ (First edition, 1924)
Epstein M (2010) The geometrical language of continuum mechanics. Cambridge University Press, UK (312 pages)
Epstein M (2014) Differential geometry (Basic notions and physical examples). Springer series on Mathematical Engineering, Dordrecht (139 pages)
Epstein M, Maugin GA (1990) The energy-momentum tensor and material uniformity in finite elasticity. Acta Mech 83:127–133
Epstein M, Maugin GA (2000) Thermomechanics of volumetric growth in uniform bodies. Int J Plast 16:951–978
Epstein M, Burton DA, Tucker R (2006) Relativistic anelaticity. Class Quantum Gravity 23:3543–3571
Ericksen JL (1960) Tensor fields. Appendix in: the classical theory of fields, Handbuch der Physik, Bd III/1, Ed. S. Flügge, Bd III/1, pp 794–858, Springer-Verlag, Berlin
Eringen AC (1971) Tensor analysis, part I. In: Eringen AC (ed) Continuum physics, vol I. Academic Press, New York, pp 1–155
Flanders H (1963) Differential forms with applications to the physical sciences. Academic Press, New York [Dover reprint, New York, 1989]
Frankel T (1997) The geometry of physics (an introduction). Cambridge University Press, UK
Gispert H, Leloup J (2009) Des patrons des mathématiques en France dans l'entre-deux guerre. Revue d'Histoire de sciences 62:39–117
Hehl FW, Von der Heyde P, Kerlick GD, Nester JL (1976) General relativity with spin and torsion: Foundations and prospects. Rev Mod Phys 48(3):393–416
Herglotz G (1911) Ueber die Mechanik des deformerbaren Korpers vom Standpunkte der Relativitätstheorie. Ann der Phys 36:493–533
Killing W (1888) Die Zusammensetzung der stetigen/endlichen Transformationsgruppen. Math Ann 31(2):252–290
Kondo K (1952) On the geometrical and physical foundations of the theory of yielding. In: Proceedings of 2nd Japanese national congress of applied mechanics, ed. Japan National committee for theoretical and applied mechanics. Science council of Japan, Tokyo, pp 41–47
Kondo K (1955) Non-Riemannian geometry of imperfect crystals from a macroscopic Viewpoint. In: Kondo K (ed) RAAG memoirs of the unifying study of basic problems in engineering and physical sciences by means of geometry, vol 1. Gakujutsu Bunken Fukyukai, Tokyo, pp 459–480
Kröner E (1955) Inneren Spannungen und der Inkompatibilitätstensor in der Elastizitätstheorie. Zeit angew Physik 7:249–257
Kröner E (1958) Kontinuumstheorie der Versetzungen und Eigenspannungen. Springer- Verlag, Berlin
Kröner E (1968, Editor) Generalized continua. Proc.IUTAM Symp. Freudenstadt1967. Springer-Verlag, Berlin
Lamé G (1859) Leçons sur les coordonnées curvilignes et leurs diverses applications. Paris
Lazar M, Maugin GA (2007) On microcontinuum field theories: the Eshelby stress tensor and incompatibility conditions. Phil Mag 87:3853–3870
Lichnerowicz A (1946) Eléments de calcul tensoriel, 1st edition (7th edition, 1964). Armand Colin, Paris [Reprint by Gabay, Paris, 2011] (English translation by J.W. Leech, "Elements of Tensor Calculus", J. Wiley & Sons, New York, and Methuen & Co Ltd, U.K., 1962)
Marsden J, Hughes T (1983) Mathematical foundations of elasticity. Prentice Hall, New York (Dover reprint, New York, 1994) [Box 4.2 in pp. 169–175 is based on works by Epstein and Segev]
Maugin GA (1993) Material inhomogeneities in elasticity. Chapman & Hall, London

Maugin GA (2003) Geometry and thermomechanics of structural rearrangements: Ekkehart Kroener's Legacy (GAMM'2002, Kroener's Lecture, Augsburg, 2002). Zeit angew Math Mech 83:75–83
Maugin GA (2011) Configurational forces. CRC/Chapman & Hall/Taylor and Francis, Boca Raton, Fl
Maugin GA (2013) Continuum mechanics through the twentieth century: a concise historical perpective. Springer, Dordrecht
Maugin GA (2014) Continuum mechanics through the eighteenth and nineteenth centuries: historical perspective from John Bernoulli (1727) to Ernst Hellinger (1914). Springer, Dordrecht
Misner CW, Thorne KS, Wheeler JA (1973) Gravitation. W.H. Freeman and Co, San Francisco
Noll W (1967) Materially uniform simple bodies with inhomogeneities. Arch Rat Mech Anal 27:1–32
Pauli W (1921) Relativitätstheorie. In: Bd. 19 of Encyclopädia der Mathematischen Wissenschaften, Teubner-Verlag, Leipzig (English translation with revisions, Pergamon Press, Oxford, 1958)
Rakotomanana LR (2003) A geometric approach to thermomechanics of dissipating continua. Birkhäuser, Boston
Ricci-Curbastro G (1887) Sulla derivazione covariante ad une forma quadratica Differenziale. Rend Accad Lincei 3(4):15–18
Ricci G, Levi-Civita T (1900) Méthodes de calcul différentiel absolu et leurs applications. Math Ann 54:125–201
Schouten JA (1954) Ricci calculus (2nd revised and enlarged edn.) Springer-Verlag, Berlin
Thiry R (1925) Eléments de calcul tensoriel: applications géométriques et mécaniques. In: P. Appell, Traité de Mécanique Rationnelle, Vol.5, pp. 1–202, Gauthier-Villars, Paris [Second edition, 1935; Gabay reprint, Paris, 1991)] (excellent but little known)
Wang CC (1967) On the geometric structure of simple bodies, or mathematical foundations for the theory of continuous distributions of dislocations. Arch Rat Mech Anal 27:33–94
Yavari A, Goriely A (2012) Riemann-Cartan geometry of nonlinear dislocation mechanics. Arch Rat Mech Anal 205(1):59–118
Zubov LM (1997) Nonlinear theory of dislocations and disclinations in elastic bodies (Translation from the Russian). Lecture Notes in Physics n.47, Springer-Verlag, Berlin, Heidelberg

Chapter 11
The Masters of Modern Continuum Mechanics

Abstract Although normal progress in a physical theory usually is a totally collective adventure within a given paradigm, it sometimes happens that a selected group of gifted and industrious individuals, working in concert or in competition, revisits an existing field and brings to it innovative and unifying ideas that tremendously accelerate the rhythm of this progress. This is what occurred with continuum mechanics in the period 1935–1980. In this specific case the main figures that clearly emerge are those of R.S. Rivlin, C.A. Truesdell, J.L. Ericksen, W. Noll, R.A. Toupin; B.D. Coleman, A.C. Eringen, A.E. Green, P.M. Naghdi, and R.D. Mindlin, and to a lesser degree but within a different political or regional background, L.I. Sedov, A.A. Ilyushin and P. Germain. Their works and definite influence through their publication of papers and books, the works of their students and direct co-workers, and their contributions to scientific life (in particular through newly created journals and scientific societies) are thoroughly perused in an objective manner.

11.1 Introduction

It is not a trivial matter that a true revival of continuum mechanics as an all embracing science[1] occurred in the second half of the twentieth century. Of course it would have been neither appropriate, nor good mannered, to have asked high-calibre scientists such as Lagrange, Cauchy or Helmholtz, or even Barré de Saint-Venant or Boussinesq, if they preferred fluid mechanics over solid mechanics, or vice versa. They were too much conscious of the common bases of these two fields of mechanics. But inevitable specialization took place, especially in the engineering community, but also in the adopted curricula, starting with the late 1800s. Nonetheless, a small group of scientists coming from various scientific or

[1]As a preliminary remark we specify that we consider only scientists who contributed to this science in this broad sense and not those, excellent and creative as they may have been, who specialized only either in solid or in fluid mechanics.

G.A. Maugin, *Continuum Mechanics through the Ages - From the Renaissance to the Twentieth Century*, Solid Mechanics and Its Applications 223,
DOI 10.1007/978-3-319-26593-3_11

engineering horizons re-kindled a global apprehending of a unified science: continuum mechanics (to be further generalized as continuum *thermo*-mechanics). This took place in the period extending roughly from 1935 to 1980. What sparked this movement is not extremely clear, although, like the Renaissance intellectuals re-evaluating to their own benefit antique Greco-Roman knowledge and vision, we can think that some of the most scientifically curious individuals also imbued with a historical vision, in some sense re-evaluated the contributions of their predecessors from the eighteenth and nineteenth centuries. This is crystal-clear in the case of Clifford Truesdell, a great admirer of nineteenth century French, Italian and German mathematicians, mechanicians and all round physicists such as, among others, Cauchy, Helmholtz, Kirchhoff, Piola, Duhem, Poincaré and the Cosserat brothers.

We have emphasized the crucial role of some well known scientists in the rebirth of continuum mechanics in the twentieth century in a recent book (Maugin 2013a), together with a sketchy biography for some of them. We will examine the career, achievements, influence and legacy of the most representative ones in greater detail herein after. What is most striking is that these scientists had extremely varied backgrounds and past experiences when they got involved in this outburst of works of a more abstract and more synthetic scope with a general purpose in continuum mechanics. They were all practically born between 1915 and 1930, so that they would have reached at least 84 years of age at the time of writing this contribution, but only a very few are still alive (e.g., Ericksen, Noll, Coleman, Toupin, Kunin), although understandably not very active in the field. Some received an initial formation in classical mechanical engineering often busy with the solution of problems involving structural members or plastic behaviour very much in the spirit expanded by the American Society of Mechanical Engineers (e.g., A.C. Eringen, R.D. Mindlin, P.M. Naghdi). Others were basically educated in the British School of applied mathematics and theoretical mechanics (e.g., A.E. Green and R.S. Rivlin). Still others were originally destined to become more or less pure mathematicians (e.g., J.L. Ericksen, P. Germain). Those educated in Russia benefited from a university system in which a close co-operation between mathematics and mechanics was cultivated (e.g., L.I. Sedov at the Lomonosov State University in Moscow, A.I. Lurie and I.A. Kunin in Leningrad/Saint Petersburg). E. Kröner was more of a mathematical physicist while R.A. Toupin came from applied physics, J. Kestin from thermal sciences and energetics, and B.D. Coleman from chemical engineering. Finally, last here but not least, C.A. Truesdell had a standard education in applied mathematics having written a doctoral thesis on classical special functions. As to W. Noll, he benefited from a mixed German education in engineering and mathematics, and a strong influence of modern abstract mathematics experienced during a stay in Paris. He obviously was the closest on the way to a global formalization of continuum mechanics, while Truesdell cultivated a love for old books and a true interest in the history of mechanics in the eighteenth and nineteenth centuries, which he combined with a practice of several European languages needed for a good appraisal of scientific publications of the past not written in English.

The selection of prominent contributors to the considered renewal and blooming may seem arbitrary to some readers and giving too much credit to American authors or to British scientists who had entertained interactions with American colleagues. But this is just an acknowledged fact. Of course, Russian authors deserve also attention for attempts at a synthetic vision that were belittled internationally for linguistic or ideological reasons. This is only slightly corrected in what follows.

Among the general trends that took place in the 1920s–1940, we must first notice that a good appraisal and modelling of behaviours such as finite-strain elasticity and non-Newtonian fluids led to a re-examination of previously devised notions (in fact introduced by the great precursors of the nineteenth century—Cauchy, G. Green, Piola, Finger, Kirchhoff, etc.) and then realizing that many elements of the kinematics of deformations were shared by both solid and fluid mechanics. Rivlin and A.E. Green are the best examples of scientists who contributed so much to this unification. The re-reading (if there ever was a real reading of these in their own time) of works by the Cosserat brothers and P. Duhem was also instrumental in this new adventure. This is where the role of Truesdell was so much important as he greatly contributed to unearthing these innovative and well thought works and directing other researchers to them. The general thermodynamic background was also very much influenced by a new appraisal of Helmholtz's, Gibbs' and Duhem's attempts at a general energetics. This is not to say that all was already done and one had only to read these masters with care and proper interpretation. Some real work was needed but nothing came completely out of the blue. That is where the analogy with the renaissance period is most obvious with a strategy involving first a good reading, next capitalizing on this acquired knowledge, and then adding new original material inspired by the work of predecessors while benefiting from recent advances in mathematics, especially in so far as formalism and the exploitation of new analytical (e.g., functionals) and geometrical (e.g., modern differential geometry) tools are concerned.

11.2 Rivlin and Truesdell

Those who have known these two scientists—always in not so friendly competition —may be surprised to find their names under the same title. They epitomized two different types of intellectuals. One, Ronald Rivlin, perfectly stands for the somewhat arrogant pontificating British scientist, a pure product of Cambridge, and the other, although formidably educated—a true pundit—and deeply influenced by old western Continental Europe, remains an American individual with all exaggeration that goes with this quality. Both were harsh critics of each other and of other scientists. They definitely believed that they owned the real truth. On the creative side we lean in favour of Rivlin with many technicalities and high analytical dexterity typical of British applied mathematics. But Truesdell was a better communicant and a forceful propagandist who succeeded through his many beautifully written pamphlets and encyclopaedic books and an astute choice of direct disciples to rally most mechanicians of the period. Indeed, he was seconded in this endeavour by bright and efficient co-workers such as Richard Toupin and Walter Noll. As to

Rivlin, his cooperation with Albert E. Green, Jerald Ericksen, and Anthony Spencer also bore fruits that influenced our whole community.

Clifford Ambrose Truesdell III (1919–2000)[2] came from a patrician family as his very name indicates. He obtained a BSc in mathematics and physics and an MSc in mathematics at the California Institute of Technology (CALTECH), respectively in 1941 and 1942. In 1943 he was granted a PhD in mathematics at Princeton University. He admits to have been much impressed by the applied mathematician Harry Bateman. His PhD thesis dealt with very classical differential equations[3] (in the line of Bateman), while his first short papers were devoted to equations issued from the applied mechanics of structural members. This does not prefigure the style and contents of his future works. But more typical of him we note that between high school and college, Truesdell took the time to travel for two years around Europe, learning then some of the languages he will need in his historical studies. On graduating he joined the U.S. Navy in its research office in 1943 before obtaining his first University position at Indiana University in Bloomington in 1950. This trajectory explains but only in part the achievements of Truesdell.

In the Navy laboratories Truesdell was kept busy with problems of fluid mechanics but he also had some time to ponder the recent developments of continuum mechanics. But as we already told in a previous volume (Maugin 2013a, pp. 62–63): "In the introduction to a long critical synthesis that was published in 1952, he tells his own story of how he got involved in the field of nonlinear continuum mechanics. He began to study the foundations of continuum mechanics in 1946 and claims—rather immodestly—that within a few months "he had set the whole field in order, to his own satisfaction" (Truesdell 1984a, b). However, an editor told him that he had underestimated the work of earlier authors. In particular, he had overlooked the then recent works by M. Reiner, R.S. Rivlin and others (e.g. Rivlin 1948). This prompted him to return to the sources cited by authors of books, and then to the sources of these sources and so on. The result of this historical search and a special effort at a synthesis was the long contribution he published in the first issue (1952) of the *Journal of the Rational Mechanics and Analysis*, published by the Graduate Institute for Applied Mathematics at Indiana University in Bloomington. This publication bears the characteristic style and aim of further encyclopaedic articles by Truesdell, a rigour allied to an obsessive mania for citations and correcting other scientists, all this in a rich but sometimes pedantic language. This opus of 175 pages, written at the latest in 1949, was adorned by an incredible list of references—with the oldest references to the seventeenth century—a large number of footnotes, and an index of cited authors spreading over six pages on two columns. Truesdell [see the reprint of the preface in Truesdell (1984a)] admitted later that he had made many mistakes and overlooked important authors. In writing this long historical review Truesdell was trying—and succeeded

[2]There exists at least one full length biography in German of Truesdell by Ignatieff and Willig (1999).

[3]This work was reviewed by Friedman (1948).

to a large degree—to supersede his great predecessors, e.g., Joseph Lagrange in his book on analytical mechanics with his historical notes, and of course Pierre Duhem in his always carefully documented various writings.

This was to provide a solid basis for the redaction of the future formidable *Handbuch* article co-authored with Richard Toupin in 1960. Alongside the publication of the *Archives for Rational Mechanics and Analysis*, this monumental contribution—that deals with the basic formulation and general principles—really is the vehicle that opened the horizon of many mechanicians to an extremely rich and fully documented framework. The historical reconstruction of the field from the seventeenth century to the late 1950s was a complete discovery to many readers, including the present writer, at a time when most lecturers rarely cited their sources and delivered courses like god-given material. Truesdell had engaged himself in a program of reconstruction and historical presentation of continuum mechanics. Pierre Duhem (1861–1916), with his complete devotion to the continuum view and his recourse to original sources, whatever the language used, probably is the model that Truesdell wanted to imitate and perhaps surpass definitely. Anyhow both Duhem and Truesdell are considered scientists of the continuum *and* historians of sciences, although Duhem specialized in the oldest texts with a keen interest in medieval science while Truesdell will devote himself to a critical edition of Euler's works in six volumes devoted to mechanics in all its aspects. Furthermore, Duhem was an advocated champion of the thermo-dynamic approach, a line that Truesdell strongly encouraged with the works of B.D. Coleman and W. Noll, when he did not himself contribute directly to this trend.

It is at this point that it seems proper to introduce **Ronald S. Rivlin** (1915–2005)[4] in greater detail since Truesdell's long contribution of 1952 reflected so much the early works of Rivlin and other "rheologists". These are the specialists of "rheology", the term coined by Eugene Bingham (1878–1945) and Markus Reiner (1886–1976) to denote the science dealing with every thing that can more or less flow, pure fluidity in the sense of Euler and strict elasticity being the two extreme behaviours (in agreement with the "continuity of states" of Noll 1955). Rivlin was very well educated in applied mathematics and physics at Cambridge (BA in Mathematics in 1937, MA in 1939). After a short stay at *General Electric Co* and two years as a Scientific Officer with the Ministry of Aircraft Production during WWII, he spent nine years (1944–1953) at the *BRPRA* (British Rubber Producers Research Association) and then the *BRRA* (British Rayon Research Association) doing both seminal theoretical and experimental works. He was naturally concerned with the behaviour of rubber and also very viscous fluids intervening in chemical engineering processes. He joined Brown University in 1954 and then switched to the Centre for the Applications of Mathematics at Lehigh University (1967–1980).

Although demonstrating then that he could also be a good experimentalist, with his initial training as a mathematician in Cambridge Rivlin was naturally led to

[4]Autobiographic notes are given in Rivlin (1996). See also Carroll and Hayes (2006).

considering a pure mathematical standpoint to formulate the expression of relevant energies for rubber like materials. This was a great achievement that provided another view to the recent results obtained by Flory, Guth and James, Treloar, and Mooney (e.g., Treloar 1946; Mooney 1940). All these scientists worked closely with the related industry of rubber and early artificial fabrics with the above mentioned *BRPRA* and *BRRA*. They all considered first an approach based on the physical description of polymers with long chains of molecules and thus necessarily exploiting arguments of statistical physics. As to Rivlin, although knowledgeable in this kind of approach, he considered a pure mathematical view of nonlinear elasticity with a postulate of an energy function depending on the correct invariants of the strain. This interest for invariants will recur in his works. This is how the celebrated *Mooney-Rivlin energy* expression for incompressible rubber-like materials was born. Such considerations required an excellent knowledge of the theory of finite deformations which Rilvin did not hesitate to revisit and develop (e.g., Rivlin 1948). This he achieved by building on previous works of F.D. Murnaghan (1873–1976; Irish mathematician established in the USA; see Murnaghan 1937, 1951) and nineteenth-century scientists such as Cauchy, G. Green, Finger, Piola, Kirchhoff, and Boussinesq. He does not seem to have been aware of the works of the Cosserats and Duhem, or to be much interested in more recent Italian contributions. This overlooking will be corrected by Truesdell.

Inevitably Rivlin had also to envisage the behaviour of some materials in time such as strongly viscous fluids, (highly viscous chemical products, paints, artificial fabrics), that is, their flow properties. In comparison with elasticity, this brings to the foreground time rates of both strains and stresses. This is rheology at its best with the notion of *non-Newtonian fluids*.[5] First true developments in this field are due to Bingham (1922) and Reiner (1945) with an attempt at a theory of visco-plastic fluids and so-called dilatancy. Characteristic effects exhibited by such media are the Poynting effect discovered in 1909 and the Weissenberg effect discovered in 1949. Several theoreticians attacked the description of these media in the period of the 1940s and 1950s.

A historical sketch of these works was given by Coleman et al. (1966) and also Rivlin (1984). The first authors, Reiner (1945) and Rivlin, followed the initial general idea of Stokes of 1845 that the stress tensor should be a tensor-valued function of the rate of deformation tensor **D**. A tensor expansion up top second order in **D** should be sufficient—and in fact correct according to the Cayley-Hamilton theorem—with coefficients still depending on the first two invariants of **D** for incompressible isotropic bodies. This provided the so-called Reiner-Rivlin model. But this does not allow for a sufficient characterization of typical flow properties of non-Newtonian fluids as shown by J.G. Oldroyd (1921–1982) in the UK with a breakthrough publication (Oldroyd 1950). This was corrected by Ericksen and Rivlin (1954) who showed that the stress should depend on higher-order time

---

[5]We have already reviewed the progress of this field in the twentieth century in Chap. 3 of Maugin (2013a).

derivatives of the deformation gradient, introducing thus the notion of Rivlin-Ericksen tensors (akin to convected or Lie derivatives in the proper geometrical invariant framework). This provided the so-called Rivlin-Ericksen model of which Green and Rivlin (1957) gave a nice definition in terms of successive time derivatives of the Cauchy-Green finite strain. In contrast, Oldroyd (1950) thought of introducing a stress constitutive equation in the form of a time integral over the history of the convected metric (or strain). This was in the line of an old proposal of hereditary integrals by L. Boltzmann. Approximation of this last modelling can be shown to provide solutions that are equivalent to those that would follow in the framework of a second-order approximation of a Rivlin-Ericksen model. From all these considerations (good kinematics of deformation, exploitation of the theory of invariants), there followed a wealth of models so that a true "industry" was sparked in the field of theoretical rheology (see Chap. 3 in Maugin 2013a, and references therein). As a result Reiner, Rivlin, Ericksen and Oldroyd can be considered the fathers of this discipline in the second half of the twentieth century. With additional works by Coleman and Noll, this will provide the backbone of the two encyclopaedia articles by Truesdell, Toupin and Noll—respectively in 1960 and 1965—that were to spread the "gospel" of modern continuum mechanics.

It is the amalgamation of nonlinear elasticity and rheology of complex materials, the necessary consideration of the common bases in kinematics, and the desire to bring some rational classification in the multi-sided behaviour of many deformable bodies that really led to the global view of modern continuum mechanics. This overall vision will be accentuated with the shared general thermodynamic background in the expert hands of Truesdell, Coleman, Noll, I. Müller, etc.

11.3 The Co-workers and Direct Disciples of Truesdell

There is hardly need to emphasize that some co-workers and direct disciples of Truesdell not only were efficient in spreading the Trusdellian spirit of continuum mechanics, but they also themselves became masters in the field demonstrating a remarkable creativity. We count among these scientists Jerald L. Ericksen, Richard A. Toupin, Walter Noll, and Bernard D. Coleman.

Jerald L. Ericksen (born 1924)[6] obtained his PhD in Mathematics at Bloomington in 1951 after serving in the US Navy. He then joined the U.S. Naval Research Laboratory (NRL) in Baltimore. He spent twenty-five years teaching in the Department of Mechanics at the Johns Hopkins University (1957–1982) still in Baltimore, before concluding his teaching career at the University of Minnesota (1982–1999). He met Truesdell in Bloomington when the latter joined the faculty in

[6]Autobiographic notes of Ericksen are to be found in Beatty and Hayes (2005). See also Ericksen (1979).

this university in 1955. They shared an interest in the theory of rods and shells on which they wrote a paper (Ericksen and Truesdell 1958). In this paper they were among the first authors to cite the ideas of the Cosserat brothers enunciated in their famous book of 1909, and subsequent works by J. Sudria (in the 1920s and 1930s).[7] This used the idea of oriented continua (a kind of generalized continuum) where internal rotations play as important a role as classical translations (the displacement). Ericksen also contributed a comprehensive introduction to tensors as an appendix to the encyclopaedic article of Truesdell and Toupin (1960). Perhaps exploiting the ideas of "directors"—that can be used to materialize the above-mentioned rotations—due to Pierre Duhem; he singlehandedly expanded a theory of anisotropic fluids that was to develop in a universally accepted theory of nematic liquid crystals (Ericksen 1960). Later on this will be generalized by Frank M. Leslie (from Scotland) and him to include dissipative effects. As already mentioned he collaborated with Rivlin on the theory of finite deformations (Ericksen and Rivlin 1954). The two had met during a stay of Rivlin at the NRL. The Rivlin-Ericksen tensors probably are the best known objects of continuum mechanics bearing their joined names. Ericksen made other very fruitful contributions to the theory of solid crystals and their phase transitions viewed in thermo-mechanics. Although close for sometime to Truesdell, Ericksen should be considered an original thinker with his name definitely attached to many objects and theorems, e.g., in addition to the Rivlin-Ericksen tensors, Rivlin-Ericksen fluids, Baker-Ericksen inequalities, Doyle-Ericksen tensor, Ericksen identity, Leslie-Ericksen theory of liquid crystals. Along his successful doctoral students we count Millard Beatty, Romesh C. Batra, Constantine Dafermos, Dominic G.B. Edelen, James Jenkins, Richard D. James and G. Zanzotto.

Richard A. Toupin (born 1926) had a different background. In his autobiographic notes (reprinted in his collected works edited by Barenblatt and Joseph 1997), Rivlin tells that he met Toupin at the National Research Laboratory in Maryland where both were visiting in 1953. Toupin was then working on a PhD with Melvin Lax at the University of Syracuse. Rivlin advised Toupin to change his research subject to the theory of deformable dielectrics where he foresaw some promising developments in the finite-strain framework of continuum mechanics, what Toupin did with the success we know (he obtained his PhD at Syracuse in 1961). The publication of his "Elastic dielectric" paper in the *Journal of Rational Mechanics and Analysis* in 1956 proved to be a true milestone (Toupin 1956).[8] This was followed by another paper about dynamics in Toupin (1963). That is when contact with Truesdell was established. This was fruitful first in the collaboration on the celebrated Encyclopaedia article of 1960. It also brought Toupin to write innovative works in the theory of oriented media, on the Saint-Venant principle, and relativity. Toupin joined the IBM Research Laboratory in 1962 after spending

[7]See Chap. 8 in Maugin (2014).

[8]We discuss this theory and its generalizations in Maugin (2013a) and our books on the electrodynamics of continua.

six years at the US NRL (1956–1962) where he produced most of his creative papers (with Rivlin, Ericksen, and Bernstein). He stayed at IBM until retirement. There he occupied different positions in research, management, consulting, and direction of a Department. He never joined a teaching position in a university save as a short-time visitor. But his Research and Development activities—in which he produced outstanding contributions such as in magnetic ink-jet printing—did not hinder his production of high-level scientific contributions (in particular with M.E. Gurtin, B.D. Coleman and E.H. Dill).

Walter Noll (born 1925)[9] still had a different background. First formed in Mathematics at the Technical University (TU) in Berlin, Noll also acquired a good knowledge of modern analysis and geometry by himself and during a short stay in Paris in 1949–1950 where he caught the virus of Bourbaki's style of mathematics. While finishing his scientific work and assisting Ivan Szabó at TU Berlin in the chair of Applied Mechanics, he was recommended to Truesdell by the well known theoretical mechanician Georg W.K. Hamel (1877–1954) who had a strong interest in the axiomatics of mechanics (cf. Hamel 1927). No doubt that Hamel's writing must have influenced the young Noll as it did with many German mechanicians. Hamel had answered a 1952 request from Truesdell, as the latter wanted a collaborator who was instilled with this rigorous and axiomatic approach. This choice could not have been better. Noll's familiarity with the formalism of modern mathematics and his knowledge of German happily complemented the historical and deep knowledge of continuum mechanics of Truesdell. Noll joined Truesdell in Bloomington, Indiana, in the fall of 1953. He defended a PhD thesis in the summer of 1954, the essential contents of which are reproduced in the history making paper of Noll (1955) on the continuity of states in continuum mechanics—see also Noll (1958). This really is the first all embracing view of this science, not yet written in the excessively formal style that will become a trademark of Noll. In addition to the formidable contribution to the *Handbuch der Physik* co-authored with Truesdell (Truesdell and Noll 1965)[10]—that deals at length more closely with the formulation of nonlinear constitutive equations defining classes of ideal materials—further works were to be published either independently or with new co-workers, primarily Bernard D. Coleman (see below). Other epoch making papers by Noll touched—and often set forth the bases of—a new approach in: linear visco-elasticity (Coleman and Noll 1962), the basic thermodynamics of continua (Coleman and Noll 1961), non-Newtonian fluids (Coleman et al. 1966), the mechanics of uniform bodies with inhomogeneities (Noll 1967),[11] the formal theory of so-called "simple"

[9]Ignatieff (1996) has written a biography of Noll. However, this text is unnecessarily hagiographic, sometimes bordering on the ridiculous. Of course, Noll is not responsible for this unfortunate outcome. More reasonably see Noll (2002). Noll claims in some autobiographic notes written in 1988 that he first got acquainted with mechanics while an assistant of I. Szabó in Berlin.

[10]Noll tells about the genesis of this volume in Noll (2002).

[11]This is where Noll introduces in some abstract way a geometric connection in the manner of Elie Cartan following previous works by E. Kröner and Bruce Bilby et al.

materials (Noll 1972), and edge interactions and surface tension (Noll and Virga 1990). He also entered a programme (cf. Noll 1987) of presenting higher algebra in an original format. This was not too well received as including too many neologisms for things called differently—but unanimously so—by "normal" people. By becoming too abstract and formal, Noll had lost many of his potential and well intended readers. Of course, this remark needs some elaboration. But we must sincerely say that Truesdell, in presenting a collection of papers of Noll (1974) exaggerates very much in saying that Noll's works are nowadays understood and applied by all engineers. This kind of unnecessary flattering is not to be put to the credit of Truesdell: the abuse of superlatives may have a counter effect.

Bernard D. Coleman (born 1930) is not exactly a co-worker of Truesdell (they co-signed only two short papers). He is more a co-worker of Noll. But it is clear that Truesdell was the most ardent propagandist of the works of Coleman and Noll as proved by the book of Truesdell (1969, 1984b), so that the names of Truesdell, Coleman and Noll are often associated as representative of a unique school of thought. Coleman was originally a chemical engineer with a PhD from Yale obtained at the early age of 24. He must have necessarily been influenced by the thermodynamic works of Josiah W. Gibbs (1839–1903),[12] the most famous scientist from Yale and one of the greatest and most original American scientists of all times. Remember that not only was Gibbs responsible for the mathematical development of physical chemistry, but he also invented dyadic algebra, one of the paths to modern vector and tensor notations. Coleman stayed from 1954 to 1957 in a Research Laboratory of the Du Pont Company. He joined Carnegie-Mellon in Pittsburgh in 1957 and remained there until 1988, becoming a professor of mathematics in 1967, and then transferring to Rutgers University, New Jersey, in 1988.

Coleman's inclination towards more abstract formulations is probably due to Truesdell's and Noll's influence after he joined their research work in the late 1950s (Note that he had obtained his BSc in Indiana in 1951). Before the 1960s, he published exclusively in the field of physical chemistry and chemical engineering. He started to publish common works with W. Noll in 1959. Then his most frequent co-workers in the domain of continuum thermo-mechanics were V.J. Mizel, H. Markovitz, M.E. Gurtin, E.H. Dill, D.R. Owen, M. Fabrizio and D.C. Newman. He also published a paper with Toupin in 1970.

Coleman's most cited published papers have certainly been those on viscoelasticity co-authored with Noll (Coleman and Noll 1961), with the principle of *fading memory*, and on the foundations of the thermodynamics of elastic materials (Coleman and Noll 1963; also Coleman 1964). It is in these papers that a real doctrine was exposed—or sometimes imposed—to our community. Its convincing

[12]Quite appropriately, Coleman obtained the title of J. Willard Gibbs Professor of Thermomechanics at Rutgers University, NJ, in July 1988, after a long career at Carnegie-Mellon University in Pittsburgh.

strength resides in its elegance and apparent simplicity. It is salient at this point to remind the reader that this consisted essentially in (a) abandoning the main idea that some notions such as thermodynamic temperature and entropy are well defined only in thermostatics, and assuming then that they also exist in full thermo-*dynamics*—i.e., outside thermal equilibrium—(b) all dependent variables of the theory should a priori depend on the same set of independent variables (this is often referred to as the principle of *equipresence*, but it is more a precautionary measure than a true principle), (c) all constitutive equations a priori are of the *functional* type over past time, (d) these are constrained by the second law of thermodynamics which, in local form, takes the expression of the so-called *Clausius-Duhem inequality*, and (e) the entropy source and entropy flux always are the ratio of the body heat source and of the heat flux to the thermodynamic temperature. First, this cannot satisfy experimentalists of thermal science and, second, some of these working hypotheses presented as rigid principles are contrary to physical reality. These hypotheses had to be relaxed in order to accommodate more complex behaviours. Furthermore, the notions of state functionals and fading memory are certainly attractive, but no so easy to implement. That is why Coleman and Gurtin (1967) introduced the notion of *internal variables* of thermodynamic state, replacing thus time functionals—whose formulation in principle requires the knowledge of the whole past behaviour—by ordinary functions of these new variables. They followed in this line original ideas expressed by Pierre Duhem and Percy Bridgman.[13] Other substantial improvements were brought in by Ingo Müller.

What we must retain from our short investigation is the world wide reception and influence of the works of the Truesdell-Toupin-Coleman-Noll School. They imported a style of presentation and writing that became the rule through their printed contributions in journals such as the *Archives of Rational Mechanics and Analysis* that they controlled and in which they generously offered so many of their influential publications. The universal diffusion—in English—of the articles in the *Handbuch der Physik* was also a determining factor in the reception and acceptance of the school ideas. But two points deserve special mention. One is about the writing style of Truesdell and is best articulated by Rivlin, albeit with this author's usual wit and exaggeration.[14] Another concerns the relationship between a priori set

[13]See Maugin and Muschik (1994) for a general discussion on the thermodynamics of continua with internal variables of state and its multiple applications.

[14]Rivlin (1984, pp. 2799–2800) wrote apropos Truesdell's style: "In his writing Truesdell evidences a strong taste for the dramatic and so there has been created a fantasy world in which various *savants* produce stream of *principles*, *fundamental* theories, *capital* results, and work of *unusual depth*. No matter that, on examination and stripped of the, often irrelevant, mathematical verbiage with which they are surrounded, they frequently turn out to be known results in a disguise, or trivial, or physically unacceptable, or mathematically unsound, or some combination of these. Nonetheless, they have been widely and uncritically reproduced in the secondary literature and have provided the starting point for many, correspondingly flawed, theses and papers." Of course, Rilvin's own style is much sober than Truesdell's somewhat grandiloquent style.

theories and experiments as expressed by Edmund Burke, a famous British politician and writer of the transition between the eighteenth and nineteenth centuries.[15]

It is impossible to list a complete list of disciples of the Truesdell-Coleman Noll School as most of the mechanicians educated in the 1960s–1970s, like the writer, all became more or less "Truesdellian", willingly or not. But we think that a few are worth mentioning. One is Morton E. Gurtin (born 1934), a former PhD student at Brown, who co-authored many creative papers with Coleman and wrote a most influential contribution on linear elasticity in the *Handbuch der Physik* (Gurtin 1973). Another one is Ingo Müller (born 1936) who much improved *rational thermodynamics* (and also was among the creators of *extended* thermodynamics). He spent a few years at Johns Hopkins in the 1960s. Finally, Kombakonan R. Rajagopal (born 1950), although a former PhD student of Roger L. Fosdick (born 1936, himself a PhD from Brown) in Minnesota, but the author of a book with Truesdell (Truesdell and Rajagopal 1999) perfectly fits in this "classification" because of his own style, his incredible production, and the formidable extent of his interests in continuum mechanics. All three scientists became influential in the field through their original works and their own students and collaborators.

Truesdell and his co-workers founded a kind of aristocratic scientific society under the name of *Society of Natural Philosophy* (SNP). This is a rather exclusive professional society with a somewhat pretentious name (copying Newton's word for "physics" and corresponding well to Truesdell's concept of his favourite selective scientific environment).

11.4 The Co-workers and Direct Disciples of Rivlin

Truesdell and Rivlin share a certain number of co-workers in common (Ericksen, Toupin). But the style and general understanding of these two scientists being so different, Rivlin remained closer to the circle of British applied mathematics and mechanics. Of course **Albert E. Green** (1912–1999), older than Rivlin, a former student (PhD 1937) of Sir Geoffrey I. Taylor at Cambridge, and with his own large experience in the elasticity of structures (shells, rods; see Green and Zerna 1954), became a natural co-worker of Rivlin in the field of nonlinear elasticity and continuum modelling (See, e.g., Green and Rivlin 1957, 1964). But Green was much less extrovert than Truesdell and Rivlin, so that his fruitful contributions to continuum mechanics are much less advertised although permeating many of its developments in the 1960s–1970s. He was also much less formal than Truesdell.

[15]"I did not vilify theory and speculation… No, whenever I speak against theory, I mean a weak, erroneous, fallacious, unfounded or imperfect theory; and one of the ways of discovering that it is a false theory is comparing it with practice".

He certainly was one of the best representatives of the British school of mathematics applied to mechanics. While teaching first in Durham and then Newcastle upon Tyne and finally for a long time at Oxford, he formed many applied mathematicians. He was twice a visiting professor at Brown University (1955/1956, 1963/1964) working there with Rivlin. Among other things he published then one of the first works on the mechanics of materials with memory (Green and Rivlin 1957), a work that influenced Coleman and Noll. After the mid-1960s he devoted himself to the thermodynamics of continua and elasto-plastic continua (in particular with Paul Naghdi; see Sect. 11.5). A close collaborator of Green has been John E. Adkins (see Green and Adkins 1960) who became a professor in Nottingham. Rather parsimonious about himself, Green provides some rare biographic information in Green (1974).[16]

Among the other co-workers of Rivlin, we must cite Antony Spencer and Michael Hayes. **Anthony J.M. Spencer** (1929–2008), an alumnus from Cambridge, started his PhD with Frank Nabarro in Birmingham and then concluded it with Ian Sneddon in Keele. He spent two years at Brown, where he started a creative co-operation with both Rivlin and Green. In co-operation with Rivlin, he became one of the most productive contributors to the theory of invariants that is of utmost importance in the formulation of constitutive equations (see, e.g. Spencer 1971). He joined the University of Nottingham to succeed John Adkins in 1965 as professor of theoretical mechanics and head of the Department until his retirement in 1994. He has done extensive research on the mechanics of fibre-reinforced materials (see, e.g., Spencer 1972). **Michael A. Hayes** (born 1936)[17] is Irish but obtained his PhD at Brown with Rivlin in 1962. They worked together on the problem of wave propagation developing a method of superposition of small motion on a static state of finite deformation ("small superimposed on large"; see Hayes and Rivlin 1961a, b). This became their favourite subject of co-operation with other works published from time to time in 1969, 1972 and 1974. A specialist of the kinematics of deformation, Hayes developed many aspects of bulk and surface wave propagation in elasticity and visco-elasticity working out more recently a theory of bi-vectors with Philippe Boulanger from Brussels. He became a professor of Mechanics at the University College in Dublin until his retirement. He mentored, among others, P. Currie, Nigel Scott, and Michel Destrade (a Frenchman now in Galway).

We note that both Green and Spencer became Fellows of the Royal Society, what was not the case of Rivlin who had become an American citizen and entertained some bad relations with some British scientists. Hayes naturally became a member of the Royal Irish Academy.

[16]Much more is to be found in the obituary published by Peter Chadwick in 2001 in the biographical Memoirs of the Fellows of the Royal Society, and the Festschrift volume edited by Naghdi, Spencer and England (1994).

[17]Biographical elements of M.A. Hayes are given by Boulanger et al. (2005).

11.5 A.E. Green and Paul Naghdi

We isolated this case because of it exemplarity of co-operation between scientists so much apart geographically but who succeeded to build a long-time extremely fruitful collaboration in various aspects of modern continuum mechanics. This was not obvious from the start because of the very different initial formations of these two scientists. Indeed, while Green was a rather typical British applied mathematician, **Paul Mansour Naghdi** (1924–1994)[18] was born in Iran but immigrated to the USA as a young man in 1943, studied in Cornell (BSc in 1946), and earned a PhD at the University of Michigan in Ann Arbor in 1951 as a standard mechanical engineer of the period. He was extremely good at solving problems of elasto-plasticity in a period where dexterity in analysis was required and only rudiments of computations were available. This is illustrated by the solution of the wedge problem that he gave in Naghdi (1957). After teaching for some years in Ann Arbor, he joined the University of California in Berkeley in 1958 where he stayed until his death in July 1994. He played there a predominant role in the establishment of a Division of Applied mechanics, while becoming quite involved in the activities of the *American Society of Mechanical Engineers* (ASME) of which he celebrated the fiftieth anniversary with a history of its Division of Applied Mechanics in Naghdi (1979).

As already mentioned Naghdi was an authority in small-strain plasticity on which he wrote a synthesis in 1960. Along with works of analytical nature, he and his co-workers conducted experiments in the field in the mid-1950s. But with time he also became an expert in finite-strain plasticity where he cultivated an alternate to the usually accepted multiplicative decomposition of the deformation gradient (cf. his critical synthesis review: Naghdi 1990). In this line he mentored J. Casey and A.R. Srinivasa. He published one of the first papers in this theory together with A.E. Green in 1965 (see also Naghdi and Trapp 1975). He was already a recognized expert in the theory of elastic shells and rods to which he contributed a beautiful critical synthesis in the *Handbuch der Physik* (Naghdi 1973). This is universally recognized as a definitive treatment of the subject. In this theory he did not hesitate to exploit a kinematic description that uses the idea of directors put forward by Duhem and then Ericksen and Truesdell (1958) in the form of "Cosserat surfaces", and on which he started to publish as early as 1963. He was in fact very much attracted by the idea of *polar materials* as initially introduced by the Cosserat brothers (1909). Some of his ideas were transmitted to his PhD student Miles Rubin. He used some of these ideas (so-called Cosserat surfaces) in a theory of sheets and jets of ideal fluids in the late 1970s and the 1980s.

Paul Naghdi, although formed as a traditional mechanical engineer, clearly was very much attracted to the fundamental questions posed by continuum mechanics. He of course read the early achievements of the Truesdellian School, but this

[18]A biographical sketch of Naghdi is given by two of his former PhD students in Casey and Crochet (1995).

tendency was strengthened with the close collaboration that he established with A.E. Green in the early 1960s and that lasted for almost thirty years. The two of them seemed to have interacted through mail but also with visits to one another as Green became a frequent visitor at the University of California, Berkeley. Their common view on finite-strain elasto-plasticity was already mentioned, but this rapidly extended to all the new directions taken by continuum mechanics in the 1960s. Practically none of the new developments escaped the attention and sagacity of the two co-workers, whether these developments dealt with generalized continuum mechanics or the theory of reacting mixtures, or new proposals regarding the foundations of the thermo-mechanics of continua (in particular with a balance of entropy with separate notions of the second law of thermodynamics; cf. Green and Naghdi 1977). Along the first line we note the theory of Cosserat surfaces and *multipolar continua* as initially proposed by Green and Rivlin. The theory of mixtures was expanded by Green and Naghdi in a series of papers in the mid-1960s. On this occasion they proposed an original energy based formulation of the fundamental laws of continuum mechanics using an argument of invariance applied to a global statement of the first law of thermodynamics (Green and Naghdi 1965). This was much elaborated upon in a series of papers published after the death of Naghdi (Green and Naghdi 1995). Their common output is a formidable one both in quantity and creativity. But the most original one certainly is the thermodynamic theory of thermodynamically irreversible continua "without dissipation" that they proposed in 1993 (Green and Naghdi 1993). The purpose of this work was to offer a possibility (but this is not the only one) to avoid the paradox of heat propagation at an infinite speed. Naghdi mentored successful contributors to continuum mechanics, among them Marcel Crochet (on non-Newtonian fluids, PhD 1966), James Casey (PhD 1978), Miles B. Rubin (PhD 1979) and A.R. Srinivasa (PhD 1991) along different lines that often opened new horizons in modern continuum mechanics.

In all, the co-operation between Green and Naghdi was exemplary and rather unique in the world of science because of its longevity and continuous good relationship, what says a lot concerning these two scientists. In most minds the two names remain associated.

11.6 A.C. Eringen and Engineering Science

Born in Turkey, A. Cemal Eringen (1921–2009)[19] immigrated to the USA after his basic formation in engineering in Istanbul. He obtained his PhD in mechanical engineering at the Brooklyn Polytechnic Institute in 1948 under the guidance of Nicholas Hoff. He was then to occupy faculty positions in three teaching

[19]We gave a long biographical sketch of Eringen and a complete list of his publications of all types in a special issue of the Int. J. Eng. Sci. (cf. Maugin 2011) dedicated to his memory.

institutions, first at the Illinois Institute of Technology (1948–1953), next at Purdue University (1953–1966), and finally at Princeton University (1966 till retirement). Although trained as an engineer Eringen always entertained a specific interest in modern mathematical physics and applied mathematics, demonstrating an insatiable curiosity for these two fields and keeping aware of all new developments. He was also very much interested in the possibility to apply irreversible thermodynamics in dissipative continuum mechanics as shown by one of the rare early publications on the subject (Eringen 1960), although most of his works during the stay in Illinois and Purdue dealt first with problems of structural mechanics (buckling, vibrations, sandwich structures, problems of impact, elastic waves of different types) and one of the first attempts to treat stochastic problems in this field (together with his first PhD student, J.C. Samuels). However, early in the 1960s he took four lines of action that were to drive all his future teaching, scientific works and editorial activities until his death in 2009. First he created a true programme of "engineering sciences" at Purdue, including a course on the foundations of nonlinear continuum mechanics. After an attempt by Prager (1961), this led to the publication of one of the first books of its kind on the subject (Eringen 1962). Furthermore, he created the *International Journal of Engineering Sciences* in 1963. This was to become a successful forum for the presentation of innovative works in continuum physics, including many of his own works and those of his students and co-workers. This created a line that was going to spread a true spirit of *engineering sciences* in the USA, in parallel, and sometimes in competition, with Truesdell journal (the A.R.M. A.). But it was generally less formal and mathematical than Truesdell School's ARMA and more open to many foreign authors. The fact that Eringen himself was of foreign origin may have been decisive in this orientation. Finally, he founded a new professional society called the *Society of Engineering Sciences* (SES) that held a yearly symposium every year (from 1963 until now) distributing several prizes and medals to the most creative engineer-scientists in the field. This move was extremely successful as it attracted many engineers who cultivated a specific taste for a recovery of the style of early nineteenth century mechanicians who combined good mathematics with physical acuity and a general interest in continuum physics.

Retrospectively, we think that Eringen's innovative book (as a teaching tool) of 1962 was a bit premature, emphasizing some trends that were not to expand so much in the future, and sometimes not selective enough, and giving useless detail. It was certainly inspired by the Truesdell-Toupin Handbuch article, but it helped many good-willed engineers to capture the re-emerging global vision of continuum mechanics. Pedagogical errors in its redaction were corrected in a more accessible, but very efficient, book (Eringen 1967) that met an immense success in the USA and abroad. As to the new journal and the new professional society, they offered the possibility to the specialized scientist or engineer—handicapped by a peculiar lack of training in fundamentals—to explore the most interesting interdisciplinary areas of cross fertilization, in which one finds the root system of science. Indeed, the new journal (IJES) immediately provided the required forum dedicated to the "re-establishment of a common tongue between engineer and scientist" (Eringen's words). It encouraged the publication of serious research works of depth and breadth at the frontier areas of science and engineering. Eringen himself will deeply participate in this advance through his many published research works (see below). The SES first met on November 4, 1963, with an attendance of about 400 engineers, physicists and mathematicians. This ecumenical spirit was maintained at the highest level with the yearly meetings of the society, sometimes in competition with established engineering societies, but now more than often in fruitful co-operation with them, notably the *American Society of Mechanical Engineers* (ASME) and the *American Society of Civil Engineers* (ASCE).

Eringen's international influence is above all to be assessed from his most creative scientific production—which we examined elsewhere in detail (cf. Maugin 2011). Indeed, in full accord with the aim and spirit of the IJES and SES, Eringen became in the 1960s one of the most active contributors with numerous works in various forms of generalized continuum mechanics (micro-morphic and micro-polar materials, non-local theory of continua), the mechanics of deformable electromagnetic materials, the theory of liquid crystals, that of mixtures, the theory of relativistic continua, the theory of defects, etc. The corresponding formidable output has had a tremendous reception all over the world. In this he was greatly helped by a selective group of PhD students and co-workers, including, to name a few, S.L. Koh, N.F. Jordan, R.C. Dixon, J.D. Ingram, J.W. Dunkin, Richard A. Grot, Charles B. Kafadar, Robert J. Twiss, W.D. Clauss Jr, T.S. Chang, James D. Lee, Hilmi Demiray, Charles B. Speziale, Patrick O'Leary, E.S. Suhubi, D.G.B. Edelen, and the present writer. Some of these co-collaborators had a brilliant career (e.g., Twiss in geology, Grot at the National Bureau of Standards, Speziale in turbulence, Lee, Demiray, Suhubi and Edelen in the academic world, and others in large engineering companies).

Among the most cited and influential works of Eringen, we note that on the introduction of the concept of micro-morphic bodies (Eringen and Suhubi 1964), that on micropolar fluids (Eringen 1966a), the one on the general thermo-mechanical description of continua (Eringen 1966b), and that on nonlocal elasticity (Eringen and Edelen 1972). Many other papers dealt with applications of these models to various media (e.g., liquid crystals, electromagnetic continua, blood flow, fibered media, and wave propagation). We refer the curious reader to the total list of publications of Eringen given in a special issue of the IJES dedicated to his memory (cf. Maugin 2011). Never at rest, Eringen continued to publish scientific papers and books until his last years. In addition to his 230 journal publications and some 60 more technical reports, he published—alone or with close collaborators—no less than 13 books on general continuum mechanics, generalized-continuum mechanics, elastodynamics, electrodynamics of continua, and he publicized continuum physics in a series of four landmark volumes that he edited and contributed (Eringen 1971/1976). In conclusion, we like to mention his lack of enthusiasm for large scientific meetings, his little interest in controversies (contrary to Truesdell and Rivlin), and his somewhat dry but crystal-clear way of exposition that is best illustrated (my personal opinion) by his presentation of the mechanics of micropolar materials in Eringen (1968). No doubt, therefore, that Eringen appears as one of the masters of the renewal of continuum mechanics and an inspiring scientist to many students and professionals all over the world.

Remark. Before examining some isolated cases in Europe, we want to mention the very special case of **Raymond D. Mindlin** (1906–1987). Contrary to Truesdell and Eringen, he contented himself with doing pure science (with more than a hundred papers in refereed journals of mechanics or applied physics) and forming students. He did not control a journal, neither did he write a real book for students nor founded a scientific society, although he was very active in the ASME. This attitude, per force, does not automatically yield an immediate world wide influence in the medium of continuum mechanics. Though, he formed excellent scientists who, together with him, greatly influenced that part of mechanics that deals with a microstructure and/or piezoelectric properties. Mindlin was formed as a standard mechanical engineer at Columbia University, New York City, (PhD 1934) where he spent most of his career, demonstrating a rather minimal mobility. He is internationally recognized for his early works in solving with dexterity beautiful problems of the strength of materials (the well-named "Mindlin"s problem") and performing initial experiments in photomechanics. These works are permeated by the excellent ASME spirit in the line of Timoshenko that we described elsewhere (cf. Maugin 2013a, Chap. 4). Then he developed an unsurpassed expertise in problems dealing with the vibrations of elastic and piezoelectric plates right at a time when this was needed in the development of electromechanical transducers in signal processing as required by the US Army and the conception of RADAR equipments. Furthermore, he efficiently contributed to the expansion of theories of granular materials, lattices and deformable media equipped with a mechanical microstructure. He was thus a pioneer in the introduction of gradient-theories of elastic materials, as also in refined theories of electro-elastic media.[20] Some of his former PhD students also reached international recognition [e.g., Harry F. Tiersten (1930–2006) and Y.-H. Pao (born 1930, PhD 1959)] especially in wave propagation and the modelling of electro-magneto-mechanical interactions. He received many honours including the Presidential Medal of Merit in 1946 and a National Medal of Science (a Presidential award) in 1979. Much less publicized than Truesdell and less versatile than Eringen, Mindlin nonetheless deserves our admiration. He stands as one of the masters of continuum mechanics in the second half of the twentieth century.

11.7 Outside the USA and the UK

Publishing in a language with no international status and in a country with closed borders like the former Soviet Union did not favour a worldwide spread of a general vision of one's own science. This was the fate of Leonid I. Sedov and

[20]Mindlin's life, complete list of publications, and detailed description of his scientific achievements (by his former students and collaborators) are given in Herrmann (1974). His collected papers are reproduced in Mindlin (1989).

Alexey A. Ilyushin, both identified as excellent technical contributors to their respective specialty (theoretical aerodynamics for Sedov, and plasticity for Ilyushin), but also scientists having a general vision and willing to ponder and shape the bases of continuum mechanics in the large. Still, we must distinguish between the former who had the chance to travel abroad on many occasions (as an official representative of Soviet mechanics) and to have the privilege to have his books translated either abroad or at home—in English and many other languages—and the later who rarely—if ever—travelled abroad and had no books translated in English.[21] The profound reasons for this last matter may be discussed.[22]

Leonid Ivanovich Sedov (1907–1999) may be considered an emblematic figure in the landscape of mechanics in Russia. He was educated in mathematics and mechanics at the (Lomonosov) Moscow State University (MSU). His first works (1934) were devoted to two-dimensional problems of hydrodynamics. He subsequently provided numerous contributions to hydro- and aero-dynamics, noticeably, those concerning the impact of bodies on water, hydroplaning, and aerodynamic forces on deformable wings. One of his favourite tools was the exploitation of dimensional analysis and so-called similarity, a subject on which he wrote a very popular monograph. He also presented a famous solution to the "blast-wave" problem, obviously of actuality then with the first nuclear explosions. But in more recent times, Sedov focused on the teaching and general formulation of continuum mechanics. His books (1962, 1971, 1973) on the subject have met an incredible success in Russian and in their many foreign-language translations by both foreign publishers and MIR Publishers in Moscow. One of his favourite general formulations is now known as "Sedov's variational principle". It is quite possible that part of the international recognition of Sedov in the 1960s–1970s was due to the fact that he was presented by Soviet authorities as the principal engineer behind the Soviet Sputnik project,[23] what helped him to become influential in the International Astronautic Union. But apart from this touch of vanity to which he did not resist, one must recognize the high value of his works, teaching and spreading of the continuum-mechanics spirit throughout the world. This was in a style very different from that of Truesdell, without any abuse of abstract notation and no recourse to Bourbakian mathematics, aiming first at efficacy.

As a matter of fact, Sedov's first book (in Russian) of 1962 may be compared to the 1962 book of Eringen. Not only dates of publication coincide, and thus some

[21]Ilyushin's book on plasticity was first translated only in French by a publishing house specialized in civil engineering (Ilyushin 1948). His general views on continuum mechanics were published only in Russian by the publishing house of the Lomonosov University (cf. Ilyushin 1971).

[22]This may be a result of the close relationship of Ilyushin with the military establishment and his friendship with the highest political authorities in the Soviet Union in Stalin's time.

[23]In a private conversation Sedov explained to the author, with a slight dose of cynicism, that all great national programmes need a well identified leader for publicity purposes. He was chosen to do the job and he accepted it. He had no visible regrets although he knew well that Sergueï Pavlovich Korolev (1907–1966) was the true driving figure in the programme.

simultaneous priority can be granted to both, but as remarked by a reviewer[24] of its English translation (Sedov 1966), this rather dense book (252 pages only in translation) conveys a similar general vision of both fluid and solid mechanics, but is less ambitious and less formal than Eringen's book. Like the latter, Sedov's book of 1962 considered finite deformations from the start. But, also less oriented toward applications of very specific behaviours, it emphasizes physical principles and develops general concepts and models applying general tensor formalism in arbitrary coordinates in a remarkably clear and to the point exposition. Thereafter, a series of books expanding this at larger length will be published (Sedov 1971, 1973) reflecting the evolution of teaching of the subject matter as a unified and synthetic presentation at the University of Moscow and some research trends. These books and the lectures of Sedov abroad have certainly opened the horizon of many engineers who then join the study of more fundamental continuum mechanics. He really gave a specific print to Russian continuum mechanics. This in turn influenced many researchers the world over. He was the recipient of many honours both in Russia and outside. He kept active until his last years in spite of a declining sight, paying special attention to the problem of inertia in space-time.

The case of **Alexey Antonovich Ilyushin** (1911–1998) is quite different. One must consult the biography given in Russian and English in a book (Kiyko et al. 2001) for a complete overview. This engineer-scientist had a long career in solid mechanics mixed with politically marked organizational roles (e.g., Rector of the University of Leningrad at a dramatic time). The strength of materials and the engineering aspects of elastoplasticity seem to have been a constant preoccupation of Ilyushin, starting with his early works in the mid 1930s. What is quite remarkable is that these most innovative and rewarding developments were made in an obviously difficult period, the 1940s, corresponding also to the strength of maturity reached by this scientist. But in perusing a large sample of his works (Ilyushin 2003–2009), we discover many contributions that were not well known outside the Soviet Union, including deeply thought works in thermo-visco-elasticity and in dynamic problems. Moreover, in his lectures at MSU published in book form (Ilyushin 1971), this author made a special effort at a general rational presentation of continuum mechanics. However, this is not at the level of C.A. Truesdell in the west, and with less international success than Sedov's courses. This matter is emphasized in Brovko (2013) who has benevolently examined the formal structure of Ilyushin's general approach. In these publications Ilyushin constantly reminds the reader that the actual state of stresses in a material body depends on the whole past history of the body. This was clearly expressed by Ilyushin in his book on plasticity of 1948. This is also continuously emphasized by him in his textbook for the Lomonosov University (Ilyushin 1971). Reception of Ilyushin's works in the West was appraised in Maugin (2013b) while the heritage of Ilyushin's works is surveyed by Brovko et al. (2011).

---

[24]See Romains' review (1966).

We can say that Ilyushin served as an inspiration for many students in Russia, so that he was one of the local successful masters of continuum mechanics. This was also the case of other scientists in Russia, among them **Anatolii Isakovich Lurie** or Lurie (pronounced Lurié; 1901–1980). Educated at the Saint-Petersburg Polytechnic Institute, Lurie became there the Head of the Department of Theoretical Mechanics before WWII and the Head of the Department of Dynamics and Strength of Machines or Department of Mechanics and Control Processes in the period 1944–1977. His numerous works span the hydrodynamics of viscous liquids as well as solid mechanics and control theory, and are the trademark of a true scientist-encyclopaedist. He proved to be quite receptive to the ideas of the Truesdellian School, as he even endeavoured to translate into Russian Truesdell's "First course in rational mechanics" of 1975. His own style is best illustrated by his most achieved book—translated into English by his son, Konstantine—published in 1980, the year of his death. This book is rather exceptional in the Russian landscape of continuum mechanics of the period, but it is unfortunately little known, even in translation, in the west. This book included the consideration of rubber as a good example of incompressible material, variational formulations with restrictions of the elasticity potential, and the problem of superimposition of small motions on a finite strain state (cf. original works of Toupin and Bernstein, and Hayes and Rivlin in previous sections). He is more "occidental" than other Russian authors in advocating the use of an intrinsic notation, does not hesitate to exploit modern techniques such as representing the angular-velocity vector by the Rodrigues-Hamilton and Cayley-Klein parameters,[25] and pre-empted the exploitation of dissipation potentials. As a consequence of his never tired activity and his ingenuity Lurie created a true Leningrad/Saint-Petersburg School of continuum mechanics the spirit of which was kept alive in the works of Pavel Andreevich Zhilin (1942–2005).This last author wrote an informative appraisal of Lurie's works in mechanics (Zhilin 2001).

Like Leonid Sedov, **Paul Germain** (1920–2009)[26] was primarily a successful and world renowned contributor to theoretical aerodynamics (transonic flow, theory of delta wings, shock structure, etc.). Among all scientists examined so far in this contribution, Germain was the one with the most solid formation in mathematics. Although admitted to the *Ecole Polytechnique*, he preferred to join the celebrated *Ecole Normale Supérieure* to specialize in mathematics. He should have become a geometer, but circumstances—the immediate post World War II period, the acknowledged need in the field with the expansion of jet airplanes, the flair of his mentor, Joseph Pérès, for promising new developing fields in mechanics—Germain started to exert his mathematical dexterity in problems of aerodynamics after a short stay in England. But he also became a professor and had to teach continuum

[25]This representation is useful in describing rotational internal degrees of freedom in generalized continuum mechanics.

[26]Germain tells his own scientific trajectory in Germain (2000) in Maugin et al. (2000). We also gave a biographical sketch of Germain in Maugin (2010). See also Chap. 7 in Maugin (2013a).

mechanics. This he did conscientiously, all the way to writing the first well structured textbook on continuum mechanics as a global science in France in 1962. This book (Germain 1962), although aimed at undergraduate students following a cursus in applied mathematics, set forth unique standards that were soon applied also in engineering schools.[27] In this book that well compares with Eringen (1962) and Sedov (1962) in timing and contents, Germain views continuum mechanics first as a whole and then gives examples in fluid and solid mechanics. The book follows a tradition exemplified by the volume devoted to continuum mechanics[28] in the formidable treatise on rational mechanics of Paul Appell (1921). It bears the print of a mathematician but remains very efficient pedagogically. A more ambitious book aimed more at graduate students, was published later (Germain 1973) incorporating sound continuum thermo-mechanics, but did not meet the same success.[29]

To fulfil needs dictated by his teaching and because of mere scientific curiosity, Germain studied in depth all new works that blossomed in general continuum mechanics in the early 1960s (especially those coming from the Truesdellian School) but remained opened to ideas of other scientists always keeping a wise critical appraisal of, and a balanced view between, different "schools" (Truesdell, Rivlin, Eringen, Sedov, etc.). He paid special attention to the thermo-mechanical formulation, often not clearly decided between the Truesdell-Coleman-Noll approach and that of more classical thermodynamicists (e.g., Joseph Kestin), each one having its own advantages and disadvantages. In truth, he was a realist who also maintained a reasonable amount of abstraction, while encouraging the use of good applied mathematics wherever needed (e.g., efficient asymptotic and numerical methods, applied functional analysis). Through his teaching at the University of Paris—and then at the Ecole Polytechnique—and the ensuing swarm in the rest of the country, he influenced a whole generation. His own "formation" in the new continuum mechanics was facilitated by the long stays he made at Brown University and CALTECH at appropriate moments when he befriended many of the American masters of this new mechanics. In France, the centralized structure of higher education and research, combined with the facts that he was in charge of the *Journal de Mécanique* (to become later on the *European Journal of Mechanics*) at its beginnings, and that he practically took the control of the Paris Academy of Sciences, helped him to build a nationwide authority in the field. The whole school of French mechanics owes its remarkable development to him in the 1970s–1980s. Internationally, he did not hesitate to travel, deliver lectures in French or English (which he had not learnt in school), and entertain the best friendly contacts with

[27]Remember that engineering schools were not integrated in French universities and admission to them was by competition. Some have been integrated in local universities only recently.

[28]We have examined in detail this treatise in Maugin (2013a, Chap. 11).

[29]We personally think that the level of teaching was ambiguous and there was an abuse of some mathematics such as convex analysis which the French have very much developed as an efficient means in the modelling of constitutive behaviour (e.g., J.-J. Moreau and his disciples).

many of his peers around the world. That, together with a gift for public communication, certainly explains why he was elected president (1984–1988) of the *International Union of Theoretical and Applied Mechanics* (IUTAM), and became member of many national academies, in both east and west. Still, in spite of this international recognition, like for the Russian authors, there remained the handicap of not writing initially in English.

11.8 Conclusion: Some Sociological Remarks

It may be thought that the above selected short list of "masters" of modern continuum mechanics is somewhat arbitrary, or too much to the own taste of the writer. It is not so. We sincerely believe in all fairness to our colleagues that the listed scientists are those who most influenced the rapid evolution in the field in the period 1940–1980. This they achieved by different means, first through their own brilliant contributions, but also through their co-workers and direct disciples yielding a true circle of devoted collaborators, the aura they gained in the world, and the benevolent but forceful control they exerted on means of diffusion and scientific publications. That is, doing an excellent scientific work clearly is not enough. This leads us to deviate from pure mathematical science to indulge in elementary sociology. Here we, and the reader, can make distinction between the different individuals. Although all expanded their global approach at about the same time, with a clear priority granted to Rivlin and Truesdell, their most active and creative period has been in the 1960s. This is not exactly a coincidence for it corresponds to a period of relative optimistic faith in the future after the drama of the Second World War and to a blooming economy in the richest parts of the world, essentially then in North America and Europe. Some of these scientists were in fact frustrated mechanicians, or engineers, who had always dreamed of becoming creative mathematicians (a real exception is P. Germain) while most of them became with age more theoretical. But to be really successful in their national or international influence, some of them understood that publishing papers—albeit excellent from a technical viewpoint—in specialized journals was not enough. They created new journals (e.g., the *ARMA*,[30]

[30]Truesdell (1987–8) tells how he came to create the *ARMA*, successor of the *JRMA*, by establishing a fruitful contact with the Editorial staff (Dr. Springer and Dr Mayer-Kaupp) at Springer in Berlin, so that the first issue of *ARMA* appeared on September 24, 1957. The journal was to work like the publications of Academies, with no review for papers published by members of the Editorial board and transmission by such a member for other papers. It favoured long articles written in the purest English. This did much to spread the Trusdellian spirit among mechanicians with a mathematical inclination. Truesdell established friendly relations with Springer, in particular with Siegfried Flügge, the Editor of the formidable Handbuch der Physik (56 volumes). Later on, he joined Flügge as editor of the volumes on mechanics. This efficient co-operation resulted also in a series of books under the title of "Springer Tracts in Natural Philosophy" which contributed further to the dissemination of the school's spirit (39 volumes published between 1964 and 1998).

the *IJES*,[31] the Journal de Mécanique),[32] founded new scientific societies (e.g., the *SNP* or the *SES*), practically controlled the publication of chapters in encyclopaedias (cf. the *Handbuch der Phyisk*), were active in some of the funding agencies (this applies to the USA in particular), or were excessively powerful in academies (e.g., Sedov in Russia or Germain in France). Of course, they also had to publish successful books at different levels (textbooks or research monographs) that would be accessible to many readers in the original language or in translation. There were some caveats in these various moves. Indeed, controlling a selective journal that becomes a mandatory inevitable channel to diffuse one's ideas may sometimes yield a true unhealthy intellectual terrorism with an expansion of some chapel-like spirit in the community of mechanics. Hubris and a definite feeling of superiority are not always good advisers. They also build blinkers that made some of these great scientists to eschew promising lines of development. The control of the editorship of a journal seems to be particularly important since we have witnessed several attempts through some intense lobbying[33] at taking the control of some well diffused journal on the death or retirement of its editor in chief.

Of course, a centralized administrative, university system and research structure favoured the emergence of (truly) scientific masters who were simultaneously endowed with a disproportionate strong power in the supervision and regulation of a field at a national level. The above scrutinized cases of Sedov and Germain are exemplary, but they are not the only ones. We have commented elsewhere (cf. Maugin 2013a, Chap. 9) on the dominant role played by Witold Nowacki (1911–1986) in the field of mechanics in Poland, reorganizing its teaching and research at a national level after the Second World War, and then in charge of the main publications (e.g., the Bulletin of the Polish Academy of Science and other specialized journals) as well as the publications of books by the Polish Scientific Publisher (P.W.N), and finally in charge of the Polish Academy of Sciences at its president. The same can be said of Heinz Parkus (1909–1982) in Austria in post WWII, with the main chair at the Technical University of Vienna, and the editorship of the influential *Acta Mechanica*. Although Germany was never truly

[31]The creation of the *IJES*, like that of the *International Journal of solids and Structures* and the *International Journal of Non-linear Mechanics* (among other journals of the same class) in the 1960s was due to the interest for science of Ian Robert Maxwell (1923–1991)—born J.L.H.B. Hoch—who had founded the *Pergamon Press* Publishing House in Oxford. Maxwell, with a flamboyant lifestyle (as witnessed by G. Herrmann and A.C. Eringen when he gathered his editors-in-chief in the UK), died at sea of unknown causes, after mismanagement of his company.

[32]The *Journal de Mécanique* (Paris; 1962–1981) was in fact founded by Joseph Pérès (1890–1962), but the latter died before publication of the first issue. Paul Germain took over the Editorship of the Journal. This was published by Gauthier-Villars Publishing Company, a much less aggressive and internationally oriented firm than Springer-Verlag.

[33]Here we do not give names in order to avoid legal pursuits for defamation.

centralized, some gifted individuals such as Ludwig Prandtl (1875–1953) and Richard von Mises (1883–1953), undoubtedly great scientists, demonstrated also some powerful insight by founding the *Gesellschaft für Mathematik and Mechanics* (for short *GAMM*) and creating an accompaying spot-on forum with the *Zeitschrift für Mathematik und Mechanik* (for short, *ZAMM*) in 1921. Their affiliation with Göttingen and Dresden, respectively, was also instrumental in their highest status among German mechanicians of the period. In the post WWII period, Horst Lippmann (1931–2008) represents also a powerful member of the German mechanical engineering community with his affiliation with T.U. Munich and the editorship of the *Archives in Applied Mechanics* (successor of *Ingenieur Archiv*) and his involvement in the *Mathematischen Forschungsinstitut* in Oberwolfach (Black Forest) and the *International Centre of Mechanical Sciences* (CISM) in Udine (Italy). To some degree, Pericles S. Teocaris (1921–1999) played a similar role in Greece with his professorship at the Technical University in Athens (NTUA) and his prevailing role in the National Academy of Sciences of Athens. In the UK, the publishing business was monopolized by Cambridge University and Oxford University. Concerning the former, some members of the community of mechanicians affiliated with the local university more or less controlled publication in the corresponding output in the Proceedings of the Royal Society and created the most successful *Journal of the Mechanics and Physics of Solids* (JMPS). This was the case of Rodney Hill, albeit a most discrete and modest scientist in spite of his breakthrough works, who was nonetheless influential in the Royal Society and having founded the JMPS in 1952. Oxford was less successful along this line and kept with Applied Mathematics while we note that one of the above examined masters, Albert Green, was not so much involved in the control of publications although himself a member of the Royal Society. Like Germany, Italy did not present a unified front, presenting rather three geographical zones of influence (the North, the South and central Rome). Influence of various masters there was essentially felt through the use of privileged text books and membership in selection committees for university positions.

Clearly, overall we have tried to honestly describe an exciting period while avoiding excessively apologetic comments in spite of our obvious enthusiasm, but we did not avoid biases and some prejudices that are naturally human.

A Gallery of Portraits

Ronald S. Rivlin (1915–2005)

Clifford A. Truesdell (1919–2000)

Richard A. Toupin (b. 1926; Photo in 2013)

Walter Noll (born 1925)

Bernard D. Coleman (Born 1930)

Paul M. Naghdi (1924–1994)

A. Cemal Eringen (1920–2009)

Raymond D. Mindlin (1906–1987)

Leonid I. Sedov (1907–1999)

Alexey A. Ilyushin (1911–1998)

Paul Germain (1920–2009)

References

Appell P (1921) Traité de mécanique rationnelle, vol III, 3rd edn. Gauthier-Villars, Paris (Facsimile reprint by Gabay, Paris, 1991)

Barenblatt GI, Joseph DD (ed) (1996) Collected papers of R.S. Rivlin, two volumes. Springer, New York

Beatty MF, Hayes MA (eds) (2005) Mechanics and mathematics of crystals. Selected papers of J. L. Ericksen. World Scientific, Singapore

Bingham EC (1922) Fluidity and plasticity. McGraw Hill, New York

Boulanger P, Knops RJ, Saccomandi G (2005) Special issue (in recognition of the 65th birthday of Michael A. Hayes). Math Mech Solids 10:132–135

Brovko GL (2013) On general principles of the theory of constitutive relations in classic continuum mechanics. J Eng Math (Special issue in the honour of A.A. Ilyushin) 78:37–53

Brovko GL, Bykov DL, Vasin GR (2011) A.A. Ilyushin's scientific heritage and development of his ideas in mechanics. Mech Solids 46:3–14

Carroll MM, Hayes MA (2006) In memory of Ronald S. Rivlin. Math Mech Solids 11:103–112

Casey J, Crochet MJ (1995) Paul M Naghdi (1924–1994). In: Casey J, Crochet MJ (eds) Theoretical, experimental, and numerical contributions to the mechanics of fluids and solids: a collection of papers in honor of Paul M Naghdi. Zeit. angew Math Phys. vol 46, Birkhäuser, Boston, pp S1–S32

Coleman BD (1964) On thermodynamics of materials with memory. Arch Rat Mech Anal 17:1–46

Coleman BD, Gurtin ME (1967) Thermodynamics with internal state variables. J Chem Phys 47:597–613
Coleman BD, Noll W (1961) Foundations of linear viscoelasticity. Rev Mod Phys 33:239–249
Coleman BD, Noll W (1963) The thermodynamics of elastic materials with heat conduction and viscosity. Arch Rat Mech Anal 13:167–178
Coleman BD, Markovitz H, Noll W (1966) Viscometric flows of non-Newtonian fluids: theory and experiment. Springer, NewYork
Coleman BD, Owen DR (1974) A mathematical foundation for thermodynamics. Arch Rat Mech Anal 54:1–104
Ericksen JL (1960) Anisotropic fluids. Arch Rat Mech Anal 4:231–237
Ericksen JL (1979) Speech of acceptance of the 1979 Timoshenko Medal, ASME. Available on the website of iMechanica (iMechanica.org/node/10396)
Ericksen JL, Rivlin RS (1954) Large elastic deformations of homogeneous anisotropic materials. J Rat Mech Anal 3:281–301
Ericksen JL, Truesdell CA (1958) Exact theory of stress and strain in rods and shells. Arch Rat Mech Anal 1:295–323
Eringen AC (1960) Irreversible thermodynamics and continuum mechanics. Phys Rev 117: 1174–1183
Eringen AC (1962) Nonlinear theory of continuous media. McGraw Hill, New York
Eringen AC (1966a) Theory of micropolar fluids. J Math Mech 16:1–18
Eringen AC (1966b) A unified theory of thermomechanical materials. Int J Eng Sci 4(2):179–202
Eringen AC (1967) Mechanics of Continua. Wiley, New York (Second and enlarged edition. 608 pp, Krieger Publishing, Melbourne, Florida, 1980)
Eringen AC (1968) Theory of micropolar elasticity. In: Liebowitz H (ed) Fracture: a treatise, vol II. Academic Press, New York, pp 621–729
Eringen AC (ed) (1971/1976) Continuum physics, vol 1. Academic Press, New York, 1971, 678 pp.; Vol. 2, 1975. 616.pp.; Vol.3, 1976, 317 pp.; Vol.4, 1976, 267 pp., Academic Press, New York (the whole translated to Chinese)
Eringen AC, Edelen DGB (1972) On nonlocal elasticity. Int J Eng Sci 10(3):233–248
Eringen AC, Suhubi ES (1964) Nonlinear theory of simple microelastic solids I. Int J Eng Sci 2:189–203
Friedman B (1948) Review: an essay towards a unified theory of special functions. Bull Amer Math Soc 56:268–269
Germain P (1962) Mécanique des milieux continus. Masson Editeurs, Paris
Germain P (1973) Cours de mécanique des milieux continus, vol I. Masson Paris
Germain P (2000) In: Maugin GA et al (eds) Autobiographical essay, pp 1–24)
Green AE (1965) Micro-materials and multipolar continuum mecanics. Int J Eng Sci 3:533–537
Green AE (1974) Speech of acceptance of the 1974 Timoshenko Medal. Available on the website of iMechanica
Green AE, Adkins JE (1960) Large elastic deformations and nonlinear continuum Mechanics. Clarendon Press, Oxford (Second edition, 1970)
Green AE, Naghdi PM (1965) A dynamical theory of interacting continua. Int J Eng Sci 3:231–241
Green AE, Naghdi PM (1977) On thermodynamics and the nature of the second law. Proc Roy Soc Lond A357:253–270
Green AE, Naghdi PM (1993) Thermoelasticity without energy dissipation. J Elasticity 31:189–208
Green AE, Naghdi PM (1995) A unified procedure for construction of theories of deformable media, Three parts. Part I Proc Roy Soc Lond 448:335–356, Part 2: 357–377, Part 3: 379–388
Green AE, Rivlin RS (1957) The mechanics of non-linear materials with memory I. Arch Rat Mech Anal 1:1–21
Green AE, Rivlin RS (1964) Multipolar continuum mechanics. Arch Rat Mech Anal 17:113–147
Green AE, Zerna W (1954) Theoretical elasticity. Clarendon Press, Oxford (Second edition, 1968)
Gurtin ME (1973) The linear theory of elasticity. In: Flügge S (ed) Handbuch der Physik, Bd VIa/2. Springer, Berlin, pp 1–295

Habib P (successor of J. Mandel as head of the LMS at Polytechnique). "Jean Mandel (1907–1982)". http://annales.org/archives/x/mandel.html
Hamel GWK (1927) Die axiome der Mechanik. In: der Physik Handbuch (ed) Geiger and Scheel, vol 5. Springer, Berlin, pp 1–42
Hayes MA, Rivlin RS (1961a) Surface waves in deformed elastic materials. Technical report Brown University and ONR No. 67, D.T.I.C
Hayes MA, Rivlin RS (1961b) Surface waves in deformed elastic materials. Arch Rat Mech Anal 8:358–380
Herrmann G (ed) (1974) R.D. Mindlin and applied mechanics. Pergamon Press, Oxford
Ignatieff YA (1996) The mathematical world of Walter Noll. Springer, Berlin
Ignatieff YA, Willig H (1999) Clifford Truesdell: Eine wissenschatliche Biographie des Dichters. Mathematikers und Naturphilosophen. Shaker Verlag, Aachen
Ilyushin AA (1948) Plasticity: elastic-plastic deformations (in Russian), Gostekhidzat, Moscow (in French: Plasticité: Déformations élastico-plastiques, translation: Popoff A, Thome P; Eyrolles Editeur, Paris, 1956)
Ilyushin AA (1971) Mekhanika sploshnoy sredy (Continuum Mechanics), Moscow State University Press, Moscow (2nd edn, 1978; 3rd edn, 1990]
Ilyushin AA (2003–2009) Selected collected Works: vol 1 (1935–1945) (in Russian), Fizmatlit: Moscow (2003); vol 2 (1946–1966), Fizmatlit: Moscow (2004); vol 3: Theory of thermoviscoelasticity, Fizmatlit: Moscow (2007); vol 4: Modelling of dynamic processes in solids and engineering applications, Fizmatlit: Moscow (2009)
Kiyko IA, Israilov MSh, Brovko GL (ed) (2001) Elasticity and anelasticity. Moscow University Press (mostly in Russian)
Lurie AI (1980) Nonlinear theory of elasticity (in Russian). Nauka, Moscow (English translation: Lurie KA), North Holland, Amsterdam, 1990
Maugin GA (2010) "Germain Paul (1920–2009)", Obituary, in: L'Archicube, 7bis, No. Spécial, Bull. de l'Association des anciens élèves, élèves et amis de l'Ecole Normale Supérieure, pp 126–129, Février 2010
Maugin GA (2011) Editorial: A.C. Eringen (1921–2009): a lifelong dedication to engineering science. In: Advances in generalized continuum mechanics (Special issue) (Guest editors: Maugin GA, Lee JD), Int J Eng Sci 49:1281–1525
Maugin GA (2013a) Continuum mechanics through the twentieth century (A concise historical perspective). Springer, Dordrecht
Maugin GA (2013b) A.A. Ilyushin's works: an appraisal from Paris. J Eng Math (Special issue in the honour of A.A. Ilyushin) 78:3–8
Maugin GA (2014) Continuum mechanics through the eighteenth and nineteenth centuries (Historical perspectives from John Bernoulli (1727) to Ernst Hellinger (1914)). Springer, Dordrecht
Maugin GA, Drouot R, Sidoroff F (ed) (2000). Continuum thermomechanics: the art and science of modeling materials' behaviour (on the occasion of P. Germain's 80th Anniversary), Kluwer, Dordrecht (see the autobiographical essay by P. Germain in pp 1–24, followed by a selection of his most representative publications by the main editor)
Maugin GA, Lee JD (eds) (2011) Advances in generalized continuum mechanics (A collection of studies in Engineering Sciences in Memory of the Late A.C. Eringen, 1921–2009), Special issue of the Int J Eng Sci 49/12:1281–1530
Maugin GA, Muschik W (1994) Thermodynamics with internal variables. I. General concepts. J Non Equilibr Thermodyn 19:217–249; II-Applications, ibid. 250–289
Mindlin RD (1989) In: Deresiewicz L, Bieniek MP, DiMaggio FL (eds) Collected papers of R.D. Mindlin, two volumes. Springer, New York
Mooney MA (1940) A theory of large elastic deformation. J Appl Phys 11:582–592
Murnaghan FD (1937) Finite deformation of an elastic solid. Amer J Math 59:235–260
Murnaghan FD (1951) Finite deformation of an elastic solid. Wiley, New York (Unfairly criticized by Truesdell in his review of this book reprinted in Truesdell, 1984, pp 148–150)

Naghdi PM (1957) Stresses and displacements in an elastic-plastic wedge. ASME Trans J Appl Mech 24:98–104
Naghdi PM (1972) The theory of shells and plates. In: Flügge S (ed) Handbuch der Physik, Bd. VIa/2. Springer, Berlin, pp 425–640 (definitive treatment of the subject)
Naghdi PM (1979) A brief history of the applied mechanics division of the ASME. J Appl Mech 46(4):723–749
Naghdi PM (1990) A critical review of the state of finite plasticity. Zeit angew Math Phys 41:315–394
Naghdi PM, Trapp JA (1975) The significance of formulating plasticity theory with reference to loading surfaces in strain space. Int J Eng Sci 13:785–797
Naghdi PM, Spencer AJM, England AH (1994) Nonlinear elasticity and theoretical mechanics. In: Honour of A.E. Green. Oxford University Press, UK
Noll W (1955) On the continuity of the solid and fluid states. J Rat Mech Anal 4:13–81
Noll W (1958) A mathematical theory of the mechanical behavior of continuous media. Arch Rat Mech Anal 27:197–226
Noll W (1967) Materially uniform simple bodies with inhomogeneities. Arch Rat Mech Anal 27:1–32
Noll W (1974) The foundations of mechanics and thermodynamics: selected papers of W. Noll. Springer, Berlin
Noll W (1987) Finite-dimensional spaces: algebra, geometry, and analysis, vol I. Martinus Nijhoff, Dordrecht
Noll W (2002) The genesis of the "non-linear field theories of mechanics" (text of a lecture given at the Meeting of the Society for Natural Philosophy in the memory of Clifford Truesdell, Pisa, Italy, Nov 2000)
Noll W, Virga EG (1990) On edge interactions and surface tension. Arch Rat Mech Anal 111:1–31
Oldroyd JG (1950) On the formulation of rheological equations of state. Proc Roy Soc Lond A200:523–541
Prager W (1961) Introduction to mechanics of continua. Ginn and Company, Boston. Adapted from a German version, Birkhäuser, Basel
Reiner M (1945) A mathematical theory of dilatancy. Amer J Math 67:350–362
Rivlin RS (1948) Large elastic deformation of isotropic materials. Part I Phil Trans Roy Soc Lond A240:459–490; Parts II, ibid, A241:379–397
Rivlin RS (1956) Solution of some problems in the exact theory of visco-elasticity. J Rat Mech Anal 5:179–188
Rivlin RS (1984) Forty years of non-linear continuum mechanics. In: Proceedings of conference at the ninth international congress of rheology. pp 2783–2811
Rivlin RS (1996) Autobiographical postcript. In: Collected papers of R.S. Rivlin, Eds. G.I. Barenblat and D.D. Joseph, p xxii, Springer, New York
Romains JE (1966) A high-level introduction (review of Sedov's book of 1966), Physics Today, Sept 1966 Issue, pp 98–100
Sedov LI (1962) Introduction to continuum mechanics (in Russian) Moscow
Sedov LI (1966) Foundations of the non-linear mechanics of continua (Translation from the Russian by Rebecca Schoenfeld-Reiner), 252 pages. Pergamon, New York
Sedov LI (1971) A course in continuum mechanics, four volumes. Wolters-Noordhoff, The Netherlands
Sedov LI (1973) Mekhanika sploshnoy sredy (Continuum Mechanics), two volumes. Nauka, Moscow (in English: Continuum mechanics, North-Holland, Amsterdam, 1971; in French: Mécanique des milieux continus, Editions MIR, Moscow, 1975)
Spencer AJM (1971) Theory of invariants. In:Eringen AC (ed) Continuum physics, vol I, Chap. 3. Academic Press, New York
Spencer AJM (1972) Deformation of fibre-reinforced materials. Clarendon Press, Oxford
Toupin RA (1956) The elastic dielectric. J Rat Mech Anal 5:849–916
Toupin RA (1963) A dynamical theory of dielectrics. Int J Eng Sci 1:101–126
Treloar LGR (1946) The elasticity of a network of long-chain molecules. Trans Faraday Soc 42:83–94

Truesdell CA (1952) The mechanical foundations of elasticity and fluid dynamics. J Rat Mech Anal 1(1):125–300
Truesdell CA (1969) Rational thermodynamics. McGraw Hill, New York
Truesdell CA (1984a) An idiot's fugitive essays on science. Springer, New York
Truesdell CA (1984b) Rational thermodynamics, second enlarged edition. Springer, New York
Truesdell CA (1987–1988) Editorial (on the occasion of the publication of the hundredth volume), Arch Rat Mech Anal 100:IX–XXII
Truesdell CA, Noll W (1965) The non-linear field theories of mechanics. In: Flügge S (ed) Handbuchder Physik, Bd.III/3. Springer, Berlin (Referred to for short as NFTM)
Truesdell CA, Toupin RA (1960) The classical field theories. In: Flügge S (ed) Handbuch der Physik, Bd. III/1. Springer, Berlin (Referred to for short as CFT)
Truesdell CA, Rajagopal KR (1999) An introduction to the mechanics of fluids. Birkhäuser, Basel
Zhilin PA (2001) A.I. Lurie—works on mechanics. In: Proceedings of the XXVIII summer school. Advanced problems in mechanics. St Petersburg, 2001, pp 1–13

Chapter 12
Epilogue

After the completion of this third volume, we have a better, if not fully complete, perspective view of the development of continuum mechanics from the Renaissance (sixteenth century) to the most recent progress early in this twenty-first century. Apart from the never ending competition between molecular approach and direct continuum vision—the latter on which we naturally focused attention—we can roughly distinguish three main periods.

The first period is one during which experiments and common sense were at work. This was mostly driven by the needs to expand hydraulic systems for everyday control of flows. This in turn led to envisaging what we now call porous media and the more or less rapid flow of liquids. Viscosity and the complex structure of observed flows planted the seeds for modern developments lastly including a comprehending of the science of flight and naturally leading to a consideration of turbulence. This was exposed in Chaps. 2, 3 and 4 in the present volume. They essentially concern what was finally put under the common umbrella of *fluid mechanics*. This was achieved mainly by civil engineers. But with time and the influence of a more theoretical and mathematical vision of Nature—that we associate with the expansion of modern science (purposefully devised experiments, introducing relationships between numbers, appearance of first deterministic notions)—these engineers mixed their approach with a new form of human activity, the newly born scientific *profession*. The transition to the second period was upheld by the essential contributions of the greatest "mechanicians" of all, namely, Galileo Galilei, Descartes, Huygens, Newton, and then the Bernoullis and Euler—only Einstein was to go much further by inventing a new paradigm in what was identified before as "classical" mechanics. With the masterpieces of such luminaries, one had at hand the basic principles on which to build a true science first of point mechanics, then the mechanics of rigid bodies, and finally continuum mechanics in what was called the framework of "*rational mechanics*". The spirit of the Enlightenment provided the general background for such a progress that inevitably yielded the second period. Noticeable contributors to that first period are not always publicized at the level of their achievements although the names of Torricelli, Pascal, Mariotte, Borda, and Bossut come immediately to mind while those of Darcy, Poiseuille, and Hagen will be representative of a closer incorporation in the second period, not to speak of Reynolds and Prandtl much later on.

G.A. Maugin, *Continuum Mechanics through the Ages - From the Renaissance to the Twentieth Century*, Solid Mechanics and Its Applications 223,
DOI 10.1007/978-3-319-26593-3_12

This second period corresponds to what we underlined as a period in which "the importance of being analytical" became a definite asset. Continuum mechanics was now clearly identified as a promising field of research sustained by obvious engineering applications. Analysis and continuum mechanics then developed hand in hand, required mathematical tools and accuracy being fostered by the need to correctly represent the newly introduced notions and to help solving practical problems on the basis of this new theoretical background. This period starts with Newton, Leibniz and the Bernoullis to gain a full momentum with Maupertuis, d'Alembert, Euler and Lagrange. Then a tremendous progress was realized by the most brilliant minds of the nineteenth century, often rightly qualified as "engineers-scientists", such as Fresnel, Cauchy, Navier, Poisson, Duhamel, Piola, Kirchhoff, Helmholtz, Lord Kelvin, Stokes, Maxwell, Barré de Saint-Venant, Boussinesq, and Voigt, the most remarkable breakthroughs being the introduction of the notion of stress tensor—that became synonymous with continuum mechanics per se -, the expansion of elasticity and viscous-flow mechanics, and the birth of a true theory of plasticity (Tresca, Saint-Venant, Lévy). This we perused with great care in the second volume (Maugin 2014) devoted to the eighteenth and nineteenth centuries.

The "Belle époque" (1880–1914) does not really constitute the third period, but it has been singled out for the following reasons. First it stands for a transition period before the dramatic events of the first half of the twentieth century, the "Belle" qualification corresponding to a period of relative peace on the European continent, an expansion through out the world, politically, economically and scientifically, but to the price of an increase in colonization of others. From the purely scientific viewpoint, this period corresponds to an often felt accomplishment and the attainment of a somewhat completeness of the field of continuum mechanics, all this in spite of the burgeoning of new fields such as relativistic studies, embryonic quantum mechanics, and biology—but Pasteur had shown the way in this last field. Also, this period still is the field of action of essentially European institutions, so that our work of appraisal of original contributions is facilitated by the sole knowledge of four languages (English, German French, and Italian). More characteristically and with deep consequences, this period was nonetheless pregnant of further progress; it provided the essential basis for a re-consideration of the field in post WWII by scientists-historians such as Truesdell. To examine this situation at the Belle époque, we decided to peruse in detail the encyclopaedic texts and lecture notes of various enlightened mechanicians or physicists, namely, Duhem, the Cosserat brothers, Caratheodory, Poincaré, Hilbert and Hellinger (see Maugin 2013a, b and the present volume), not necessarily the most creative contributors to our field, but perspicacious examiners. We find in these remarkable timely syntheses—that give a state-of-the-art view at that crucial moment—all the material that was to provide the main ideas in the new construct and rational re-organization of the field in post WWII. Among this material we emphasize the formulation of basic balance laws (often still on the basis of the principle of virtual power), the seeds of good elements of thermo-mechanics as a unified science, some rich views on field discontinuities, elements of singular behaviours to be dealt with in greater detail in the future, and suggestions for a finer mechanical description of complex materials (those with a potent microstructure), all these to be dutifully exploited during the third period.

We envisage as third period that one which succeeded the Second World War, although some definite signs had shown in the 1920s and 1930s, in particular concerning behaviours such as visco-plasticity (Bingham, Reiner), plasticity (Nadai, von Mises, Hencky, Prandtl, Reuss), dislocations (Taylor, Polanyi, Orowan), and finite-strain elasticity (Murnagham, Signorini et al.). This period starts to show the emergence and powerful rise of research in North America that suggest a further globalization of research in the field and supremacy of English as a *lingua franca* permeating all of science. Although we still perceived some national schools and styles—in particular in Maugin (2013a, b) (cf. US, British, French, Polish, German, and Russian schools) -, the internationalization of the field became an obvious fact. We clearly emphasized the prevailing role played by a selected group of brilliant individuals whom we called the "masters of modern continuum mechanics" (e.g., Truesdell, Rivlin, etc.), while we also naturally had to account for more dilute contributions from a wealth of scientists, in particular those who were in direct contact with these masters. For obvious reasons of generational relations, we are much better informed about all these people, especially if one adopted a more cosmopolitan approach to science than before (as was the case of the present author). After that of analysis, the role of modern differential geometry in the solid basis of continuum mechanics with its delivery of new useful technical tools in describing complex mechanical behaviours of materials has been appraised at its just value. Simultaneously, the synthetic thermo-mechanical framework was practically adopted everywhere, while applications of modern functional analysis (functional spaces, variational techniques) allowed rigorous unambiguous statements of many theorems of existence, uniqueness, and regularity of solutions. This was a required prerequisite for efficient well designed applications of numerical schemes allowed by developments in computer science (finite-element method, finite-volume technique, mixed analytical-numerical approaches, multi-scale approaches, mathematically justified homogenization techniques).

These underlined new developments offered the unique occasion to cast the successes of the master scientists in their proper socio-economical background as science itself of necessity became part of this general landscape, essentially with established networks of contact, influence, communication, and diffusion. This is indeed what fostered the present era That is why on a more scientific level we gave a conspectus of different specialized fields which, in our opinion, shaped most of the true accomplishments in the period of interest, and our own contributions in particular. Although this often reflects a personally biased interest and expertise, in all, the domains of the viscoelasticity of solids, plasticity, Fracture, generalized continuum mechanics in its different avatars, configurational mechanics, electro-magneto-mechanical interactions, and relativistic continuum mechanics were analyzed more or less thoroughly in Maugin (2013a, b) and the present volume. To be complete, one should add to these the contents of several historical reviews published in other places (Maugin 2011, 2012, 2013a, b, 2015a, b). The best conclusion and prospective view of the near future were given at the end of Maugin (2013a, b), emphasizing the fostering role played by new experimental equipments and computational means, the interactions between scales, and the increasing importance of a good geometrical framework, but simultaneously regretting the ever increasing demand of multi-sided knowledge from the most concerned scientists. Perhaps that we are now experiencing a period quite similar to that of the "Belle époque" with an overestimated confidence in our own prowess, i.e., some exaggerated hubris.

References

Maugin GA (2011) Solitons in elastic solids (1938–2010). Invited "historical review". Mech Res Commun 38:341–349 [Also available on: Virtual special issue on Historical reviews in Mechanics, MRC, Springer, 2015]

Maugin GA (2012) The principle of virtual power: from eliminating metaphysical forces to providing an efficient modelling tool. Cont Mech Thermodynam 25:127–146 (Special issue in the honor of G. Del Piero)

Maugin GA (2013a) Continuum mechanics through the twentieth century: a concise historical perspective. Springer, Dordrecht

Maugin GA (2013b) Sixty years of configurational mechanics (1950–2010). Invited "historical review". Mech Res Commun 50:39–49 [Also available on: Virtual special issue on Historical reviews in Mechanics, MRC, Springer, 2015]

Maugin GA (2014) Continuum mechanics through the eighteenth and nineteenth centuries: historical perspectives from John Bernoulli (1727) to Ernst Hellinger (1914). Springer, Dordrecht

Maugin GA (2015a) The saga of internal variables of state in continuum thermo-mechanics (1893–2013). Invited "historical review". Mech Res Commun 69:79–86

Maugin GA (2015b) Some remarks on generalized continuum mechanics. Math Mech Solids 20:280–291 (Special issue dedicated to K.R. Rajagopal)

Index

G.A. Maugin, *Continuum Mechanics through the Ages - From the Renaissance to the Twentieth Century*, Solid Mechanics and Its Applications 223,
DOI 10.1007/978-3-319-26593-3

MIX
Papier aus verantwortungsvollen Quellen
Paper from responsible sources
FSC® C105338
FSC
www.fsc.org

If you have any concerns about our products,
you can contact us on
ProductSafety@springernature.com

In case Publisher is established outside the EU,
the EU authorized representative is:
Springer Nature Customer Service Center GmbH
Europaplatz 3, 69115 Heidelberg, Germany

Printed by Libri Plureos GmbH
in Hamburg, Germany